Zebrafish

The Practical Approach Series

Related **Practical Approach** Series Titles

Please see the **Practical Approach** series website at

http://www.oup.com/pas

for full contents lists of all Practical Approach titles.

No. 261

Zebrafish

A Practical Approach

Edited by

Christiane Nüsslein-Volhard

and

Ralf Dahm

Max-Planck-Institut für Entwicklungsbiologie,
Abteilung Genetik, Spemannstr. 35,
72076 Tübingen, Germany

OXFORD

UNIVERSITY PRESS

OXFORD
UNIVERSITY PRESS

Great Clarendon Street, Oxford OX2 6DP

Oxford University Press is a department of the University of Oxford.
It furthers the University's objective of excellence in research, scholarship,
and education by publishing worldwide in

Oxford New York

Auckland Bangkok Buenos Aires Cape Town Chennai
Dar es Salaam Delhi Hong Kong Istanbul Karachi Kolkata
Kuala Lumpur Madrid Melbourne Mexico City Mumbai Nairobi
São Paulo Shanghai Taipei Tokyo Toronto

Oxford is a registered trade mark of Oxford University Press
in the UK and in certain other countries

Published in the United States
by Oxford University Press Inc., New York

A catalogue record for this title is available from
the British Library

Library of Congress Cataloging in Publication Data
(Data available)

ISBN 0 19 963809 8 (Hbk)
ISBN 0 19 963808 X (Pbk)

10 9 8 7 6 5 4 3 2

Typeset by Footnote Graphics
Printed in Great Britain
on acid-free paper by CPI Bath

Preface

In recent years, the zebrafish (*Danio rerio*) has become one of the most important model organisms to study biological processes *in vivo*. This is due to a combination of advantages making it an ideal organism for researchers interested in many aspects of embryonic development, physiology and disease. As a vertebrate, it has many of the strengths of invertebrate model systems, such as a small size, a large number of offspring and a short generation time. Moreover, its very rapid and synchronous embryonic development greatly facilitates phenotypic analysis and large-scale experimental approaches. Its transparent, easily accessible embryos as well as their robustness make it ideally suited to micromanipulation and *in vivo* observations. Thus, the zebrafish allows using many if not most of the approaches and techniques used in other model organisms, bridging the gaps between them. The important role the zebrafish will play in the future of biomedical research is indicated by the many initiatives looking at gene function in vertebrates, which already have chosen the zebrafish as their model organism.

With the community of zebrafish researchers growing worldwide, new techniques and methodologies are becoming available at an ever-increasing rate. This book is intended to provide researchers with a background as well as an up-to-date and comprehensive set of tools for working with the zebrafish. It will allow those new to the field to establish zebrafish in their laboratory. It also gives a broad overview of commonly used methods and a complete collection of protocols describing the most powerful techniques for those already working with the zebrafish. The book includes a list of all published zebrafish mutants including brief phenotypic descriptions and references as well as chapters on morphological development necessary for the staging and analysis of embryos, juvenile and adult fish.

This book would not have been possible without the help of numerous people and the editors would like to thank all those who have contributed to this volume. In particular, we would like to thank all authors for their time and effort in preparing their chapters and for sharing their expertise to ensure that this volume will be an invaluable source of information for those working with zebrafish. We would also like to thank all members of the zebrafish group in Tübingen for proofreading texts at various stages of development, for very productive feedback and for sharing unpublished protocols and information with us. We are also grateful to the Graphics Department at our Institut for their help and advice in preparing the figures. Last but not least, we would like to thank all staff at Oxford University Press for their technical advice and help in preparing the manuscript.

Tübingen   Christiane Nüsslein-Volhard
June 2002   and Ralf Dahm

Contents

Protocol list

Abbreviations

ALL	anterior lateral line
A-P	anterior-posterior
APS	ammonium persulphate
ATPase	adenosine triphosphatase
BAC	bacterial artificial chromosome
BB/BA	benzyl benzoate/benzyl alcohol (2:1)
BCIP	5-bromo-4-chloro-3-indolyl-phosphate
BM Purple	Boehringer Mannheim Purple (precipitating substrate)
BSA	bovine serum albumin
cfu	colony forming units
DAB	3,3'-diamino-benzidine
DASPEI	2-(4-dimethyl-aminostyryl)-N-ethyl pyridinium iodide
ddH$_2$O	double-distilled water
DEPC	diethyl pyrocarbonate
DGGE	denaturing gradient gel electrophoresis
dH$_2$O	distilled water
DIC	differential interference contrast (*syn.* Nomarsky)
DIG	digoxigenin
DMSO	dimethyl sulphoxide
dpf	days post-fertilization
DRG	dorsal root ganglia
D-V	dorso-ventral
E3	embryonic medium (see Chapter 1)
EDTA	ethylenediaminetetraacetic acid
EM	electron microscope or electron microscopy
ENU	N-ethyl-N-nitrosourea
EP	early pressure
ES	embryonic stem (cell)
ESTs	expressed sequence tags
EtOH	ethanol
FITC	fluorescein isothiocyanate
GFP	green fluorescent protein
hpf	hours post-fertilization
HRP	horseradish peroxidase
HS	heat shock
HTA	head-trunk angle

IAA isoamyl alcohol
IPTG isopropyl-β-D-thiogalactopyranoside
IVF *in vitro* fertilization
MBT mid-blastula transition
MeOH methanol
MESAB ethyl-*m*-aminobenzoate methanesulphonate
mpf minutes post-fertilization
NA numerical aperture
NaAc sodium acetate
NBT 4-nitroblue tetrazolium chloride
NCBI National Center of Biotechnology Information
NGS normal goat serum
OD outer diameter
OVL otic vesicle length (refers to the number of times the otic vesicle could be placed in the space between the eye and the otic vesicle)
PAC P1-derived artificial chromosome
PBS phosphate-buffered saline
PCR polymerase chain reaction
PFGE pulsed-field gel electrophoresis
PLL posterior lateral line
PNS peripheral nervous system
PTU 1-phenyl-2-thiourea
RAPD random amplified polymorphic DNA
RH radiation hybrid
RT-PCR reverse transcriptase polymerase chain reaction
RZPD Resource Center of the German Human Genome Project
SCC solitary chemosensory cell
SDS sodium dodecyl sulphate
SNP single nucleotide polymorphism
SSC saline sodium citrate
SSCP single-strand conformational polymorphism
SSLPs simple sequence length polymorphisms
TE Tris, EDTA
TEMED *N,N,N,N*,-tetramethylene diamine
TRITC tetramethylrhodamine isothiocyanate
Tü Tübingen
UTRs untranslated regions
UV ultraviolet
YAC yeast artificial chromosome
YSL yolk syncytial layer

Introduction: zebrafish as a system to study development and organogenesis

Christiane Nüsslein-Volhard, Darren T. Gilmour, and Ralf Dahm

Max-Planck-Institut für Entwicklungsbiologie, Abteilung Genetik, Spemannstr. 35, 72076 Tübingen, Germany

1 History

Until the beginning of the twentieth century classical zoology focused on the discovery and description of a large number of new species. The mode of early development, and the shape of larval stages, provided crucial parameters for the assessment of evolutionary descent and relationship to other species, more so than adult morphologies and traits. With time, zoologists interested in embryonic development *per se* displayed an increasing tendency to focus on particular organisms that were chosen for traits that made them amenable to experimental manipulation. Each of these organisms allowed a particular approach to be taken and together they gave rise to all of the major concepts in developmental biology. For example, from several marine organisms, such as sea urchins, eggs could be obtained in incredible numbers. Species with large robust embryos, such as amphibia and some insects, were amenable to the transplantation and ablation experiments that lead to the notion of organizing centres in embryos. Early understanding of cell lineage was gained from work on animals with invariant cell lineages that could be traced throughout development (*Ascaris* spp.). Some organisms were chosen for reasons that at first glance might seem insignificant. For instance, the presence of cortical pigment granules in egg cytoplasm of ascidians gave rise to the concept of cytoplasmic inheritance. Despite these advances, the incredible diversity of phenomena observed in the different systems made embryology an increasingly bewildering topic. Generalizations, or even useful comparisons, were often difficult, even within phyla. In particular, the molecules underlying these phenomena remained elusive, despite considerable efforts.

In the 1950s, molecular biology was born. The first breakthroughs in molecular understanding came from research on the simplest living beings, namely bacteria and their viruses. These studies were subsequently extended to the field of developmental biology, when questions concerning the regulation of gene activity in multicellular organisms were addressed with studies on chromatin, ribosomal RNA synthesis, and eukaryotic transcription. It was at this time that most of the model organisms still popular with developmental biologists today were established. These include the slime mould *Dictyostelium*, the coelenterate *Hydra*, the fruitfly *Drosophila*, the clawed frog *Xenopus*, the chicken, and the mouse. It was, of course, not a single feature that made researchers focus on these 'model' organisms, rather it was a combination of useful traits. *Drosophila*, for example, had a long tradition as a genetic organism, through which most of the principles governing the transmission of genes from one generation to the next were discovered. The ease with which mutants could be induced and the existence of giant chromosomes, allowing the physical visualization of genes and their activities, promoted the use of the fruitfly. Developmental genetics in *Drosophila* began with the investigations into the developmental potential of imaginal discs. The first *Drosophila* developmental mutants, such as *bithorax* and *engrailed*, had been discovered as early as the 1920s, and by the 1940s mutants affecting the morphology of embryos had been isolated.

When compared to many other insects, *Drosophila* is not particularly suited for embryological manipulation. However, any embryological disadvantages *Drosophila* may have had were overcome by the power of its genetics. The possibility to perform saturating mutagenesis screens for embryonic mutants and the discovery of P element-mediated transfomation meant that almost any question in development, no matter how daunting, could be tackled using this little fly. This has led to *Drosophila* becoming perhaps the most widely used genetic model organism. This popularity is, at least in part, due to the fact that many key regulators of development first discovered in *Drosophila*, most prominently the *Hox* genes, were subsequently shown to play analogous roles throughout the animal kingdom.

Likewise, important experimental advances have been made with other model organisms. In the case of the mouse, the existence of large collections of mutant strains together with embryonic stem (ES) cell-based reverse genetic techniques, such as knockouts, has made them the model system of choice for mammalian development. Other vertebrates, such as *Xenopus* and chicken, with their large, easy-to-manipulate embryos, also have advantages as experimental animals. For example, the technique of RNA injection into *Xenopus* oocytes provides a powerful expression system to test rapidly the potency of a large number of molecules *in vivo*.

It is clear that each model organism has its particular strengths and weaknesses. The advantages of each system, in a feedback mechanism, has meant that they are suited to the study of particular aspects of development, such that the results cannot easily be compared with those obtained for other systems. Therefore, while establishing a body of sophisticated knowledge for a single organism, there are considerable limitations in allowing general conclusions. For example, the understanding of *Drosophila* embryology is dominated by findings based on genetics, while in the frog, the transplantation and injection experiments prevail. Therefore, research in animals allowing the application of genetic approaches in addition to embryological methods would help solve this problem.

Two such model organisms were established in the 1970s—the roundworm *Caenorhabditis elegans* and the vertebrate zebrafish, *Danio rerio*. Sydney Brenner and George Streisinger had studied much simpler living systems before. Both had been members of the Phage Group, unravelling the nuts and bolts of transcription and protein synthesis in *Escherichia coli*, using bacteriophage genetics. Thus for both of them, the possibility of applying genetics to new model organisms was of prime importance. Sydney Brenner was in search of a simple multicellular animal to which genetic methods could be applied, with the aim of understanding cell lineage, neuronal function and behaviour.

Streisinger, who tragically died in 1984, chose the zebrafish, which is easy to keep and breed and whose embryos are transparent and develop rapidly, as an animal model for vertebrate development. He exploited the possibilities of genetically manipulating his chosen organism, laid out in his seminal paper published in 1981 (1). These include *in vitro* fertilization, a method to induce haploid development, and the establishment of homozygous lines. The paper also describes the first zebrafish mutation, *golden*, which gives a homozygous viable pigment phenotype, and was used in many of these pioneering studies. In the fish group at the University of Oregon at Eugene, the embryology of zebrafish soon was described in detail and the first developmental mutants, *cyclops* and *spadetail* were isolated. Later, methods for high-efficiency *N*-ethyl-*N*-nitrosourea (ENU) mutagenesis were developed and large-scale screens resulting in close to 2000 characterized mutants were carried out (2). Since then, the zebrafish community has increased rapidly and to date nearly 300 laboratories are using this organism in their studies. The public database Zfin lists over 3000 zebrafish publications, 406 of which appeared in 2000 alone.

2 Phylogeny

Zebrafish is a teleost fish of the cyprinid family in the class of ray-finned fishes (Actinopterygii). The lineages leading to the cyprinids and mammals split about 450 million years ago. The teleosts include other model organisms, such as medaka (*Oryzias latipes*), the pufferfish (*Takifugu rubripes, Tetraodon nigroviridis*), and the three-spined stickleback (*Gasterosteus aculeatus*). The latter three species are located on opposite sides of the lineage tree with respect to the zebrafish, and the last common ancestor lived about 100 million years ago (a similar distance as that between humans and the mouse). The phylogenetic position of the zebrafish in the vertebrate lineage is illustrated in *Figure 1*.

3 The zebrafish genome

The zebrafish genome is 1.7 gigabases in size, which is just more than half that of tetrapods, such as human and mouse. It is divided up into 25 chromosomes ($1n$). The absolute number of genes in zebrafish is currently not known; however, a large set of genes has been cloned and analysed to date. Comparisons of these genes between tetrapods and fishes has shown that in zebrafish (and other teleosts) there are often two homologues of the mammalian equivalent. This suggests that there has been an additional genome duplication shortly before the teleost radiation. This duplication must have been either partial, or followed by rapid gene loss, as gene duplication is estimated to account for only about 20% of the zebrafish genes examined. In all cases that have been investigated, the paralogues in zebrafish had highly diverged with respect to both spatial and temporal expression. This suggests that in the teleosts the function of ancestral genes may now be divided up between two genes, each having a more restricted function than the original gene. A consequence of this could be that in many instances the function of individual genes might be less complex in fish than in tetrapods and therefore easier to study in the former.

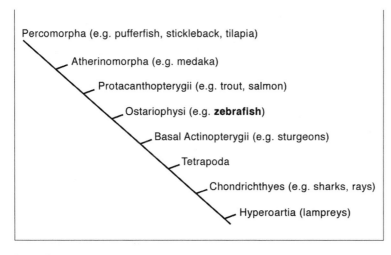

Figure 1 Schematic phylogenetic tree of the vertebrate lineage.

Recently, a genomics initiative with the aim of completely sequencing the zebrafish genome has been started by the Sanger Centre, UK in collaboration with groups in Tübingen, Germany and Utrecht in The Netherlands. The complete annotated sequence is expected to be available by 2004.

4 Advantages of the zebrafish as a model system

It is the combination between excellent embryology and the powers of genetic manipulation that make the fish one of the most important vertebrate model organisms for biological research. The zebrafish embryo has many characteristics that make it ideal for studying development. Fertilization and subsequent embryonic development are external and occur synchronously in large clutches. The embryos are relatively large and their development can be observed easily through the chorion. During the first 24 hours of development, the embryos are completely transparent, allowing the visualization of developing organs, even deep inside living embryos. Moreover, embryonic development is rapid. After about 2 days all common vertebrate specific body features can be seen, including a compartmentalized brain, eyes, ears, and all internal organs. In comparison to higher vertebrates, the organs in zebrafish are like a minimalist version, using far fewer cells to fulfil the equivalent function in the organism. For example, the kidney of the larval zebrafish consists of a single glomerulus which runs most of the body length, in comparison to the tens of thousands of glomeruli in the mammalian kidney. The zebrafish larvae have hatched and are able to swim and search for food as soon as 5 days after fertilization.

The second crucial advantage of zebrafish is its usefulness for gene identification by mutations. It has a number of unique features that facilitate genetic analysis, inbreeding schemes, and stock keeping. Compared to other vertebrates, the generation time of 2–4 months is not particularly short. However, the large number of progeny greatly facilitates genetic analyses, for example collecting mutant or recombinant individuals for mapping purposes.

As with other model organisms, the number of methods and tools that were developed for the zebrafish has increased considerably over the past few years and has allowed new experimental approaches to be undertaken.

5 Purpose of this book

This book contains a collection of chapters that describe the most commonly used techniques and approaches for studying zebrafish. Chapter 1 is a beginners guide to setting up a fish lab. It describes how to raise and keep zebrafish and gives an overview of different tank and fish facility designs.

The morphology of wild-type fish and ways to stage embryos and larvae are covered in Chapters 2 and 3. Both also give protocols on how to stain fish embryos for different structures. Appendix 2 is a guide of how to stage zebrafish embryos.

The external development of the zebrafish allows a wide range of experimental manipulations and Chapter 4 covers many of these embryological techniques, such as cell labelling and transplantations.

One of the strengths of the zebrafish is that it is relatively easy to manipulate gene expression. Chapter 5 explains in detail how to generate transgenic fish lines that allow the stable expression of specific gene products. Further, commonly used techniques, like gene misexpression by RNA injection and gene knockdown by morpholinos are also covered.

Chapter 6 gives a step-by-step guide to generating mutants from mutagenesis protocols to maintaining mutant stocks. It also describes specific strategies, such as allele and haploid screens, and those allowing the isolation of maternally required or dominant genes.

To date, mutagenesis screens have produced thousands of mutants with visible phenotypes, many of which have been mapped and cloned. Appendix 3 gives an up-to-date list of all published mutants. It also lists the 60 or so mutants that have been cloned to date.

Such molecular cloning would not have been possible without the help of the several molecular linkage maps of the zebrafish genome that have been constructed. However, most of these were isolated due to their similarity to known genes in other species. If the zebrafish is to have an impact on the future of biomedical research, it is essential that those mutations that fall in novel genes can also be isolated using positional cloning strategies. To this end, higher-resolution physical maps have to be constructed. Chapter 7 provides guidelines to assigning mutations to both physical and genetic maps, as well as describing the state of the art of zebrafish gene mapping.

This little compendium is aimed primarily at those new to working with zebrafish. However, we also hope that it will also provide a useful source of up-to-date protocols for the more experienced.

We would also like to point out that this is by no means the first time zebrafish methods have been described in print. First and foremost, the *The zebrafish book* by Monte Westerfield (available online at http://zfin.org/zf_info/zfbook/zfbk.html), now in its second edition, is a very valuable source of zebrafish protocols. More recently, two Methods in Cell Biology volumes have been published: *The zebrafish: biology* and *The zebrafish: genetics and genomics* (volumes 59 and 60, respectively; Academic Press, 1999), both edited by H. W. Detrich, M. Westerfield and L. I. Zon. Finally, the zebrafish community would not be the same without the www database ZFIN (http://www.zfin.org), maintained and regularly updated by Pat Edwards, Monte Westerfield, and colleagues at the University of Oregon.

Acknowledgements

We would like to thank all members of the Tübingen zebrafish lab for generously providing and testing many of the protocols included in this book and for proof-reading the text. We are grateful to Karl-Heinz Nill for his assistance with many of the figures to be found throughout the book. We are also indebted to Kai-Erik Witte for help with phylogeny.

References

1. Streisinger, G., Walker, C., Dower, N., Knauber, D., and Singer F. (1981). *Nature*, **291**, 293.
2. Zebrafish Issue (1996). *Development*, **123**.

Chapter 1
Keeping and raising zebrafish

Michael Brand

Max-Planck-Institute for Molecular Cell Biology and Genetics, Pfotenhauerstr. 108, 01307 Dresden, Germany

Michael Granato

Department of Cell and Developmental Biology, University of Pennsylvania, 1210 BRB II/III, 421 Curie Blvrd, Philadelphia, PA 19104, USA

Christiane Nüsslein-Volhard

Max-Planck-Institut für Entwicklungsbiologie, Abteilung Genetik, Spemannstr. 35, 72076 Tübingen, Germany

1 Introduction

Over the past decade, the zebrafish has become an important model organism for biological research. In recent years, more than 1800 zebrafish mutants have been identified, providing a rich resource for the study of embryonic development in vertebrates. An increasing number of laboratories are now starting to use this unique collection of zebrafish mutants for their research.

Raising and maintaining zebrafish is more demanding than keeping stocks of invertebrates, but less so than keeping stocks of mammals. This chapter is aimed at giving a background on how to raise mutant and laboratory wild-type stocks and how to maintain them in good breeding conditions. We describe aquarium systems and water conditions as well as procedures to raise and breed zebrafish. While this chapter is aimed primarily at investigators entering the zebrafish field, the recipes and protocols may also provide useful information for experienced zebrafish researchers.

2 Aquaria systems and water conditions

2.1 Aquaria systems

The major demands to aquaria systems are posed by the problem of keeping the water in good condition. If the water is not constantly kept clean, the fish will suffer from the toxicity of their own excretions and derivatives thereof. The simplest way to keep the water clean is to keep fish at very low densities and to replace the water at regular intervals. A more elaborate method is to equip the aquaria with biological filters through which the water is circulated and purified. The filters contain material with a large surface area on which bacteria can grow, which degrade toxic ammonium to less toxic nitrates. In larger systems, water from several (often more than 1000) tanks may be recycled through one large common filter (recirculating systems).

The exact type of facility required depends very much on the actual work to be done. For experiments involving only wild-type embryos and one or two mutant strains, a few large tanks are sufficient. However, genetic experiments with zebrafish often need facilities with up to several thousand tanks. When setting up such a system, the space requirements have to be considered, as well as the costs of maintenance and the overall safety of the system. In several laboratories, large-scale aquaria systems that are tailored to a variety of experimental needs have been built. The principles of such systems are described below.

2.1.1 Systems without filtering

On the one hand, non-circulating systems avoid all costs of establishing and servicing filters and are therefore relatively cheap to set up. On the other hand, they are very space-demanding, because fish have to be kept at low densities. Moreover, they require high maintenance and, in practice, they work only in areas where water of very high quality can be produced at a low cost.

Adult fish can be maintained without feeding and water exchange for up to 10 days at low densities (no more than two fish per litre). Keeping fish for longer periods requires feeding and regular water exchange. This can be done on a daily basis or continuously. In flow-through systems the tanks are typically large (20–200 litres) and each has its own adjustable inlet and an overflow (for example, a hole, equipped with a sieve, drilled just below the upper rim of the tank). The rate of freshwater supply should be about 0.3 tank volumes per day, allowing fish densities of one fish in 1–3 litres, depending on the feeding regimen. The tap water should be charcoal filtered and run through ultraviolet (UV) lamps before entering the aquaria. As the charcoal removes oxygen from the water, it is necessary to aerate the tanks using bubble stones. Care has to be taken to avoid overfeeding of fish, as surplus food quickly rots and thus decreases the water quality. The tanks have to be cleaned daily to remove faeces and surplus food after the last feeding. To avoid spread of infections between tanks, it is routine to handle fish only with sterilized or disposable items. However, it is not easy to avoid spreading infections in genetic experiments, where it is often necessary to mix fish from different tanks and different rooms.

2.1.2 Recirculation systems

With biological filters, the water is purified, cleaned, and reused. A principal advantage of recirculation systems is that they provide high-quality water and a high water exchange rate. For small-scale experiments, fully equipped tanks with an appropriate filter, illumination, heater, and aeration can be purchased at local pet stores. Gravel and plants are not required. The tanks should be placed on a dark surface (because the fish like those better than light surfaces) and not in the vicinity of windows, in order to avoid growth of algae. With an appropriate filter, which is serviced regularly, and bi-weekly exchange of one-third of the water, fish can be kept at densities of about one fish per litre with little maintenance.

In large-scale recirculating systems, the water of many tanks is cleaned and regenerated through a common filter unit. Such systems therefore permit the raising and maintaining of very large numbers of fish in a comparatively small space. As hundreds, or more, tanks can be serviced by one filter, very high-quality water is provided with little maintenance. On the other hand, diseases can spread between the interconnected tanks in recirculation systems. Sterilizing the water by UV irradiation before redistribution can reduce this problem.

A large-scale recirculation system suited for genetic screening purposes was developed for the 1993 Tübingen mutagenesis screen. This system has proved to be very reliable and effective, and is therefore described here in detail (*Figure 1*). Smaller-scale versions of this system, with minor modifications, are now in use in many zebrafish laboratories throughout the world, including the authors' laboratories. The system was built by an experienced company (Schwarz, Germany), but similar systems are now also available from other companies (e.g. Müller & Pfleger, Germany or Marine Biotech Inc., USA).

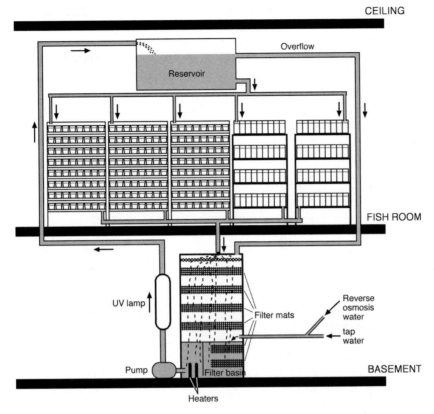

Figure 1 Large-scale recirculation system. The upper room contains the aquaria (serial tanks, mouse cages, and/or single boxes). Water is distributed to the aquaria by gravity flow from a common reservoir located above the aquaria. The water from the aquaria is collected in a common pipe and recycled through a large filter unit in the room below. From the filter basin, the water is pumped back up into the reservoir. On its way, it passes through a UV filter unit.

For the purpose of an inbreeding genetic screen, a large number of relatively small fish containers is needed. Two types of aquaria are in use, which will be described below: serial tanks and overflow containers (see Sections 2.2.2 and 2.2.3). The water from a room holding between 200 and 1500 aquaria is collected and recycled through a big common filter. From the filter basin, the water is pumped up into a large reservoir located above the racks of aquaria in the fish room (*Figure 1*). From the reservoir it is distributed to the aquaria by gravity flow. The flow rate can be adjusted for each row of aquaria. For adult fish, an exchange rate of about three tank volumes per hour is recommended. This high flow rate is the major determinant to allow the fish to be kept at high densities. The water is kept in constant motion and debris is generally automatically flushed out of the tanks, reducing the necessity of regular cleaning.

Such systems can also be divided into several smaller units with individual filters, e.g. one for each rack of fish containers. This can be done to further reduce the possibility of spreading diseases. However, each filter unit has to be serviced individually, thus increasing the maintenance work. Stand-alone units with just a single rack of fish containers and an integrated biological filter, pump, UV sterilizer, and particle filter, etc. are also available (*Figure 2*; e.g. from Marine Biotech Inc. or Schwarz). Several different sizes of fish containers can be used in such units and several units can be combined into a bigger system.

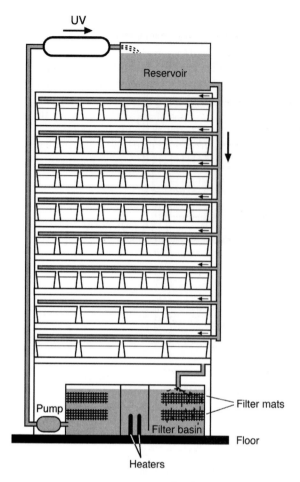

Figure 2 Stand-alone unit. A single rack of fish containers (shown here for single boxes and mouse cages) is equipped with an integrated water reservoir, a biological and particle filter, a pump, and a UV sterilizer.

2.2 Aquaria

2.2.1 Tanks

Aquaria should be made of high-quality glass to minimize the risk of the glass breaking. As zebrafish are surface-living fish, the height of the aquaria does not have to exceed 25 cm, provided the tanks have a relatively large surface area. Aquaria made of plastic may also be used; however, they do not keep as clean as glass tanks and get scratched easily over time. Commercial mouse or rat cages (Ehret, Germany) are available in sizes of 2.5, 5, and 12 litres and also make good fish containers. Lids should be made of clear plastic and fit very tightly because fish can escape through even relatively small slits. It is advisable to use extra strong tape to secure one end of the lid to the container. Lids should have a small hole, no larger than 1 cm in diameter, to allow feeding with a squirt bottle without having to open the lid. The tubing carrying the water, outlets, etc. should be non-transparent silicone or green polypropylene, in order to avoid algal growth.

In large-scale recirculating systems, two types of aquaria are in use: serial tanks and overflow containers.

2.2.2 Serial tanks

Serial tanks are long glass aquaria that are divided transversely by partitions to provide a number of individual compartments for keeping adult fish (*Figure 3a*). The aquaria are about 120 cm long, 60 cm deep, and 22 cm high. They are subdivided by glass plates that leave a 1.5 mm slit at the bottom, resulting in a series of 8–10 interconnected compartments of about 12 litres each. This volume is just right to keep one family of 60–70 fish. Each compartment has a separate lid. Water in a row of tanks flows into the compartment at one end and runs through the slits from one tank to the next, carrying dirt particles with it. The water is drained from the last compartment at the other end of the row by an overflow pipe, which

(a) SERIAL TANK

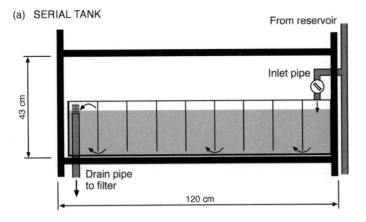

(b) OVERFLOW CONTAINER

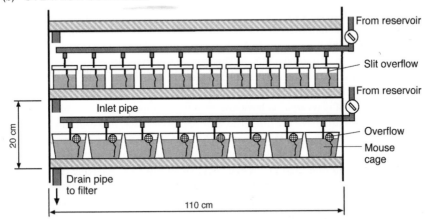

Figure 3 (a) A row of serial tanks. Serial tanks consist of a long glass aquarium which is divided into individual compartments by glass partitions. The glass partitions leave a 1.5 mm slit at the bottom, resulting in a series of interconnected compartments. Water in a row of tanks flows into the compartment at one end and runs through the slits from one tank to the next. It is drained from the last compartment by an overflow pipe. The overflow is equipped with a grid to prevent fish from escaping. (b) Overflow containers. Individual containers (shown here as single boxes and mouse cages) are placed on to a rimmed glass shelf. Water enters into each container through a small outlet from a pipe that runs along each shelf. From each container, it is drained on to the shelf through an overflow. From the shelf it is collected by a common drainage pipe that takes the water to the filter unit.

determines the water level. The overflow must be equipped with a grid (we use hair curlers) to keep fish from escaping from the tanks. The advantage of this type of tank is that it has only one inlet and one outlet for 12 compartments, which is cost effective and also reduces the time that has to be spent checking the water flow. Moreover, the water passes through the serial compartments at the bottom, thus flushing out any dirt that has sunk down. The system works only for juvenile or adult fish that are large enough not to pass through the slits. We also use small serial tank units, in which each compartment holds about 2.5 litres, to keep identified carrier fish for long periods of time, with hardly any maintenance work needed.

2.2.3 Overflow containers

In this system, individual containers are placed in a row on to a rimmed glass shelf. Water enters into each container through a little outlet and exits through an overflow on to the shelf, from where it is drained into the communal filter (*Figure 3b*). We mostly use 2.5-litre mouse cages for raising larvae and 1-litre plastic boxes for keeping single fish in this set-up.

Mouse cages are equipped with an outflow at the top margin. This is a hole of 3.5 cm diameter into which a grid is inserted. Grids of three different mesh sizes (0.3 mm, 1 mm, and 2 mm), depending on the age of the fish, are used to prevent fish from escaping through the overflow. For 1-litre boxes, a 2-cm long slit with a width of about 2 mm is cut into the top edge of the box. The water enters into these containers via a common pipe with a series of outlets equipped with silicone tubing. Cut-off Eppendorf pipette tips are stuck into the ends of the silicone tubes. The outlets are spaced such that each container is supplied individually with running water (*Figure 3b*). The water flow in these outlets does not have to be adjustable individually, as the pressure resulting from the reservoir located above the shelves is sufficient to provide an even flow of water.

At one end of the shelf, the water from all containers on the shelf is collected and drained into the filter system. The system holds several sizes of containers and is therefore very versatile. For example, rat cages with a volume of 12 litres can be used to raise families of fish. Mouse cages or rat cages may be further subdivided by plastic partitions with slits on the bottom, like those in the serial tank system, to ensure a flow of water from one compartment to the other. Compared to small boxes for individual fish, the 'hybrid' compartments stay cleaner and work well for small numbers of fish. This type of container also ensures a more efficient use of space. For instance, when males and females of the same genotype need to be kept separated, they can still be kept in the same mouse cage.

Overflow systems made of glass aquaria have also recently become available (Schwarz or Müller & Pfleger). These aquaria have the dimensions of the serial tanks; however, the compartments are completely separated from each other. A slit at the back allows the water to enter a drainage canal running along this side. Each tank has an individual water inlet at the front. At the overflow a small filter mat keeps even very small fish from escaping. Juvenile fish (about 1 month old) can be transferred into these tanks. Although this arrangement has the advantage of complete separation of individual containers, it is not as easy to keep clean as the serial tanks. Moreover, it is more expensive to set up and service.

3 Water

3.1 Fresh fish water

The mineral content and buffering capacity of the water is important for the health and fecundity of the fish. In general, elevated salt concentrations have a positive effect on fish health because they reduce bacterial growth and the pH is more stable when the quality of the water changes. However, lower salt concentrations promote fecundity of the fish.

3.1.1 Tap water and charcoal filtering

In many laboratories, tap water can be used without further processing. However, as a minimal precaution, the water should be passed through a charcoal filter of an appropriate capacity in order to remove possible organic contaminants (e.g. pesticides) and chlorine.

The capacity of the charcoal depends on the individual brand of charcoal. It is usually given as the capacity to absorb a certain concentration of a compound to be removed (e.g. chlorine) and the total amount of water that can be filtered per volume element of charcoal. From these numbers and the amount of water entering the system per day, the total lifetime of the charcoal filter cartridge can be calculated.

3.1.2 Diluted tap water

If the water is too rich in calcium carbonate, it should be mixed with deionized water. Tübingen tap water, for instance, is rich in calcium (73.3 mg/l) and hydrogen carbonate (195 mg/l), and has a correspondingly high electrical conductivity (approximately 700 μS) and a pH of 8 or higher. The tap water is mixed with reverse osmosis water, to give a conductivity of about 300 μS and a pH of 7.5. This seems to be a good compromise between sufficiently low conductivity, which stimulates egg laying, and a sufficiently high carbonate buffering capacity at an acceptable pH.

3.1.3 Conditioned water

In most areas, it is advisable to desalt the water completely using reverse osmosis. Following this treatment, salts must be added such that the water has a pH of about 7 and a conductivity of approximately 180 μS.

It is essential to keep records of the fish water quality in the fish facility. The different parameters should be measured at least once a week.

Protocol 1

Conditioned water

Equipment and reagents

- Reverse osmosis system
- Thermometer
- pH meter
- Conductivity meter (1–1000 μS range)
- Nitrite dip sticks (e.g. Aqua Lab III Nitrite Test Strips, That Fish Place, USA)
- Chlorine test strips (e.g. Aquaculture Supply, USA)
- Sodium bicarbonate
- Sea salt (Instant Ocean, Aquaculture Supply, USA)
- Calcium sulphate

Method

1 Run tap water through charcoal filters and subsequently through a reverse osmosis system.

2 To generate conditioned water, add per 1000 litres of reverse osmosis water: 75 g sodium bicarbonate, 18 g sea salt, 8.4 g calcium sulphate.

3 10 to 20% of the total fish water volume should be exchanged with conditioned water on a daily basis.

4 Measure values daily and keep records:

   (a) Temperatures: 26–28 °C (water), 27–29 °C (air) to prevent condensation of water.

   (b) pH: about 6.8–7.5, not lower than 6, not higher than 8 (use bicarbonate or HCl, respectively, to adjust pH).

(c) Conductivity: 180–350 µS.

(d) Nitrate/nitrite: nitrite should not be detectable with dip sticks. Moderate levels of nitrate (between 100 and 200 mg/l) are easily tolerated by the fish. If necessary, increase the amount of conditioned fish water added daily to reduce the levels of nitrate and nitrite.

(e) Chlorine: should not be detectable with dip sticks. Caution: Dip sticks are not very useful as the detection levels are already above those toxic for young fish larvae.

3.1.4 Problems with the water

If the water does not have a sufficient buffering capacity (or due to other imbalances), the pH can drop below 6. As a low pH has toxic effects, bicarbonate should be added to the water to increase the pH. If the biological filters are not yet working, or if their capacity is too low, the pH may rise to values above 8. This prevents the growth of denitrifying bacteria in the filter and causes a build-up of toxic ammonium compounds in the fish water. In combination with excessive feeding, this may cause the water to become turbid due to growth of other bacterial species. Frequent water changes and adjustments of the pH by careful addition of HCl can solve the problem.

All biological filters require a start-up phase until the denitrifying bacteria have colonized the filter mats. Usually this takes at least several weeks. This period can be shortened if the filter mats are inoculated with a sample of water from an established biological filter. Alternatively, preparations of denitrifying bacteria are available in aquarium stores (e.g. from Tetra Germany). During the start-up phase, the water should be checked daily for ammonium compounds (with Merck test strips). The levels of nitrite should be very low or undetectable. If they rise, the water should be exchanged immediately. It is wise to populate a new fish facility gradually over a period of 2 months to allow the biological filters to equilibrate.

Some researchers also aerate the water. With a dripping filter such as those used in the large-scale recirculating systems (*Figure 1*) this is not necessary. However, it should be kept in mind that the denitrifying bacteria require oxygen for their task, therefore when filters are placed in individual tanks, aeration of the filters improves the water quality. If fish water is prepared from reverse osmosis water, or after charcoal filtering, aeration also is required as the charcoal absorbs oxygen.

Care should be taken to ensure that the materials used in setting up the aquarium systems, such as pipes, plastic connections, tubing, siphons, pumps, etc., do not leak toxic compounds into the water. If contamination is suspected, charcoal filters may help to remove toxic compounds.

3.2 Filtering

3.2.1 Biological filters

The simplest commercial filters consist essentially of a box holding gravel, foam, or other material with a large surface area, which is placed into the aquarium. Bubbling air through the filter causes a constant stream of water. More elaborate filters are placed outside the aquarium and water from the aquarium is pumped and recycled using appropriate connecting tubing. Small-scale filters of the type available in every aquarium store will do for individual tanks.

In large-scale recirculating systems, the water is filtered collectively through biological filters with filter mats (approximately 1000 litres of water per m² of foamed plastic with a 10–30 ppi pore size, 3 cm thickness). We have found that such filter mats are more effective in denitrification than several other commercially available filter materials. The filter mats act as coarse mechanical filters, but most importantly they provide a large surface area for aerobic denitrifying bacteria (*Nitrosomonas* and *Nitrobacter*, see ref. 1), which degrade ammonium compounds to nitrite and finally nitrate. Ammonium and nitrite are very toxic to the fish even at low levels, whereas nitrate is tolerable below about 200 mg/l. Since the surface area of the filters must be large, mats of foamed plastic are arranged in a series of shelves, with increasing density of foam, through which the water drips into the filter basin. The filter basin contains additional foam mats. A top layer with disposable synthetic filter material can be used to collect coarse debris. This should be checked twice a week for clogging and exchanged if necessary. The filter mats themselves are cleaned about once a month by rinsing them in clean water or in a commercial washing machine without adding any washing powder. Mat cleaning should be scheduled such that no more than half of the mats from a given filter are washed at the same time.

The production of nitrate leads to a lowering of the pH. Normally 5% of the water is exchanged with fresh water on a daily basis. This is sufficient to keep the nitrate at a tolerable level. Water is exchanged using a simple timer-operated valve available in gardening supply stores. If the pH drops and/or if the nitrate increase is rapid (e.g. caused by excessive feeding), more fresh water has to be added. In addition, the filter basins contain a device (a float like the ones used in toilet tanks) that causes automatic refilling with fresh water in case the water level drops below a set limit. Filtered water is pumped from the bottom reservoir up into the top reservoir that is located above the aquaria (*Figure 1*). There are two pumps per filter unit that run on alternate 12-hour cycles. In the case of one breaking down, the other one automatically takes over. The top reservoir has an overflow pipe, allowing water to flow into the bottom reservoir to prevent spillage of water from the top reservoir in case it is too full.

3.2.2 Mechanical filtering and UV sterilization

The biological filters serve as coarse mechanical filters and to degrade organic compounds biologically at the same time. However, water that has passed through these filters still contains a significant amount of small debris that can cause clogging, particularly of the fine water outlets in the mouse-cage system. The amount of debris depends on the amount of food brought into the system and general cleanliness of the system. Some debris collects in the glass trays of the mouse-cage system, which should be removed with a commercial water vacuum cleaner. Additional filter devices can be installed to suppress clogging problems if necessary, e.g. a sand filter or a pleated cartridge filter (Aquaculture Supply, Inc.).

Installation of UV lamps in the circulation is advisable, because the UV radiation significantly reduces microbial growth and hence the transmission of diseases. For different microorganisms, the lethal dose varies over a 100-fold range. In order to be effective, UV lamps should deliver a dose of at least 10 000 W/s/cm^2, which kills 99% of all *Mycobacterium tuberculosis*, a major fish pathogen. The lamps need to be cleaned and exchanged on a regular basis, e.g. about once a year, depending on the individual bulbs used and their run time.

4 Room conditions and maintenance

4.1 Room conditions

A start-up facility does not require much space, e.g. 30 m^2 for about 1000 containers, including bench and shelf space for setting up crosses. The major requirements for a suitable room are that it should have a water supply line, ideally an additional one with reverse osmosis water, a sink, a floor drain, several power lines and programmable room light. Care should be taken that the floor construction is suited to support the weight of the tanks. All furniture and equipment in the room (racks, shelves, table tops, etc.) should be made of material that tolerates the high humidity and the frequent exposure to water.

In large facilities, the filters and reverse osmosis may be in a separate room below the fish rooms (*Figure 1*). However, in smaller facilities, it is often more convenient to place them in the same room with the aquaria, e.g. underneath one of the racks of tanks (*Figure 2*).

4.2 Temperature

Zebrafish require water temperatures ranging between 25 °C and 28 °C. We normally adjust the temperature to around 26 °C, using several heaters placed into the filter basin. The room temperature should be set slightly higher (e.g. at 27 °C), which prevents condensation of water and growth of mould on the walls of the rooms. Higher temperatures are uncomfortable for people working in the fish rooms and might also reduce the life span of the fish. Further, the higher the temperature, the lower the oxygen content of the water. Since there are several heaters for each basin, none of which on its own is sufficient to heat up above tolerable values, there is no danger of overheating. Temperatures dropping to room temperature by failure of the heaters are not dangerous for the fish.

4.3 Illumination

Zebrafish are usually kept on a 14-hour light–10-hour dark cycle, using appropriate timers. For convenience, lamps may be installed above every row of tanks, although a brighter illumination will also result in increased growth of algae. Although fish also do well in dim light, bright illumination of the room is generally desirable because it allows researchers to work more easily. Red-light lamps may also be installed in the fish rooms, allowing researchers to perform tasks while the fish are 'sleeping'.

Zebrafish spawn when the light is turned on in the morning. If embryos at particular stages of development are needed at specific times of the day, the parent fish should be kept on a shifted time schedule in special light cabinets. Light cabinets consist of a normal tank system within a light-tight box. This box is equipped with a programmable light source. Having several such cabinets allows the generation of embryos of many different developmental stages in parallel, e.g. for transplantation/injection purposes. Feeding at inconvenient times can be done by cheap automatic feeding devices that are available in all aquarium stores.

4.4 Cleaning

The cleanliness of the aquaria and filters is a most important feature of keeping fish in healthy breeding conditions. To avoid the spreading of diseases, all containers and tools with which the fish may come into contact should be kept clean. Debris that collects at the bottom of tanks should be siphoned out. The overflow containers stay clean only if there are enough fish to sufficiently stir up the water so that the dirt is carried out of the overflow. With only a few adult fish in a large volume and careless feeding, the containers may get very dirty and have to be cleaned regularly. Letting the water run at higher speed does help to flush out dirt, but there is a limit as the reservoirs may run dry. In the serial tanks, dirt often collects in the last compartment, which should be cleaned regularly.

In general, there are few mechanical problems with serial tanks. However, clogging of the outlets, grids, and slits can occur in the overflow containers. It is therefore necessary to check that the water is running in all aquaria on a daily basis. If blocked, the outlets and grids have to be cleaned (e.g. with a toothbrush), as fish being kept at high densities may die overnight if the water is not running. Clogging may be further suppressed by mechanical filtering devices, such as sand filters or removable pleated cartridge filters.

Another problem may be algae that grow on the walls of the tanks, which should be removed at regular intervals. Razor blades or a scrubbing sponge (Scotch brite) are suitable for this task. As a precaution against algal growth, many researchers keep fish together with snails. Florida freshwater snails (*Planorbella* spp.) clean the walls of algae and eat any surplus food. In general, they have a very positive effect on the water quality and also reduce turbidity that might occur at times of heavy feeding. Adult snails are sometimes killed and baby snails are eaten by adult zebrafish, so it is necessary to regularly resupply snails to the fish tanks. If one or two adult snails are added to a tank of larval fish, snails propagate readily. Even though snails introduce a small amount of extra work, they are very helpful in keeping the tanks clean. A tank without snails is easily spotted due to the fact that its walls are overgrown by algae and the fish inside can no longer be seen. Care must be taken that snails are introduced from disease-free colonies, otherwise they may be sources of infection. Snail spawn may be bleached using the recipe described for fish embryos (see *Protocol 5*).

All containers and aquaria should be cleaned thoroughly before resupplying them with live fish. For small boxes, such as mating containers, a dishwasher is useful. Care must be taken to rinse away all washing soap before putting fish into a container. For mouse cages and other overflow containers, scrubbing is necessary, as commercial mouse-cage washing machines are not sufficient. Mouse cages may be autoclaved or dipped in disinfectant (see *Protocol 12*). The interconnected serial tanks should be disinfected before they are filled with a new batch of young fish to be grown up. We use hydrogen peroxide as a disinfectant.

Protocol 2
Tank disinfection

Reagents

- 5% acetic acid solution: dilute 100 ml of acetic acid in 2 litres of water
- Hydrogen peroxide/sodium hydroxide: add 20 ml of 10% NaOH to 200 ml of 30% hydrogen peroxide and fill to 2 litres with tap water. Final concentration is 3% hydrogen peroxide in 0.1% NaOH

Method

1 Clean the aquaria: remove snails and other solid waste and brush the walls of the tank to remove the worst of the dirt using the water that is in the tanks. Remove the outflow tube to allow water to empty into the system. Allow the tanks to dry at this stage for at least 24 h, as this is thought to reduce the number of infectious bacteria.

2 Replace overflow pipe firmly into outflow. From this point on until the end it is essential that no water/cleaning solution enters the system. Please make sure that the pipe is placed firmly into the outflow and that the water level never exceeds the height of this pipe.

3 Make 2 litres of 5% acetic acid solution. **Do not pour into tanks.** Instead soak a sponge and wipe the walls and floor with acetic acid solution. Wipe above the usual water mark to remove chalk deposits. When all tanks have been wiped clean, rinse with cold tap water using a hose. Be sure to spray down the walls. Siphon water into the drain (**not into the system**) using plastic tubing. Alternatively, a water vacuum cleaner can be used to suck out the liquid. Any remaining acid will be neutralized by the next step.

4 3% hydrogen peroxide in 0.1% NaOH. Once again, soak a sponge in the solution and use it to wipe **all inside surfaces** of the tanks. Wipe down the tanks twice with this solution before rinsing.

Protocol 2 continued

Do not get this solution on the skin (it burns) or clothes (it bleaches the colour). Wear rubber gloves. If it comes into contact with skin, the NaOH can be neutralized using a weak acid solution such as 5% acetic acid or tap water.

5   Rinse the tanks carefully using cold tap water through a hose until the tanks are two-thirds full. Spray the walls down with water at high pressure to ensure that they are properly rinsed. Siphon water out of tanks and empty the remainder using a vacuum cleaner if necessary. **Repeat** this rinse step **3 times** to ensure removal of all wash solution.

6   Start the flow of system water and allow the water to run overnight before adding fish. It is recommended that the tanks are tested first by placing some 'tester' fish in them for a day or two before filling all of the tanks with fish.

Nets are sterilized by placing them into a pot of boiling water for at least 20 min. For each fish collection, a separate net should be used.

Used plastic Petri dishes can be recycled by collecting them in a large cylindrical container filled with tap water and bleach (50 ml bleach in 50 litres of water). After bleaching, the dishes should be rinsed thoroughly with tap water and dried at 50 °C.

4.5 Safety

When designing an aquarium system, it is important to minimize the number of maintenance and control items. In the systems described above, a number of safety considerations have been observed. In case of a breakdown of the heating control, the water cannot be overheated, and dropping to room temperature is no problem for the fish. There are two pumps per filter unit, which automatically replace each other in case of a breakdown. If both pumps stop working at the same time, the room will be flooded, but the tanks cannot run dry. The tanks are constructed in a such way that they stay relatively clean without extra effort. Regular control of the pH and nitrate levels of the water has to be carried out, and the outlets and inlets of the overflow containers have to be checked on a daily basis. With controls taking place regularly, only rare coincidences will result in a disaster. Alarm systems can be installed for pumps, temperature, the water levels in the reservoirs above the aquaria, or the reservoir tank for deionized water, as well as for the ventilation of the rooms. For smaller facilities, regular inspections are quite sufficient, although an alarm system for the water level in the reservoirs above the aquaria is recommended.

5 Fish care

5.1 Keeping adult fish

5.1.1 Fish densities

The maximum fish density in the aquaria depends on the efficiency of the filters, the age of the fish, and the amount of feeding. As mentioned above, in systems without filtering, the fish must be kept at low densities to keep them in good breeding condition. For fish kept in quarantine, it is not convenient to use filters. Instead, they can be kept with complete daily exchanges of water and moderate feeding at densities of about three fish per litre in individual containers. As mentioned above, pairs of fish or single fish can be kept without feeding and water changes in 1-litre boxes for up to 10 days.

In large-scale recirculating systems, families of sibling adult fish are kept in serial tanks at densities of five adult fish per litre (60 fish/12 litres). They are raised at densities of up to 60 fish in 2.5-litre mouse cages in the overflow system. Adolescent fish should be transferred into larger tanks when they are big enough not to pass through the 1.5mm slits, e.g. after about 2–3 months, otherwise they might be retarded in their growth. The fish do well at these high densities and the water in the tanks remains clean as it is kept in constant motion and dirt is carried out via the slits and the overflow.

Zebrafish tend to be aggressive if few fish are kept together in small volumes of water. This can be compensated by adding plastic 'grass' (Tetra) to the containers to give them an opportunity to hide from one another.

Fish that are no longer kept are anaesthetized and killed by placing them on ice.

5.1.2 Feeding

Beyond about 1 month of age, zebrafish can be fed with standard dry food flakes. It is advisable to use different kinds of flakes, as they might differ in quality and nutritional value. A typical feeding regimen is to feed adult fish tanks twice a day (once on weekend days). Dry food can be mixed into water and fed with a laboratory squirt bottle. When feeding, it is important to take the number of fish in a tank into account and not to overfeed the fish. It is good practice to check whether all food has been eaten within about 10min. Feeding too much can spoil the water, because rotting food is a burden for the biological filters and a breeding ground for bacteria. Adult fish that have to be kept for longer periods of time without breeding require very little feeding (e.g. twice a week, preferably live food). Two weeks of rich feeding will bring them back into breeding condition again.

Dry food alone is not sufficient to keep fish in good breeding conditions. Therefore it is necessary to supplement it with live or frozen food. The most commonly used additional live food is *Artemia nauplia*. These should be fed at least twice a week, or daily for 'high performance' breeding fish. Alternatively, or in addition to *Artemia*, *Drosophila* larvae or different types of frozen food that are available from aquaculture supply stores can be used. Live or frozen food (e.g. tubifex, *Daphnia* and *Chironomus* larvae) that has been harvested from freshwater systems that also harbour fish, should be avoided, as it may be a source of pathogens. On the other hand, salt-water-dwelling articulates are safe (e.g. frozen adult *Artemia* and krill).

5.2 Raising fish

5.2.1 Collecting eggs

Zebrafish mate and spawn at dawn. In order to recover the eggs, it is necessary to provide a means of protecting them from being eaten by the parents. A traditional, simple means is to cover the bottom of the tank with a layer of glass marbles and to siphon out the eggs. Breeding traps, consisting essentially of two plastic containers that fit tightly into one another, are more convenient. The inner container holds the fish. Its base has been replaced by a grid, through which the eggs fall into the outer container where they are protected from the fish (*Figure 4*). Such breeding traps may be placed directly into the aquarium or, for more efficient egg collection, the fish are taken out of the tank and placed into these containers for mating. For the collection of eggs from groups of fish, e.g. for injection experiments, mating containers built from small-sized mouse cages are convenient. For controlled breeding, eggs are collected from single-pair matings.

5.2.2 Setting up pair crosses

Mating crosses are best set up in the afternoon or early evening using the same conditions (e.g. water quality, light–dark cycle) under which the fish are otherwise maintained. Mating containers are available in different sizes and designs. In our laboratories we use mating containers consisting of 1-litre acrylic boxes equipped with a removable inner container (sieve)

Mating box, side view

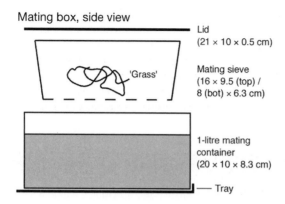

Lid
(21 × 10 × 0.5 cm)

Mating sieve
(16 × 9.5 (top) /
8 (bot) × 6.3 cm)

1-litre mating
container
(20 × 10 × 8.3 cm)

— Tray

Tray with five mating containers, front view

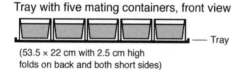

— Tray

(53.5 × 22 cm with 2.5 cm high
folds on back and both short sides)

Stack of 3 trays with 15 mating containers

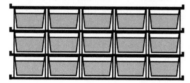

Figure 4 Mating boxes consist of two plastic containers that fit tightly into one another. The bottom of the inner container holding the fish has been replaced by a grid through which the eggs fall into the outer container. In order to prevent fish from jumping out of the mating boxes, the boxes are covered with a lid. Several mating boxes can be placed on a tray and several trays can be stacked to save bench space.

that holds the fish (*Figure 4*). A Rubber Maid food container with a hanging breeding trap (Aquaculture Supply, USA) is an alternative. When setting up pair maitings, we add a small bundle of green plastic 'grass' (Tetra), which serves as a 'barrier' between the two fish, to reduce aggression.

Zebrafish can jump, and it is therefore important to keep the mating containers covered at all times. Five breeding containers fit on to one tray with a bent-down front lip. The trays can be stacked, which allows convenient handling of large numbers of crosses and saves space in the setting-up area. Because the outer mating containers are made of clear plastic, mating boxes containing eggs can be spotted easily.

Usually, not all pairs that are set up will lay the next morning. About 12% will lay only on the second day. Average laying rates can be as high as 50% of the pairs set up, with a large variation between individual families.

5.2.3 Collecting, sorting, and bleaching of embryos

The containers are checked for successful matings before noon the following day. Mating containers in which eggs are visible on the bottom are labelled, the embryos are collected (by pouring the water from the outer container through a fine mesh sieve), and are transferred into a Petri dish in E3 medium (see *Protocol 3* and 4). A good clutch consists of between 70 and 300 eggs, of which at least 80% are fertilized. Unfertilized eggs provide a source of nutrients for growth of bacteria and mould, which may quickly spoil the water. It is therefore necessary to transfer the fertilized eggs into a new dish with fresh E3 medium soon after they are laid. To allow sufficient aeration, no more than 70 eggs should be kept in a 95-mm Petri dish, and dead embryos or empty chorions should be removed regularly using a disposable plastic pipette.

Protocol 3

Setting up pair matings, collecting and sorting embryos

Equipment

- Fish nets
- 90-mm Petri dishes
- Plastic tea strainer
- Tape
- Flexible plastic pipette

Mating containers (see also *Figure 4*)

A 1-litre acrylic box (Semadeni, Switzerland) is equipped with a removable mating sieve. The sieve is made from a tightly fitting Tupperware plastic container, the bottom of which has been replaced by a mesh of stainless steel (mesh size: 2 mm). There is a space of about 2 cm between the mesh net and the bottom of the outer box. Both inner and outer box are level at the top, to prevent fish from escaping into the outer box, and covered with a lid. This design allows the eggs to sink to the bottom but prevents the adults from eating them. The mating containers also contain some plastic grass, allowing fish to hide from each other (reduces aggression).

Method

1 Five mating containers are placed next to each other on a plastic tray and are filled with water from the system. When fish are netted from a tank, it is convenient to use specially made nets that fit tightly into the serial tanks. Care should be taken to avoid injuring the fish. Fish are released into a mouse cage filled with fish water and are transferred to the setting-up area. One female and one male each are caught using small nets and placed carefully into the mating boxes. Once the fish are inside, the boxes are covered with a lid or with another tray in order to prevent the fish from jumping out of the mating boxes (see *Figure 4*).

2 Stack the mating boxes from one F_1 stock as shown in *Figure 4*. Do not mix crosses from different stocks in the same tray or stacks of trays, in order to avoid mix-ups between stocks. Label each stack clearly with a label indicating the genotype and tank origin of the fish.

3 Check the mating containers for eggs late on the morning of the following day. For those crosses that laid, place two labels, indicating the genotypes of the parents, on to the mating container. In case of several crosses with the same parental genotypes, include a running number to distinguish each pair. Leave the crosses that did not lay and recheck them the next day.

4 The adults along with one label are transferred into a lidded 1-litre container filled with system water. This container can be connected to the water circulation system or it can be left disconnected for up to 10 days without feeding.

5 To collect embryos, place a 90-mm Petri dish in front of each mating container with eggs. Pour the water from the mating container through a plastic tea strainer. Using a squeeze bottle filled

Protocol 3 continued

with E3 medium, rinse the eggs from the tea strainer into the Petri dish and stick the second label with the genotypes of the parents onto the lid of the Petri dish.

6   Sort 60 embryos that develop normally from the dish using a flexible plastic pipette under a dissecting scope and transfer them into a fresh 90-mm Petri dish which is approximately ¾ filled with E3. Do not overfill such that the lid contacts the E3, as this might cause insufficient gas exchange between the E3 and the surrounding air and hatched embryos might get stuck to the lid. A fresh Petri dish ensures a minimum of contamination and the best survival rates for the embryos. Be aware that embryos can get stuck in your pipette, so check pipettes when starting to sort a new dish. Sorting eggs a few hours after they were laid facilitates the distinction between fertilized and unfertilized eggs.

7   The dishes with the embryos are kept in an incubator at 28 °C.

Protocol 4

Embryo media

Reagents

- NaCl
- KCl
- CaCl$_2$
- MgSO$_4$
- Methylene Blue

- KH$_2$PO$_4$
- Na$_2$HPO$_4$
- NaHCO$_3$
- Penicillin
- Streptomycin

E3 (for standard work with embryos)

1   5 mM NaCl, 0.17 mM KCl, 0.33 mM CaCl$_2$, 0.33 mM MgSO$_4$, 10^{-5} % Methylene Blue.

2   E3 can be made up as a 60× stock (without Methylene Blue). The 1× medium keeps under non-sterile conditions at room temperature for over a week.

3   For 10 litres of a 60× stock solution, mix 172 g NaCl, 7.6 g KCl, 29 g CaCl$_2$.2H$_2$O and 49 g MgSO$_4$.7H$_2$O. Store stock in fridge.

4   Dilute 160 ml of 60× stock in distilled H$_2$O (dH$_2$O) to make up 10 litres of E3 (1×), and add 30 ml of 0.01% Methylene Blue as a fungicide.

E2 Pen/Strep (to raise stocks that are not very healthy)

1   17.5 g NaCl, 0.75 g KCl, 2.4 g MgSO$_4$, 0.41 g KH$_2$PO$_4$, 0.12 g Na$_2$HPO$_4$ to 1 litre of water to make a 20× E2 stock. Autoclave. Add 7.25 g CaCl$_2$ to 100 ml of water and autoclave. Add 3 g NaHCO$_3$ to 100 ml of water and autoclave. Keep stocks refrigerated.

2   To make 1 litre of 1×E2, combine 50 ml 20× E2 stock, 2 ml of the CaCl$_2$ solution, 2 ml of the NaHCO$_3$ solution and add distilled water to 1 litre.

3   E2 Pen/Strep: make up a 60 mg/ml penicillin, 100 mg/ml streptomycin stock, aliquot and store at −20 °C. Dilute 1:500 in 1× E2 for use.

In order to decrease the risk of spreading diseases, we routinely treat eggs to be raised with bleach. This should be done after epiboly is completed, but before 28 hours of development. After 28 hours of development, bleaching can damage the embryo, since the chorion is already partially degraded by hatching enzymes. As the bleaching procedure interferes with hatching of the embryos, a small amount of pronase has to be added after bleaching to ensure hatching.

Protocol 5

Bleaching embryos for raising

Equipment and reagents

- 70% ethanol
- 5 plastic containers (each approximately 1 litre)
- Flexible plastic pipettes
- Plastic tea strainers (sterile)
- Timer

- 90-mm Petri dishes
- 10–13% NaOCl solution (Sigma)
- E3 medium
- Pronase (Roche Molecular Biology, 30 mg/ml in E3; aliquot and store at −20 °C)

Method

Embryos should be bleached after tail bud stage (approximately 10 hpf) and before they have reached 28 hpf (hours post-fertilization).

1 Clean the bleaching area (with 70% ethanol).

2 Set up five containers, in the following order: bleach bath 1, water bath 1, bleach bath 2, water bath 2, water bath 3. Prepare two bleach baths by adding 380 µl NaOCl per 1 litre E3 (from a 10–13% NaOCl solution, Sigma). Use tap water.

3 Transfer embryos to clean tea strainers and stick corresponding label with genotype on to the handle of the tea strainer.

4 Transfer the tea strainer with the embryos to the first bleach bath and incubate for 5 minutes (make sure embryos are submerged in the bath).

5 Transfer the tea strainer with the embryos successively into the following baths, submerge and incubate for 5 min each.

6 Wash the embryos into a new Petri dish with E3 medium.

7 Transfer the labels from the tea strainer handles to the lids of the corresponding Petri dishes.

8 Add 10 µl pronase (30 mg/ml in E3) to each Petri dish and place dish into a 28 °C incubator. Embryos will not hatch without addition of pronase!

9 Next day: check that all embryos have hatched and help them out with clean forceps if necessary.

After hatching, the larvae remain in a Petri dish until day 5 of development. At this stage most of the larvae have inflated their swim bladders (except for incrosses, where 25% of the embryos display the mutant phenotype, which very often is associated with non-inflated swim bladders). To achieve the highest survival rate and a homogeneously growing population of fish, it is important to transfer and feed the larvae within 48 hours after their swim bladders are inflated.

5.2.4 Raising of zebrafish larvae

During the first period of feeding, (day 6 to day 15), there is no water exchange and 60 fish larvae are kept in about 1 litre of water. At about 2 weeks of age, water is added to 2 litres and exchanged continuously by connecting the tank to the water system. At the age of 2 months the fish are transferred to the serial tanks. They will be ready to breed at the age of 3–4 months. The critical period for raising larvae is the first period, during which the water quality and the feeding regimen have to be very carefully controlled.

Protocol 6

Raising larvae

Equipment and reagents

- Mouse cages with small grids
- Fish water
- Sea salt (optional)
- Tetra-AZ 200 (Tetra Germany) or Fry food Kyona A 250 (keep stock closely sealed in a refrigerator, aliquot in the fish room; food may clog when it gets moist)
- Food dispenser (e.g. plastic baby spoon, bottom replaced with stainless steel mesh and lid)

Method

1 Within 24–48 hours after the larvae inflate their swim bladders, feeding should be started (around days 5–8 of development). Remove remaining chorions and any dead or abnormally developed embryos under a dissecting microscope.

2 Fill mouse cages with 1 litre of fish water. If conditioned water prepared by reverse osmosis is used, the water must be aerated. A 20-litre container is filled with fish water and bubbled with air using a simple aquarium pump for 24 hours before being used. Some laboratories also add salts to the water used for raising fish larvae (e.g. Coral Reef Red Sea Salt, final concentration: 3 g/l). Salt is thought to inhibit bacterial growth and increases the buffering capacity.

3 Slowly and carefully pour the larvae into the mouse cage and put the lid on. Transfer the label from the Petri dish to the mouse cage. At this stage, do not yet connect the mouse cages to the fish water supply.

4 Use baby dry food to feed the larvae twice a day from day 6 to day 16. Take a little bit of powder with the fingertips and sprinkle it on to the water surface, or use a food dispenser as described above. Fill the dispenser with powder and cover tightly. A slight tap at the rim of the mouse cage should release just the right amount of powder. There is no need to cover the entire surface with food, the fish larvae will find it. Apply the powder such that most of it stays on the surface and does not sink down. This is necessary as the larvae only feed on the dry food provided at the surface. Within minutes, the larvae will start eating the food from the surface. Dry food that sinks to the bottom is not consumed by the fish and diminishes the water quality, which can reduce the survival rate dramatically. Caution: Do not overfeed the larvae at this stage. As the containers are not yet connected to the water supply, rotting excess food will cause a rapid deterioration of the water quality. On the other hand, underfeeding may be just as harmful, as baby fish easily die of starvation or are retarded in their growth.

5 If fed appropriately, there is no need to add or exchange water until day 16. Occasionally a layer of scum can form on the surface, preventing the larvae from reaching the food. Carefully remove the scum using a brush. With time, and appropriate feeding, the water gets cloudy but this is no cause for alarm.

Protocol 6 continued

6 Start feeding babies with *Artemia* at day 16. Once the babies are big enough to be fed *Artemia*, the mouse cages have to be connected to the water system. Insert a grid with the smallest mesh size into the outlet hole. Check the inflow tip for dirt.

7 Adjust the dripping rate of the incoming water to 1–2 drops per second for the next 7 days. The 2-week-old fish are tiny at this point and can easily be flushed out of the mouse cage if the water flow rate is too high. The larvae should never be swept around by the water current in the mouse cage.

8 Start feeding small amounts of *Artemia* to the larvae and check 5 minutes later if they have eaten all the *Artemia* (see *Protocol 8* for *Artemia* production). Make sure the *Artemia* being fed are freshly hatched and move around. Small fish do not eat dead *Artemia*, which rapidly sink to the bottom of the mouse cage. Only after the fish have finished eating all *Artemia*, add more *Artemia* and check again 5 minutes later. Avoid adding too many *Artemia* at a time as they may clog up the mouse-cage outflow grid. Well-fed larvae have a round red belly after each meal. Feed *Artemia* every day, three times a day at approximately 4-hourly intervals. Don't add snails to the mouse cages as long as the fish are still very young (optional).

9 Check the flow rate daily for the first 3 days and check daily for clogged overflow nets. As the fish grow, replace the outflow nets with nets of a larger mesh size and increase the amount of *Artemia* being fed. At the age of about 5 weeks, start feeding dry food in addition to *Artemia*. As the fish reach 2 months of age, they need about 5–10 times more *Artemia* than they were fed on day 16. Once fish reach 2–3 months of age, they should be transferred into 12-litre aquariums. They should reach sexual maturity within the next month.

The best food for baby fish is live infusoria, e.g. paramecia, or rotifers. Suspensions of the collected and washed infusoria are added to the water in which the fish swim.

Protocol 7

Raising infusoria

A number of live food sources for fish larvae younger than 3 weeks have been reported, including paramecia, *Tetrahymena*, and rotifers. An incomplete list of references for various protocols with detailed instructions for raising paramecia, *Tetrahymena*, and rotifers is given below.

A. Rotifers
Reagents

- Culture Selco food (CS, Inve Aquaculture, Belgium)
- Sea salt (e.g. Instant Ocean or Forty Fathoms, Aquaculture Supply, USA)
- 2.5% glutaraldehyde solution
- *Brachionus* spp. (Drs P. Dhert and P. Sorgeloos, University of Ghent, Belgium)

Method

1 Grow *Brachionus* spp. rotifers in 25- or 50-litre round plastic tanks with 15 g/l sea salt in deionized water at about 25 °C with slow-bubbling aeration. Rotifers are inoculated at a density of 100–150 rotifers/ml and are fed on instant Culture Selco (CS) food.

2   Depending on the density of the culture, feed increasing amounts of CS: 6 g when set up and approximately 6, 7, 8.5, 10, 11, and 12 g, respectively, on subsequent days.

3   After 7 days the culture has over 500 rotifers/ml and is harvested through a fine net. The rotifers are resuspended in 14 litres of sea water (15 g/l sea salt), of which about a quarter is used to re-inoculate the culture.

4   The remainder can be stored in the refrigerator for up to a week and is used for feeding fish larvae (2 ×12 ml/mouse cage/day). The rotifer density is determined by withdrawing a few millilitres with a pipette from the medial water column of the tank, of which 10 aliquots of 100 μl are spotted on to a Petri dish. After adding 5 μl of a 2.5% glutaraldehyde solution to immobilize the rotifers, they can be counted and the density can be calculated.

B. Paramecia

Equipment and reagents

- Paramecia starter culture (obtain from another zebrafish lab or from Carolina Biological Supply Co.)
- 2-litre laboratory glass bottles
- Wheat grains, 20 per bottle

Method

1   Autoclave 1 litre of tap water in each 2-litre bottle and, separately, a supply of wheat grains.

2   After cooling, add about 20 wheat grains per bottle and shake.

3   Add 50 ml of starter culture per bottle and grow for 10 days in the dark at 27–28°C (e.g. in a cabinet in the fish room). Do not screw the lid tight so that air can get in. Keep at least two cultures in parallel in case one of them crashes.

4   For 7 days, feed about 2 ml of this culture to each mouse cage with zebrafish larvae, four times a day. From the number of zebrafish you want to raise, calculate the amount of paramecia bottles you need.

5   After having used up half of the culture, add autoclaved tap water back to a volume of 1 litre. After 2 months, also exchange the wheat grains. Continue the culture as in point 3.

Other paramecia protocols

Kimmel, C. (1994). *The Zebrafish Science Monitor*, **3**(3).
Kavumpurath, S. and Aleström, P. (1992). *The Zebrafish Science Monitor*, **2**(1).

Tetrahymena protocols

Speksnijder, J. E. and Bijmolt, E. (1995). *The Zebrafish Science Monitor*, **3**(4).
Gerson M. and Stainier D. (1995) *The Zebrafish Science Monitor*, **3**(5).

Infusoria are difficult to produce in large quantities. Therefore, the larvae are fed baby fish dry food (see *Protocol 6*). Although the nutritional value of the dry food is excellent, the problem is that surplus food rots quickly and spoils the water. Of course, underfeeding is also dangerous, because it prolongs the time until babies can eat *Artemia* and may cause death by starvation or unequal growth of fish.

Protocol 8

Artemia nauplia

Equipment and reagents

- *Artemia* hatcheries (e.g. from Aquaculture Supply, USA or Marine Biotech Inc., USA) or inverted plastic bottles with cut-off bottoms (*Figure 5*)
- Air pump with bubble stone
- Submerged heater (optional)
- Filter cloth or a mesh filter bag (120 μm mesh size, e.g. from Aquaculture Supply, USA)
- Sea salt (e.g. Instant Ocean or Forty Fathoms, Aquaculture Supply, USA)
- *Artemia* cysts (available from pet stores or from suppliers such as Aquaculture Supply, USA)
- Household bleach

Method

1. For convenience, use commercially available *Artemia* hatcheries. Alternatively, hatcheries may be constructed from inverted plastic bottles with cut-off bottoms, the neck of which is fitted with an outlet. Bubble air through the salt water to ensure vigorous mixing.

2. To 14 litres of tap water add 100 ml sea salt and aerate vigorously. Add 65 ml of shrimp eggs and use a small submerged heater to keep the temperature at 28 °C (necessary only when room temperature is below 20 °C).

3. After 24 hours most of the *Artemia* should be hatched. When feeding zebrafish larvae, do not incubate the *Artemia* for much over 30 hours (up to 2 days for feeding adults), as they rapidly lose most of their nutritional value. Add nettle powder or dried algal powder to feed *Artemia* to be kept longer.

4. For harvesting, stop aeration to allow hatched *Artemia* to sink and shells to float. Do not let *Artemia* sit for longer than 20 min without aeration, as they will die.

5. Use a filter cloth or a mesh filter bag to collect hatched *Artemia*. Avoid shells that float. If necessary, shells can also be removed later with the aid of a suction pump. Wash briefly with fish water and suspend *Artemia* in water. Unhatched *Artemia* cysts will rapidly sink to the bottom and the live *Artemia* should be separated from them by decanting in a plastic laboratory squeeze bottle.

6. Always feed the youngest fish larvae first. When feeding takes long, aeration of the collected *Artemia* or stirring with a magnetic stirrer keeps them alive for longer periods of time.

7. Clean hatchery thoroughly after each use. Once a month, bleach the hatchery, heater, and air bubbler for 12 hours with 1 ml bleach per litre of tap water. Rinse several times with tap water. If *Artemia* die immediately after hatching, it is usually caused by a dirty hatchery. If they do not hatch in time, it may be too cold. If only a small proportion of the *Artemia* hatch even under favourable conditions, change the supply company.

Fish at low densities grow faster, probably because for them food is not limiting. When trying to shorten the generation time by raising fish at lower densities, it should be taken into account, however, that densities influence the male to female ratios, low densities favouring female development. Feeding all fish more intensely helps shorten the generation time. On the other hand, it may result in severe problems with cleanliness and water quality, as the capacity of the filtering system might become limiting.

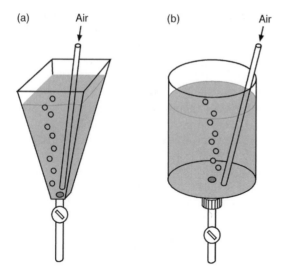

(a)   Air   (b)   Air

Figure 5 *Artemia* hatcheries. Hatcheries can be made of, for example, inverted plastic bottles or barrels with cut-off bottoms (their size depending on the amount of *Artemia* being raised). Air is bubbled through the water by a pipe connected to an air pump. At the bottom of the hatchery, a valve attached to a pipe allows the partial or complete drainage of the tank.

6 Maintaining stocks

6.1 Maintaining wild-type stocks

A number of robust and healthy wild-type laboratory strains have been established, including AB* (2), TU (Tübingen) (3), TL, HK (4), and WIK (5). In addition, two additional isogenic but less robust lines, SJD (6) and C32 (1), are available. These strains can be obtained either from the Tübingen stock centre (TU, TL, and WIK, or the stock centre of the University of Oregon (AB*, C32, and SJD). A number of homozygous viable pigmentation mutants, such as *golden* or *albino* (7) are useful for observations in live fish as well as antibody and *in situ* stainings at developmental stages when pigmentation may interfere with the visualization of the staining.

When propagating wild-type stocks or homozygous viable mutants, care should be taken not to inbreed too much. Although in theory inbred lines are very desirable for genetic experiments, in practice they may be too much trouble, because of low vitality and fecundity, as well as their susceptibility to diseases. Deriving the strain from many parents ensures a large gene pool and counteracts the negative effects of inbreeding. To prevent the propagation of spontaneous mutations in wild-type stocks, 50 embryos each from 10–15 individual pair matings should be sorted separately and inspected for abnormal embryos at 5 days of development. Dishes containing mutant phenotypes should be discarded. All larvae from the remaining dishes are combined and from this mixture the desired number of containers with 60 larvae each are raised (*Protocol 6*).

6.2 Maintaining mutant stocks

For a short description of published zebrafish mutants see Appendix 3. A more detailed list in searchable format is available through ZFIN (www.zfin.org). Many mutants identified in the various mutagenesis screens can be obtained through the Tübingen or Eugene stock centres. Most of the zebrafish mutants available are caused by recessive, zygotic lethal mutations.

Lethality is apparent before the homozygous mutants reach the reproductive age and therefore the mutant strains must be propagated through adult, heterozygous mutant carriers. Propagation of mutant stocks includes: (1) incrossing or preferably outcrossing of heterozygous mutant carriers; (2) raising embryos to adulthood; and (3) identifying adult heterozygous mutant carriers.

The fraction of mutant carriers is higher after an incross. Nevertheless, we generally prefer to propagate mutant stocks by outcrossing since repeated incrossing often significantly reduces growth and fertility due to inbreeding. For each mutant stock to be maintained, we recommend outcrossing two heterozygous fish (see *Protocol 3*) and raising 60 progeny from each cross separately (*Protocol 6*). If future mapping of the mutant locus or genotyping of individual embryos is being considered, mutant stocks should be maintained in the same genetic background in which the mutation was introduced. Mapping of the mutant locus and genotyping of individual embryos requires a mapping cross between the mutant strain and a polymorphic strain, such as WIK or SJD (for details on mapping and genotyping, see Chapter 7).

6.3 Identifying and maintaining adult heterozygous mutant carriers

To identify carriers for a heterozygous, recessive mutation, eggs are collected from crosses between sibling males and females obtained from an outcross. Because ½ of the fish in an outcross tank are heterozygous for the mutation, only ¼ of all successful crosses will produce mutant progeny. As not all of the crosses will give eggs, we recommend setting up as many crosses as possible from a family (see *Protocol 3*).

Those fish which produced eggs are kept as pairs in 1-litre containers until their embryos have been screened for the presence of the mutant phenotype. If this takes much longer than a week, the boxes should be connected to the water system and the fish should be fed. After having screened the embryos, the pair(s) which produced embryos with the expected mutant phenotype are kept as *identified* fish (ID fish), e.g. in 1-litre single boxes or in 2-litre mouse cages with two separate compartments, and are re-connected to the water supply and fed. Those pairs which did not produce mutant embryos are returned to the stock tank. As long as the heterozygous carriers are kept as separated male and female individuals, they can be successfully mated once every week, although mating the fish once every 2 weeks generally results in higher numbers of eggs.

Genotyping tail fin tissue clipped from individual adults, or genotyping embryos with appropriate green fluorescent protein (GFP) inserts linked to the mutant locus before raising them (10), might become a more feasible alternative to sibling crosses as more mutant loci are mapped or cloned.

Protocol 9

Preparation of fin-clip DNA for PCR screening

Equipment and reagents

- DNA extraction buffer: 10 mM Tris pH 8.2, 10 mM EDTA, 200 mM NaCl, 0.5% SDS, and 200 μg/ml proteinase K
- 0.16% Tricaine® in H_2O stock solution (Tricaine®; 3-aminobenzoic acid ethyl ester, Cat. No. A 5040, Sigma)

- Microcentrifuge
- Clean razor blades/scalpels

Protocol 9 continued

Method

1 Anaesthetize F_1 fish by placing them into 0.16% Tricaine® stock solution diluted appropriately in fish water. The amount of Tricaine® to use varies with water volume and number of fish, but should be used sparingly so as not to kill the fish.

2 Cut off ½ to ⅔ of the caudal fin using a sterile razor blade/scalpel and place the fin clip into a separate Eppendorf tube containing 100 μl of DNA extraction buffer. Return fish into separate mouse cages.

3 Vortex tubes thoroughly and incubate at 55 °C for 2–3 hours. Vortex tubes periodically to ensure thorough mixing. After incubation, centrifuge at 14 000 rpm for 20 min at room temperature to spin down debris. Dilute 100 nl to 1 μl for PCR screening (see *Protocol 6*). If desired, multiple DNA samples can be pooled to minimize the number of PCR reactions.

Identified fish can be maintained for years with little feeding, although fertility will decline. Generally they are kept until carriers from the next generation are identified and no longer than 2 years.

6.4 Maintaining mutant stocks as frozen sperm

Mutant strains can be kept as sperm samples frozen in liquid nitrogen. This form of stock keeping reduces the space required, the work load, and the costs when compared with the maintenance of live stocks. In addition, frozen sperm samples are an excellent 'backup' of stocks kept as live fish in the facility. To derive frozen sperm samples from mutant stocks, heterozygous males are squeezed to obtain sperm. Alternatively, the testes can be dissected out of adult males. The sperm is gradually cooled and stored in liquid nitrogen.

Protocol 10

Freezing zebrafish sperm for long-term storage

A. Isolating sperm from live zebrafish

Equipment and reagents

- Styrofoam box with lid, filled with dry ice
- Dewar with liquid nitrogen
- 10 μl pipette and tips
- 50 ml Falcon tubes
- Eppendorf tubes
- 1.8 ml cryovials
- Flat forceps (Millipore)
- 10 μl capillaries (should fit 1.8 ml cryovial)
- 0.02% Tricaine® in fish water
- Ice bucket, filled with dry ice
- Round sponges, approximately 2.5 cm in diameter, 3 cm high; make a cut (approximately 1 cm deep) on one of the flat sides

- Ginzburg Ringer: 6.5 g NaCl, 0.25 g KCl, 0.3 g $CaCl_2$, 0.2 g $NaHCO_3$ per 1 litre double-distilled water (ddH_2O); can be stored in frozen aliquots at −20 °C, place on ice when thawed
- Freezing medium: 9.0 ml Ginzburg Ringer, 1.0 ml methanol, 1.5 g Carnation non-fat milk powder. Vortex for 10–15 min (it is very important that all milk powder is dissolved in the Ringers, no precipitates). Place on ice, foam should be removed from the top, so that it does not get mixed with the sperm sample.

Protocol 10 continued

Method

1 Pre-chill 1.5 ml Eppendorf tubes in ice bucket. Pipette 10 μl of ice-cold freezing medium to each tube, keep on ice. Label 1.8 ml cryovials with date and genotype of sperm (no tape, only pen). Record the volumes, dates, times, genotype, freezing medium, etc.

2 Anaesthetize adult male in 0.02% Tricaine®. Rinse fish briefly after Tricaine® treatment and dry briefly on paper towel. Place the fish belly up in the cut, slightly moist sponge and position it under a dissecting microscope. Using flat forceps, push the anal fins to the side to expose the anus. Blot excess water from the anal fin area.

3 Using the forceps gently squeeze the sides of the fish at a point just anterior to the anal fins, collecting the sperm with the capillary tube, attached to a mouth pipette (like the one used for loading sequencing gels). Collect sperm in the capillary at the calibrated end, volumes may range from 0.5 μl to 2.0 μl for each fish. Although 2–3 fish can be used to reach a higher volume, the faster the sample is processed into the dry ice, the better. Sperm should be cloudy and white, clear samples are too dilute. Use the mouth pipette to add the sperm (0.5 μl to a maximum of 2 μl) to a pre-cooled 10 μl aliquot of freezing medium (avoid bubbles). Mix by pipetting up and down 1–2 times, then draw all freezing medium with sperm into capillary.

4 Transfer capillary into an appropriately labelled 1.8 ml cryovial, and place the cryovial in a 50 ml Falcon tube. The cryovials and Falcon tubes must be pre-labelled and pre-chilled on dry ice. Put the male into a container with fresh water to revive.

5 Submerge Falcon tube for 30 min horizontally in dry ice (in a styrofoam box). Then quickly transfer the cryovial from the Falcon tube into liquid nitrogen. After all males are squeezed, transfer the cryovials into a liquid nitrogen storage box (e.g. from Nalgene), record the position within the box (number), the level of box within the rack (bottom is one), and rack number, e.g. B24-L2-R3. We typically freeze at least three vials of sperm for each mutant.

For a protocol describing the collection of sperm for in vitro fertilization, see Protocols 6 and 7 in Chapter 6.

B. Isolating sperm from dissected testes

The method described below needs experience before it can be employed successfully. Therefore it should be practised with dispensable fish until a fertilization rate of at least 10% per *in vitro* fertilization (IVF) is obtained routinely. Under optimal conditions, 5 min are needed per testis preparation and 2 min per IVF to the step where 750 μl of fish water is added (see below). If well trained, a fertilization rate of 10–50% per IVF can be obtained.

Equipment

- Polystyrene box (at least 30 × 40 × 30 cm) for dry ice
- Ice bucket
- Transport bucket holding 1–3 litres of liquid nitrogen
- 50 ml Falcon tubes
- 2 ml cryovials with screw caps
- 10 μl capillaries 3 cm long (Brand or Hirschmann)
- Sterile 1.5 ml Eppendorf tubes
- 5 cm Petri dishes
- Scalpel
- 10 μl, 200 μl and 1 ml micropipettes
- Crystal tips (Eppendorf): 200 μl tips and 1000 μl tips
- Mouthpiece and capillary adaptor for a hose to fill the capillaries
- Two pointed pairs of forceps and one blunt pair of forceps
- Small microscopy scissors

Protocol 10 continued

- 30 cm pair of forceps to place cryovials into liquid nitrogen
- Vortex apparatus
- Binocular microscope
- Timer

Please note that all equipment and material used to isolate sperm has to be dry as water activates the sperm!

Reagents

- Tricaine® solution: 0.4% Tricaine® (3-aminobenzoic acid ethyl ester, Cat. No. A 5040, Sigma) + 1% $Na_2HPO_4.2H_2O$, pH 7.2). Use 6 ml Tricaine® solution added to 100 ml fish water.
- Ginzburg Ringer solution (GRS): 6.5 g NaCl, 0.25 g KCl, 0.4 g $CaCl_2.2H_2O$ in 1 litre ddH_2O add 0.2 g $NaHCO_3$. Store in 4.5 ml aliquots at −20 °C.
- Freezing medium: 4.5 ml Ginzburg Ringer solution + 750 mg fat-free milk powder (Carnation milk, Nestle) + 0.75 ml methanol. The milk powder has to be completely dissolved in the GRS by vortexing the solution for 15 min. Then add the methanol. Prepare fresh on the day of the sperm isolation. Keep cold on ice.

Method

1 Prepare the freezing medium.

2 Fill the polystyrene box (at least 10 cm high) with crushed dry ice and place six Falcon tubes up to the screw cap into the dry ice. Do not take these tubes out of the ice during the experiment.

3 Label the cryovials and cool them in the ice bucket on water ice.

4 Label the Eppendorf tubes, fill 20 ml of ice-cold freezing medium into the tubes and keep them on ice.

5 Anaesthetize a male fish in the Tricaine® solution until the gills stop moving.

6 Take the fish out of the solution and dry it very carefully with a paper tissue.

7 Put the fish into a Petri dish and decapitate it with a scalpel.

8 Using the scalpel, dissect the male. First, make a cut transversely at a level immediately posterior to the pectoral fins. Subsequently cut longitudinally along the ventral side from the plane of the first section to the anal region.

9 Open up the body cavity and grip the testis with the pointed forceps below the swim bladder. Pull the testis out.

10 Transfer the testis into the corresponding Eppendorf tube. Release the sperm by gently stirring and pipetting the testis with the help of a crystal tip (cut off 3 mm from the tip before usage).

11 Take the sperm solution up into capillaries: $2 \times 10\,\mu l$ or $4 \times 5\,\mu l$.

12 Transfer the capillaries into the corresponding cryovials (on ice). Be careful when tightening the screw cap: first gently put the screw cap on to the vials, turn them upside down so that the capillaries fall into the screw cap and then tighten the cap.

13 Transfer the cryovials into the Falcon tubes, which are buried in dry ice. Do not take the Falcon tubes out of the dry ice. Freeze the vials for 30 min on dry ice.

14 Transfer the cryovials into liquid nitrogen storage (transport bucket).

15 Sort the cryovials into a nitrogen container.

Protocol 10 continued

Please note that a patent application covering this method has been filed by Artemis Pharmaceuticals GmbH. If you want to use this method for **commercial** purposes, please contact Artemis (http://www.artemis-pharmaceuticals.de). The use of this method for non-commercial purposes is not restricted.

An alternative method for freezing sperm obtained from adult testes has recently been described (11). Although this method provides a much greater volume of sperm (about 100 μl per fish) and would therefore be the method of choice, it does not yet give consistently high fertilization rates. However, due to the great promise this method holds, the interested reader is referred to discussions in the zebrafish newsgroup (see ZFIN).

To recover a live stock from frozen sperm, females are squeezed to obtain eggs, which are fertilized *in vitro* with the thawed sperm (*Protocol 11*, see also Chapter 6). For this procedure, it is advisable to use females from strains that have been selected for large clutch sizes and 'easy-to-squeeze' females, e.g. AB, or golden Tübingen. The *in vitro* fertilization rate can be as high as 80%.

Protocol 11

In vitro fertilization using frozen sperm

For a protocol describing the stripping of eggs from females and the use of non-frozen sperm for *in vitro* fertilization, see *Protocols 8* and *10* in Chapter 6, respectively.

Equipment and reagents

- Set up six AB* females and one male in a 2-litre container (e.g. mouse cage with lid) the night before the *in vitro* fertilization. Remove the male the next morning as soon as the lights are turned on (the AB* strain has been selected over many generations for large clutch size, 'easy-to-squeeze' females)
- Ice bucket
- Sperm vials in liquid nitrogen
- Capillary pipettor
- 5 cm Petri dish
- 200 μl and 1000 μl pipettor with tips
- Timer
- Fine paintbrush
- 0.02% Tricaine® in fish water

- Dissecting microscope
- Fructose-egg water: 0.5% fructose in egg water at 28 °C
- Hank's (final): 9.9 ml Hank's premix, 0.1 ml Stock 6.
- Hank's premix. Combine the following in order: 10.0 ml Solution 1, 1.0 ml Solution 2, 1.0 ml Solution 4, 86.0 ml ddH$_2$O, 1.0 ml Solution 5. Solution 1: 8.0 g NaCl, 0.4 g KCl in 100 ml ddH$_2$O. Solution 2: 0.358 g Na$_2$HPO$_4$ anhydrous, 0.60 g K$_2$H$_2$PO$_4$ in 100 ml ddH$_2$O. Solution 4: 0.72 g CaCl$_2$ in 50 ml ddH$_2$O. Solution 5: 1.23 g MgSO$_4$.7H$_2$O in 50 ml ddH$_2$O. Solution 6: 0.35 g NaHCO$_3$, 10.0 ml ddH$_2$O. Store Hank's premix in the refrigerator along with the Hank's solutions

Method

1 Anaesthetize females in 0.02% Tricaine®. Rinse fish briefly after Tricaine® treatment and dry briefly on a paper towel.

2 Place the female on her side into a 5 cm Petri dish and place one finger of one hand on the dorsal side of the female. Using one finger of the other hand press out the eggs by gently pressing on to the ventral side of the fish, starting just behind the pectoral fins and moving towards

Protocol 11 continued

the tail. Only gentle pressure is needed—if the fish has eggs they will come out easily. If gentle pressure fails to produce eggs, do not continue to squeeze harder. Extra squeezing may injure the fish.

3 Carefully move the eggs away from the body of the female with the fine paintbrush. Cover the dish with its lid and inspect the eggs using a dissecting scope: good eggs look slightly yellowish with a granular appearance. Only use large clutches (approx. 200 eggs) of good-looking eggs with your precious sperm samples. Put the female into a recovery container to revive. Work quickly.

4 Place 70 µl (for every sperm sample) of ice-cold Hank's (final) next to the eggs. Remove sperm capillary from liquid nitrogen, attach capillary to capillary pipettor (e.g. mouth pipette) and carefully expel the sperm into the 70 µl of Hank's (avoid bubbles).

5 Mix the sperm in Hank's carefully with the eggs and let sit for 30 seconds. Add 750 µl of fructose-egg water and let sit for 2 minutes. Add more fructose water to bring the level up to half the volume of the dish. Record the volume of sperm and genotype on the lid and incubate the dish at 28 °C.

7 Fish health

7.1 Fish diseases

Zebrafish are generally very hardy fish. Nevertheless, a number of diseases that can severely affect the entire fish colony are known, most commonly velvet disease, fish tuberculosis or TB (mycobacteriosis), and nematode infection (intestinal capillariasis). There are clear symptoms for each of these diseases, and for velvet and capillariasis efficient treatments are available.

The best strategy to maintain a healthy colony is to avoid introducing fish from the outside without rigid quarantine procedures, and to discard any sick fish as quickly as possible. All fish parasites and infectious agents survive well only if sufficiently moist, therefore infections via the air or thoroughly dried materials is unlikely. Fish tuberculosis is a disease that exists in the background in many systems. It affects stressed or injured fish most frequently. Genetically inbred fish are also more susceptible to the disease. TB is probably transferred via contacts with nets, tank walls, sibling fish, and other items. Fish die slowly from this disease, therefore to prevent spreading, it is important to spot them early and remove them. Spreading of the disease to 'downstream' tanks in the serial system has sometimes been observed. However, as the recirculated water runs through UV lamps, there is less chance of the disease spreading to other containers within the same filter system. It is probable that infection and spread of the disease takes place during handling. We take routine measures to reduce spreading the disease between rooms or systems. The properties of the system that are relevant for health in the colony have already been mentioned, and include UV lamps, sand filters, and general construction features that allow cleaning and sterilization of the tanks.

Protocol 12

Treatment of capillariasis

Reagents

• Fluke tabs (active components: mebendazole and trichlorfon)

Protocol 12 continued

Method

Symptoms of nematode infections are a skinny, arched body or a failure to come upon feeding. Fish will stop eating and not mate any more. Individual fish may be cured with fluke tablets dissolved in water according to the manufacturer's instructions, by treating them twice for 2 days each, interrupted by a period of 7 days. If the infection is spread widely, it is possible to add the medicine to an entire system. Fluke tablets will kill snails, which must therefore be removed. They do not harm the bacteria that colonize the filters.

7.2 Fish handling in systems with latent TB

Keep the system as clean as possible. Handle fish very gently, avoiding injuries of the skin. Keep track of the age of the fish in each aquarium and avoid keeping fish that are older than 2 years. Check tanks daily. Remove dead fish and fish that display symptoms of disease or just generally look unhealthy, and keep records of the tanks in which sick fish were observed. Common symptoms of a TB infection are raised scales, open lesions, and wound spots. Try to discard the entire tank as soon as possible.

In order to avoid contaminating other fish, disinfect fish nets between handling different fish by keeping them in boiling water. Autoclave containers in which larvae are to be raised and wash mating boxes with hot tap water and soap in a dishwasher each time after having set up fish. Since many of the available mating containers do not withstand the high temperatures during autoclaving, they should be sterilized on a regular basis in a Clidox® bath (Pharmacal Research Labs, Naugatuck, USA). Clidox® is a chlorine dioxide-based, high-level disinfectant, which kills vegetative bacteria, including *Mycobacterium bovis*, *Pseudomonas aeruginosa*, *Staphylococcus aureus*, and *Salmonella choleraesuis*, and pathogenic fungi and viruses (see *Protocol 13*). Disinfect fish tanks before refilling them (see *Protocol 2*). To avoid the transfer of spills from one room to another, people should walk through troughs containing disinfectant when entering a fish room. If more than one water system is available, it might be advisable to remove all fish from one system and thoroughly clean and disinfect it from time to time. Also, embryos should be bleached before growing them up, in order to avoid a vertical transmission of the disease.

Protocol 13

Decontamination of mating containers and small systems using Clidox®

Equipment and reagents

- Base MSDS (Pharmacal Research Labs, Naugatuck, USA)
- Activator MSDS (Pharmacal Research Labs, Naugatuck, USA)
- Protective plastic or rubber gloves
- Goggles for eye protection
- Tap water
- Large container (for disinfecting smaller containers)
- Large net to submerge mating containers in Clidox® bath

Method

1 Use a brush to remove the worst dirt from containers or tanks.

Protocol 13 continued

2 For disinfecting smaller containers, dilute Clidox® to disinfectant level in a large container: 1:18:1 (one part base plus 18 parts water plus 1 part activator). Mix and let sit for 15 minutes before using it. Add items to be disinfected and close the lid. In a closed container diluted Clidox® is stable for 14 days.

3 To disinfect systems, add the appropriate amounts of base and activator to the water to achieve a 1:18:1 dilution. At 1:18:1 Clidox® kills vegetative bacteria, including *Mycobacterium bovis*, *Pseudomonas aeruginosa*, *Staphylococcus aureus*, and *Salmonella choleraesuis*, and pathogenic fungi and viruses in 5 min at 20°C when used on pre-cleaned surfaces. Cover tanks – the vapour pressure of Clidox® is similar to that of water.

4 Leave containers for 1 hour in bath or leave in tanks for 1 hour.

5 Rinse several times with water.

For disinfections of a large system, SanAqua (That Fish Place) may be used.

7.3 Quarantine

When obtaining stocks from stock centres or other laboratories, they usually come as larvae from eggs that were bleached. Nevertheless, as a routine, these fish should be grown in quarantine to adulthood. The next generation can be raised from bleached eggs in the normal facility. Particular care has to be taken when introducing adult fish from pet shops or from the wild. When acquiring fish from pet shops, it is advisable to check the general hygiene condition in the shop as well as the health and vitality of its fish. When in doubt, go to another shop. Pet shop fish are frequently infected with exoparasitic diseases such as *Ichthyophthirius*. There is an efficient treatment that should be routinely applied to pet shop fish while kept in quarantine.

Protocol 14

Preventive treatment of exoparasites (Ichtyo)

Reagents

- Malachite green
- K_2(Cu-EDTA)
- Hexamethylpararosanilin chloride

- Stock available as *Fauna special* from Zoo Fachring (Germany)

Method

1 Prepare stock solution of 15 mg malachite green, 7.5 mg K_2(Cu-EDTA), 40 mg hexamethyl-pararosanilin chloride in 100 ml water.

2 Add two drops of stock to 1 litre of water, aerate vigorously. Keep fish in solution for 3 days. Change water. If fish are still infected, repeat treatment.

In quarantine, it is not advisable to use filters. Instead, the water should be frequently exchanged with fresh system water. A convenient system is to place individual containers with overflows in a rack equipped with inflows and rimmed shelves, such as that used for raising baby fish. Once a day all the water is exchanged with fresh water. Particular care has to be taken to avoid transfers of water between containers (e.g. while feeding with a squirt bottle) and only disposable or sterilized items should be used in the quarantine unit. Obviously, no material should be exchanged between the quarantine room and the rest of the facility.

Generally, it is not advisable to cure fish from diseases, but rather to discard them quickly. The Oregon stock centre has a fish pathology section to which diseased fish can be sent for diagnosis and advice for further treatment.

References

Zebrafish information database, ZFIN: http://www.zfin.org

1. Krause, H. J. (1992). *Handbuch Aquarientechnik.* bede Verlag, Kollnburg, Germany.
2. Streisinger, G., Walker, C., Dower, N., Knauber, D., and Singer, F. (1981). *Nature,* **291**, 293–6.
3. Mullins, M. C. and Nüsslein-Volhard, C. (1993). *Curr. Opin. Genet. Dev.,* **3**, 648–54.
4. Knapik, E. W., Goodman, A., Ekker, M., Chevrette, M., Delgado, J., Neuhauss, S., Shimoda, N., Driever, W., Fishman, M. C., and Jacob, H. J. (1998). *Nature Genet.,* **18**, 338–43.
5. Rauch, G. J., Granato, M., and Haffter, P. (1997). *TIGS-Technical Tips Online,* **13**, 461.
6. Johnson, S. L., Africa, D., Horne, S., and Postlethwait, J. H. (1995). *Genetics,* **139**, 1727–35.
7. Streisinger, G., Singer, F., Walker, G., Knauber, D., and Dower, N. (1986). *Genetics,* **112**, 311–19.
8. Haffter, P., Granato, M., Brand, M., Mullins, M. C., Hammerschmidt, M., Kane, D. A., Odenthal, J., van Eeden, F. J. M., Jiang, Y.-J., Heisenberg, C.-P., Kelsh, R. N., Furutani-Seiki, M., Vogelsang, E., Beuchle, D., Schach, U., Fabian, C., and Nüsslein-Volhard, C. (1996). *Development,* **123**, 1–36.
9. Westerfield, M. (1993). *The zebrafish book.* University of Oregon Press. Available online (http://zfish.uoregon.edu/zf_info/zfbook).
10. Kawakami, K. and Hopkins, N. (1996). *Trends Genet.,* **12**, 9–10.
11. Ransom, D. G. and Zon, L. I. (1999). Collection, storage and use of zebrafish sperm. *Methods Cell Biol.,* **60**, 365–72.

Chapter 2
Looking at embryos

Stefan Schulte-Merker

Artemis Pharmaceuticals GmbH, Spemannstr. 35, 72076 Tübingen, Germany

1 Introduction

The many advantages of zebrafish, such as its powerful genetics (Chapters 6 and 7) and experimental accessibility (Chapter 4), are described throughout this book and need no further elaboration. In this chapter we would like to focus on the one feature of zebrafish that brings all other advantages to full appreciation: this is the fact that zebrafish embryos are transparent, and that from the moment of fertilization onwards cellular and developmental processes, as well as organogenesis, can be observed in great detail. This is an extremely important asset and immensely facilitates the studies of development in this small teleost. Its transparency is also the reason why zebrafish are so successful as a system for performing genetic screens: while *Drosophila* larvae provide a cuticle pattern that allows a great readout for morphogenetic processes that occurred earlier in development, zebrafish are equally suited for screens. They do not have the epithelial hallmarks of a cuticle, but all cells and organs are easily visible, which results in a comparable ease of analysis.

In this chapter we describe some of the methods required for observing embryos, for analysing gene expression, and for documenting data, both in life as well as in fixed material. This chapter provides the basic principles of performing *in situ* hybridization, either with or without simultaneous detection of a protein antigen in the same specimen, and describes performing antibody staining in embryos and larvae. We then cover preparing *in situ* or antibody-stained specimens for microscopy, either without or with prior sectioning. Observing live embryos is described, as well as a protocol that prevents pigmentation in embryos. While staining protocols for cartilage and bone are contained in Chapter 3, the procedure for highlighting the vasculature of larvae is described in this chapter.

2 *In situ* hybridization

In studying the expression of genes of interest, there are a number of possibilities. First, one can produce reporter constructs driving the expression of β-galactosidase of a promoter sequence, as described in Chapter 5. Secondly, one can generate an antibody and use it to uncover the expression of the protein product, as described below. Thirdly, and most easily, one can generate an epitope-tagged antisense RNA probe directed against the gene/mRNA of interest. *In situ* hybridization has the advantage over other methods of being cheap, fast, and requiring nothing but a fragment of the gene of interest in hand (no antibody production necessary).

In tissue samples of small size, such as zebrafish embryos, hybridization and detection can usually be performed in the whole mount, permitting analysis of the result in three dimensions.

It is always possible to section after whole-mount hybridization, but it is very cumbersome to do sections first and then having to reconstruct the three-dimensional image from sections!

Ever since its first demonstration in zebrafish, whole-mount *in situ* hybridization has proved to be one of the most robust and informative methods for analysing gene expression. Over the years, many labs have contributed to improving the method, in both sensitivity as well as speed. Also, protocols have evolved allowing detection of multiple probes and combining the method with antibody staining. Examples of these are given below.

All protocols listed use RNA probes, rather than DNA probes. While DNA probes work as *in situ* probes, and are equally straightforward to synthesize, they are hardly ever used because the yield from the *in vitro* synthesis differs by a factor of five in favour of the RNA probes.

Many researchers, when using *in situ* hybridization for the first time, express worries about working with RNA probes, and are extremely careful about minimizing the possible influence of RNases at the bench. While we certainly do support working carefully, we would like to point out that there are only very few steps during the procedure where RNases could possibly interfere with the successful outcome of the experiment: most of the time, the embryos are either in formaldehyde, varying concentrations of methanol, or in 50% formamide at 55°C, all of which are conditions not particularly favourable to RNase activity. Even for preparing PBST (see *Protocol 1*), we simply use autoclaved phosphate-buffered saline (PBS) (no diethyl pyrocarbonate (DEPC) treatment), and take no further precautions. The protocol is robust enough to use it in undergraduate lab classes.

2.1 Single probe

The most routine method used for *in situ* hybridization involves fixing the material, digesting with low amounts of proteinase to partially free the target mRNA from proteins, pre-hybridizing to block non-specific binding of probe to the fixed material, hybridizing with the probe, and detecting the probe enzymatically. The probe is labelled through incorporation of an epitope-tagged nucleotide, usually digoxigenin (DIG)- or fluorescine-tagged uracil (*Figure 1*, see also *Plate 1*). There are many variations to this scheme, and some of these will be discussed.

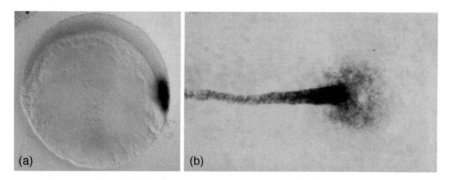

(a) (b)

Figure 1 (see Plate 1) *In situ* hybridization using a single probe. Embryos were processed for *in situ* hybridization as described in *Protocol 1*. (a) 60% epiboly embryo, dorsal to the right. The embryo was hybridized with a digoxigenin-labelled probe against the *chordino* gene (9). Detection of the probe was carried out using alkaline phosphatase and BM Purple as a substrate. The embryo was mounted in a viewing chamber (Section 4.1) using glycerol as a mounting medium. (b) Tailbud embryo, flat-mount. Anterior to the left, dorsal view. The embryo was hybridized with a digoxigenin-labelled probe against the *no tail* gene (10) and the probe detected as described in (a). Mounting was in benzyl benzoate/benzyl alcohol. Please note that while the stained cells of the notochord and the tailbud are very apparent (even the 'negative' nuclei in the anterior part of the notochord are visible), the remainder of the embryo cannot be seen, due to the strong clearing effect of benzyl benzoate/benzyl alcohol.

Protocol 1

In situ hybridization using a single digoxigenin-labelled probe

Equipment and reagents

- Water bath or hybridization oven at 55–65 °C
- RNase-free linearized DNA construct containing an appropriate RNA polymerase promoter sequence to allow generation of antisense RNA probes
- DIG-labelling kit (Roche)
- G50 Sephadex in DEPC-treated, autoclaved double-distilled water (ddH$_2$O)
- PBST: PBS plus 0.1% Tween
- SSCT: saline sodium citrate (SSC) plus 0.1% Tween
- HYB$^-$: 50% formamide, 5 × SSC, 0.1% Tween-20
- HYB$^+$: HYB$^-$, 5 mg/ml torula (yeast) RNA, 50 μg/ml heparin. The torula RNA is prepared by dissolving in DEPC-treated water. HYB$^-$ and HYB$^+$ should be kept at −20 °C.
- Blocking solution (final concentration: 150 mM maleic acid, 100 mM NaCl, 2% blocking reagent (Roche), pH 7.5, 0.1% Tween-20). The blocking reagent takes some time to dissolve

and is best prepared as a 5× stock solution that is stored at −20 °C.
- Anti-digoxigenin Fab fragments, coupled to a detection enzyme such as alkaline phosphatase.
- Staining buffer: 100 mM Tris pH 9.5, 50 mM MgCl$_2$, 100 mM NaCl, 0.1% Tween-20, 1 mM levamisol (add fresh). Levamisol inhibits acidic phosphatases, which can, depending on the stage of the fixed material, lead to staining in tissues expressing phosphatases, such as retina and intestinal tissues.
- NBT (4-nitroblue tetrazolium chloride), 75 mg/ml in 70% dimethylformamide
- BCIP (5-bromo-4-chloro-3-indolyl-phosphate), 50 mg/ml in dimethylformamide
- An alternative chromogenic substrate that we recommend is the BM Purple substrate from Roche (use undiluted or mix 1:1 with staining buffer, warm up to room temperature before adding to samples, and mix well before removing from bottle).

Method

All steps are carried out at room temperature, unless stated otherwise.

1. Fix embryos with 4% paraformaldehyde in PBS overnight at 4 °C. Use fresh paraformaldehyde (or paraformaldehyde frozen right after production), otherwise background staining is likely to form. Fixation is best done in glass vials or 2 ml Eppendorf tubes. It does not matter whether the embryos are dechorionated or not, and sometimes it is easier to remove the chorions enzymatically prior to fixation rather than using watchmaker forceps as described below.

2. Wash embryos twice in PBS, 5 min each.

3. Transfer embryos into a glass Petri dish and dechorionate them using watchmaker forceps.

4. Transfer embryos to glass vials or 2 ml Eppendorf tubes containing 50% methanol. Replace 50% methanol with 100% methanol after 5 min, replace with fresh 100% methanol after a further 5 min, and store at −20 °C. Use methanol and not ethanol, since ethanol causes background staining. From this point on, the embryos remain in the same vial until they are ready for pre-hybridization. Embryos can be stored in methanol for months; excellent results have been obtained with embryos that had been in methanol for 6 months. The methanol step is necessary for permeabilization of embryos even if they are not stored.

5. Rehydrate embryos through incubation in 50% methanol, in 30% methanol/PBST, and in PBST (twice for 5 min each). Embryos can also be dechorionated at this point, but chorions tend to be sticky after having been in methanol.

6. Fix for 20 min in 4% paraformaldehyde in PBST.

Protocol 1 continued

7   Wash twice with PBST, 5 min each.

8   Digest with proteinase K (10 µg/ml in PBST) for 5–12 min at room temperature (the time span depends on the stage of the embryos, with younger stages being more sensitive; the optimal time for digestion also depends on the particular batch of the enzyme, which needs to be tested when using a fresh batch). The proteinase K digestion increases the sensitivity of the procedure significantly, probably by digesting away proteins from the mRNA, thereby allowing better access of the probe, and by allowing better penetration of the antibody.

9   Rinse briefly in PBST to dilute the proteinase, then wash once with PBST for 5 min.

10 Fix for 20 min with 4% paraformaldehyde in PBST, then wash twice with PBST (5 min each). Transfer the embryos (up to 40) into small Eppendorf tubes (0.8 ml) containing 300 µl HYB⁻. This is best achieved by using a glass Pasteur pipette, cut off with a diamond knife and flame polished.

11 Incubate for 5 min at 55 °C. Afterwards, replace HYB⁻ carefully with 200 µl HYB⁺ and pre-hybridize at 55–68 °C for 1 hour (minimum) or overnight in HYB⁺.

12 RNA probes are prepared according to the manufacturer's instructions (Roche). Roche provides upon request free brochures (*DIG user's guide* and *In situ hybridization manual*) which contain all general information required. About 5–10 µg of digoxigenin-labelled probe are transcribed from 1 µg of a linearized plasmid. Probes do not need to be hydrolysed (1) unless they are longer than 3–4 kb, but care should be taken to remove unincorporated nucleotides as these are extremely sticky and will result in high background staining unless removed from the probe. The easiest way to remove mononucleotides is by using spin columns filled with Sephadex. After spin purification, use 1 µl of the eluate to measure the concentration of the synthesized probe. Precipitate the probe with 0.1 volumes 3 M ammonium acetate and 2 volumes ethanol for 20 min on dry ice, spin down at 4 °C for 30 min at 14 000 rpm, and wash once with 70% ethanol (use DEPC-treated water). Dissolve the probe directly in HYB⁺ and store at −20 °C.

13 Take off 100 µl of the HYB⁺ without letting the embryos touch air. Add 1 or 2 µl of probe to the samples, containing 20–100 ng RNA probe. For some probes, signal intensity decreases below 10 ng. Amounts higher than 100 ng are usually not required. The amount of probe added depends on the amount of RNA you want to detect and the accessibility of the target mRNA. The amount of probe required to obtain best results can vary between probes and needs to be tested individually.

14 Incubate overnight at 55–68 °C. The hybridization temperature depends on the required stringency: while in most cases 55 °C is satisfactory, in some cases 68 °C is required because the probe proves to be sticky (e.g. the probe might contain repetitive elements or a common protein motive such as a homeobox). Remove probe, but take care not to let the embryos touch air. Most probes can be reused up to two times; using them more often may result in weaker signals. Probes in HYB⁺ are stable for at least 6 months at −20 °C.

15 Wash twice at 55–65 °C in 50% formamide/2× SSCT (20 min each minimum), once at 55–65 °C in 2× SSCT (15 min), and twice at 55–65 °C in 0.2× SSCT (20 min each). All solutions should be pre-warmed, in order to ensure that the required temperatures are reached quickly during the washing procedure.

16 Transfer embryos into PBST (5 min), and subsequently into blocking solution.

17 Block for at least 1 hour at room temperature (better overnight at 4 °C).

18 Incubate with anti-digoxigenin Fab–alkaline phosphatase at a 5000-fold dilution in blocking solution and rock gently for 4 hours at room temperature or overnight at 4 °C.

Protocol 1 continued

19 Wash 4 times (25 min each) with blocking solution, then 3 times (5 min each) in staining buffer.

20 Incubate in staining buffer with 4.5 μl NBT and 3.5 μl BCIP per ml.

21 Stain for 30 min to overnight in the dark, until staining intensity increases no further or until background staining starts to develop. Stop developing by removing staining solution and by washing extensively in PBST.

22 Keep samples in the dark as much as possible after removing from staining solution, or background staining will develop (particularly in the yolk). Embryos can be stored at this point in the dark, preferably in 4% paraformaldehyde in PBST at 4 °C.

23 Mount as described below.

If using alkaline phosphatase as a detection enzyme, the NBT/BCIP precipitate (and the BM Purple precipitate) will fade somewhat in anhydrous solutions. Fixing the embryos after having stopped the colour reaction is necessary if you want to clear the embryos in alcohol. After fixation, even weaker signals are reasonably stable in alcohol. Occasionally, we also overstain with careful monitoring of the dehydration and clearing process. Another alternative is to photograph weak signals immediately after transferring the specimen to alcohol or to clear them in glycerol.

2.2 Detecting two differently labelled probes *in situ*

In many cases, it is very useful to be able to compare the expression patterns of different genes in the same specimen. This can help to determine which cells express the gene of interest by comparing its expression with the expression of a marker gene. Alternatively, one might want to compare the expression patterns of genes that are known to interact with each other, and to determine to what extent the expression domains of these genes overlap or how the temporal aspects of the respective expression patterns compare. While this is possible by using probes labelled with the same epitope, the task is greatly facilitated by using two different epitopes to label the respective probes, allowing detection of the expression domains in different colours.

With the emergence of ribonucleotides that are linked to fluorescein, it has become possible to examine the expression of genes by monitoring both expression patterns at a time in the same specimen. It is, of course, possible to use the fluorescence of the probe as the primary mode in which to detect the signal; however, the sensitivity of doing so is considerably lower than amplifying the signal by using an anti-fluorescein antibody with subsequent enzymatic detection.

Protocol 2

In situ hybridization using digoxigenin-labelled and fluorescein-labelled probes within the same specimen

Reagents

- 100 mM glycine, pH 2.2, 0.1% Tween-20
- Anti-fluorescein Fab fragments coupled to alkaline phosphatase
- Staining buffer:

100 mM Tris, pH 9.5
100 mM NaCl
50 mM MgCl$_2$
0.1% Tween-20

- Fast Red buffer:
  100 mM Tris, pH 8.2
  100 mM NaCl
  50 mM MgCl$_2$
  0.1% Tween-20
- Fast Red staining solution: dissolve 1 tablet
  Fast Red (Roche) in 4 ml Fast Red staining

buffer; spin down undissolved particles and transfer the supernatant to a new tube. This solution must be used within 30 min and cannot be stored. Try to take ½ of a tablet or even ¼ if you need less.

Method

Follow the *Protocol 1* up to and including step 11. All steps are carried out at room temperature unless otherwise stated.

12 RNA probes are prepared according to the instructions of the supplier (2, 3) as described in *Protocol 1*, step 12.

13 Add appropriate amounts of both probes to the embryos in HYB$^+$.

14 Incubate overnight at 55–68 °C. See comments in *Protocol 1*.

15 Wash twice at 55–65 °C in 50% formamide/2× SSCT (20 min each minimum), once at 55–65 °C in 2× SSCT (15 min), and twice at 55–65 °C in 0.2× SSCT (20 min each). All solutions should be pre-warmed, in order to ensure that the required temperatures are reached quickly during the washing procedure.

16 Transfer embryos into PBST (5 min), and subsequently into blocking solution.

17 Block for at least 1 hour at room temperature (better overnight at 4 °C).

18 Incubate with anti-digoxigenin Fab–alkaline phosphatase at a 5000-fold dilution in blocking solution and rock gently for 4 hours at room temperature or overnight at 4 °C.

19 Wash 4 times (25 min each) with blocking solution, then 3 times (5 min each) in staining buffer.

20 Incubate in staining buffer with 4.5 µl NBT and 3.5 µl BCIP per ml added.

21 Stain for 30 min to overnight in the dark until staining intensity increases no further or until background staining starts to develop. Stop staining reaction by removing staining solution and by washing a few times in PBST.

22 Incubate in 100 mM glycine for 30 min. This efficiently destroys any enzymatic activity from the alkaline phosphatase bound to the digoxigenin-labelled probe. Then wash in PBST, 3 × 5 min each.

23 Block for at least 1 hour in blocking solution (better overnight at 4 °C).

24 Incubate with anti-fluorescein Fab–alkaline phosphatase at a 5000-fold dilution in blocking solution and rock gently for 4 hours at room temperature or overnight at 4 °C.

25 Wash 4 × 25 min each with blocking solution, then 3 × 5 min each in Fast Red buffer.

26 Develop stain in Fast Red staining solution in the dark. As the staining solution is orange and the Fast Red precipitate pink/red, it is not always easy to detect the signal within the staining solution. Stop staining reaction by washing excessively in PBST.

27 Analyse and document as soon as possible, as the Fast Red precipitate is not as stable as the NBT–BCIP precipitate or BM Purple precipitate.

It is advisable to use the darker substrate first: the reason for doing so is that while any remaining enzymatic activity from the first alkaline phosphatases might lead to a small amount of Fast Red precipitate, this will hardly be noticeable among the dark-blue precipitate from the first staining reaction. If the order of staining substrates is reversed, however, this would have the undesired effect of having the red precipitate turn blue.

We recommend that the probe expected to give a weaker signal is detected with the blue substrate, as it results in a more pronounced signal than the Fast Red staining.

3 Antibody staining

Performing *in situ* hybridization using antisense ribonucleotide probes, as described above, is a widely used method, due to its simplicity and due to the fact that sequence information is abundant. With the help of reverse transcriptase polymerase chain reaction (RT-PCR) virtually any expressed sequence tag can be converted into a suitable expression vector that allows the generation of an antisense probe. To obtain antibodies involves considerably more work, but once obtained, an antibody gives information at a level of resolution that can not be gained with an RNA probe: for example, the precise localization of a particular protein and its intracellular localization (nuclear protein versus cytoplasmic, for example), its time window of expression (the protein might still be there long after the mRNA is no longer detectable), and possibly, in the case of secreted proteins, its range of distribution within a tissue.

A plethora of fixatives and fixation methods are available (4), and the same can be said about secondary antibodies used to detect the primary antibody. Commercially available secondary antibodies can be obtained that are labelled with fluorochromes or different enzymes, such as horseradish peroxidase or alkaline phosphatases. The intention of this section is to provide the most commonly used and robust staining procedures. We have selected two protocols as a guideline for antibody staining: the first protocol deals with early embryos and a biotinylated secondary antibody that is subsequently detected by a complex of avidin and biotinylated peroxidase. The second protocol describes staining of cartilage matrix proteins in 4-day-old larvae that require additional treatment for good penetration of reagents.

3.1 Detecting antigens in early embryos using biotinylated secondary antibodies

Embryos younger then 24 hours are easy to permeabilize, and in most cases the measures outlined below are sufficient to achieve good penetration of the primary and secondary antibodies. However, when using a particular fixation regime for the first time, it is a good idea to section your stained material and to check whether staining is homogeneous or whether, for example, the distal part of the stained structure stains more intensely than the proximal part. In such cases, it is necessary to alter the fixation and penetration procedure.

Protocol 3

Detecting antigens in early embryos

Reagents

- Blocking solution as in *Protocol 1*
- Biotinylated secondary antibody (available from a number of suppliers, e.g. Vecta)
- ABC Vecta Stain kit, containing avidin (component A) and biotinylated horseradish
- peroxidase (component B)
- DAB (3,3'-diamino-benzidine) stock at 20 mg/ml in 50 mM Tris, pH 7.5
- 0.3% H_2O_2. Make fresh from 30% stock with ddH_2O

Protocol 3 continued

Method

All steps are carried out at room temperature unless otherwise stated.

1 Fix embryos using 4% paraformaldehyde in PBS overnight at 4°C. Paraformaldehyde is a comparatively mild fixative, and most epitopes tolerate this fixative. However, in some cases epitopes are sensitive to formaldehyde, and other fixatives have to be used, such as tri-chloracetic acid (4) or picric acid (4). Fixation is best done in glass vials or 2 ml Eppendorf tubes.

2 Wash three times for 5–10 min in PBST to remove the paraformaldehyde, and dechorionate embryos if necessary (use watchmaker forceps while embryos are in a glass dish).

3 The embryos are permeabilized by immersing the material in solutions with increasing amounts of methanol, for 5 min each (30% methanol, followed by 50%, 70%, 90%, and twice 100%), or by immersing in ddH_2O (5 min), pre-cooled acetone (7 min at −20°C), and ddH_2O (5 min). An alternative method to achieve permeabilization is to use Nonidet P-40 at 0.5–1% in cases where epitopes are sensitive to anhydrous solutions. In this case, omit steps 3 and 4 and add Nonidet P-40 to all solutions up to and including step 12.

4 Rehydrate by taking the material through a descending methanol series (5 min each in 70% methanol, 50% methanol in PBST, 30% methanol in PBST); then wash twice in PBST for 3 min.

5 Block any potentially adhesive sites that could non-specifically bind to the primary or secondary antibody by incubating in blocking solution for a minimum of 60 min at room temperature or overnight at 4°C. Gently rock the specimen to ensure better perfusion.

6 Incubate in primary antibody diluted in blocking solution overnight at 4°C. The optimal titre of both primary and secondary antibody, as well as the time for incubation (and subsequent washes), can vary considerably depending on the antibodies, the age of specimen, fixation procedure, and method of permeabilization, and therefore needs to be determined empirically.

7 Wash four times for 25–30 min in blocking solution.

8 Incubate with biotinylated secondary antibody for 4 hours at room temperature or overnight at 4°C. For example, if your primary antibody was raised in a rabbit, use a secondary anti-rabbit antibody diluted in blocking solution.

9 Wash four times for 25–30 min in blocking solution.

10 During the last wash, gently mix 5 μl solution A with 1 ml blocking solution, add 5 μl B, mix gently and let sit for 30 min without shaking or rocking. Avidin and biotinylated peroxidase (components A and B from the standard ABC Vector Kit) form a complex during this step.

11 Incubate embryos for 45 min in AB complex at room temperature. Do not incubate longer than 1 hour maximum.

12 Wash three times (30 min each) in blocking solution and once with PBST (30 min).

13 Add 0.1 ml 20× DAB (20 mg/ml) to 1.9 ml PBST, and incubate the embryos for 15 min in this solution. DAB is carcinogenic and extreme care must be taken when working with it.

14 Add 2 μl 0.3% H_2O_2 to 1 ml DAB–PBST solution, mix well and stain the embryos, while carefully monitoring the colour reaction. If the staining reaction is very slow, it is possible to increase the amount of H_2O_2 by a factor of 5.

Protocol 3 continued

15 Stop the colour reaction by washing with PBST several times. Make sure that DAB is inactivated appropriately.

16 The precipitate is stable in alcohol. Therefore the specimen can be stored in 4% formaldehyde in PBST, or in methanol (take the specimen stepwise through an increasing methanol series of 30%, 50%, 70%, 90%, and 100% methanol).

The avidin/biotinylated peroxidase system has an amplification effect as it allows for binding of a number of peroxidase molecules per epitope. An even more significant amplification can be achieved by using the Elite kit from the same supplier, but this is only required if it is difficult to detect the antigen .

For strongly expressed proteins an alternative (and cheaper) method is to use a secondary antibody that is linked directly to the detection enzyme of choice (usually horseradish peroxidase or alkaline phosphatases). Of course this also includes using secondary antibodies covalently linked to fluorochromes such as rhodamine and fluorescein. When using these at step 8 of the above protocol, steps 10 through 16 are superfluous.

3.2 Antibody staining of zebrafish larvae

Early embryos can be permeabilized with ease and therefore require no special treatment for antibody staining other than the measures outlined above. However, embryos that are older than 24 hours, and particularly ones that are older than 4 days, have an epithelial skin layer and extracellular matrix components that are resistant to methanol or acetone treatment. Hence, additional measures are necessary to obtain satisfactory results when staining older larvae. The following protocol works well even for epitopes expressed in chondrocytes, which are completely surrounded by extracellular matrix and therefore difficult to stain.

Protocol 4

Antibody staining of zebrafish larvae (cartilage proteins)

Reagents

- Blocking solution as in *Protocol 1*
- 10 mg trypsin/10 ml PBST/1 mM EDTA (ethylenediaminetetraacetic acid)
- 50 mg hyaluronidase/10 ml PBST
- Staining buffer: 100 mM Tris pH 9.5, 50 mM MgCl$_2$, 100 mM NaCl, 0.1% Tween-20, 1 mM levamisol (add fresh). Levamisol inhibits

acidic phosphatases, which can, depending on the stage of the fixed material, lead to staining in tissues expressing phosphatases, such as retina and intestinal tissues.
- NBT, 75 mg/ml in 70% dimethylformamide
- BCIP, 50 mg/ml in dimethylformamide

Method

All steps are carried out at room temperature unless stated otherwise.

1 Fix anaesthetized larvae in 4% paraformaldehyde for 1–4 hours.

2 For storage, rinse twice with PBST and place fish in methanol for at least 10 min at room temperature, or for longer periods at −20 °C.

3 Transfer fish for 5 min to acetone (this only applies if the epitope is stable in acetone).

Protocol 4 continued

4 Rinse 3 × 5 min with PBST.

5 Incubate fish in 0.1% trypsin in 1 mM EDTA in PBS (30 min at 37 °C).

6 Rinse 2 × 2 min with PBST.

7 Incubate in 0.5% hyaluronidase for 30–60 min at 37 °C.

8 Rinse 2 × 2 min with PBST.

9 Wash 60 min in blocking solution.

10 Add primary antibody at the required dilution in blocking solution.

11 Incubate overnight at 4 °C or 4 hours at room temperature.

12 Rinse 2 × 5 min in blocking solution and wash 2 × 20, then 2 × 30 min in blocking solution.

13 Add secondary antibody (coupled to alkaline phosphatase) at the required dilution in blocking solution.

14 Incubate overnight at 4 °C or 4 h at room temperature.

15 Rinse 2 × 5 min in blocking solution and wash 2 × 20, 1 × 30, and 1 × 60 min in blocking solution.

16 Incubate embryos in staining buffer (3 washes, 5 min each).

17 Proceed with staining as described in *Protocol 1* (steps 20–22).

3.3 Double *in situ* hybridization combined with antibody staining (triple stain)

In some cases, it will be necessary to combine double *in situ* hybridization with the detection of a protein (*Figures 2* and *3*, see also *Plates 2* and *3*). The main task here is to minimize the time required to carry out all incubations and wash steps, and to minimize the background staining which, of course, will build up with every additional staining procedure.

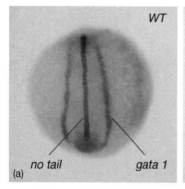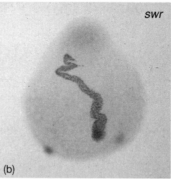

Figure 2 (see Plate 2) *In situ* hybridization combined with antibody staining. Embryos were processed according to (a) Five-somite-stage embryo (dorsal view) hybridized with the a digoxigenin-labelled probe against the *gata1* gene, depicted in blue. Subsequently, staining for detection of the No-tail protein was carried out, using a biotinylated secondary antibody and the avidin/biotinylated horseradish peroxidase system as the detection method. DAB was used as a substrate. (b) Mutant *swirl* embryo, processed as in (a). Note the complete lack of background staining, achieved through titration of all components involved in the antibody procedure.

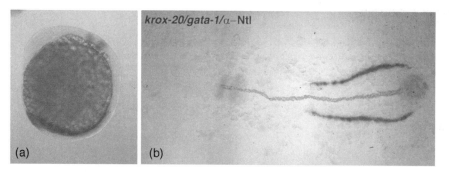

Figure 3 (see Plate 3) Detection of three different gene products within the same embryo (11). A 5-somite-stage embryo was processed according to *Protocol 5*. Following hybridization with antisense probes that were labelled with digoxigenin (myoD, depicted in blue) and fluorescein, respectively (Krox–20, depicted in red), the staining reactions were carried out in that order. Subsequently, the protein distribution of the No-tail protein product was visualized by using an anti-No-tail antibody that was detected by a horseradish peroxidase reaction. The embryo was mounted for photography as described in *Protocol 8*, using araldite as a mounting medium.

Protocol 5

Detecting three different gene products via antibody staining followed by *in situ* hybridization using two differently labelled *in situ* probes (triple staining)

Reagents

- Diamino-benzidine at 20 mg/ml in 50 mM Tris pH 7.2
- 0.3% H_2O_2, freshly diluted from 30% stock with ddH_2O

- All other reagents as in *Protocol 2*

Method

Follow *Protocol 2* up to and including step 17, unless the antigen you intend to detect via antibody staining is sensitive to proteinase K digestion. This needs to be tested for each epitope and antibody individually. Most antigens will not withstand proteinase K digestion, so steps 8, 9, and 10 will need to be omitted in these cases.

18 Add anti-digoxigenin Fab–alkaline phosphatases at a 5000-fold dilution in blocking solution, as well as the primary antibody of choice, and shake for 4 hours at room temperature or overnight at 4 °C.

19 Wash 4 × 20 min each with blocking solution.

20 Add secondary antibody (peroxidase coupled) in the suggested dilution in blocking solution and shake for 4 hours at room temperature or overnight at 4 °C.

21 Wash three times (20 min each) with blocking solution, and twice with PBST (15 min each).

22 Add 0.1 ml DAB stock (20 mg/ml) to 1.9 ml PBST; incubate the embryos for 15 min in this solution to equilibrate them. Then add 2 µl 0.3% H_2O_2 to 1 ml DAB–PBST and stain the embryos in this solution.

23 Stop the colour reaction by adding PBST several times; wash carefully.

24 Wash three times (5 min each) in staining buffer.

Protocol 5 continued

25 Incubate in staining buffer with 4.5 μl NBT and 3.5 μl BCIP per ml. Alternatively, use the BM Purple substrate from Roche (use undiluted or diluted 1:1 with staining buffer, warm up to room temperature before adding to samples, and mix well).

26 Stain for 30 min to overnight in the dark, then stop the reaction by washing excessively in PBST.

27 Incubate for 10 min in 100 mM glycine pH 2.2 (this is a crucial step that inactivates the first alkaline phosphatase).

28 Wash 4 times (10 min each) in PBST.

29 Block for at least 1 h with blocking solution.

30 Add anti-fluorescein antibody at a 1:5000 dilution in blocking solution and shake for 4 hours at room temperature or overnight at 4 °C.

31 Wash 4 times (25 min each) with blocking solution, then 3 times (5 min each) in Fast Red staining buffer.

32 Stain for 1–3 hours, then stop reaction by washing excessively in PBST.

33 Fix for 1 hour in 4% paraformaldehyde in PBST.

34 Transfer material stepwise into 87% glycerol, and avoid anhydrous mounting methods: the Fast Red signal quickly fades away otherwise.

3.4 Pre-absorption of antibodies

Invariably, even when taking good care during all steps of the aforementioned procedures, unspecific background staining will develop in same cases. This might be due to a number of causes. One scenario, where unincorporated mononucleotides within the probe are the problem, has been discussed (see above). Other reasons can stem from the antibody that is used to detect the epitope. The concentration of the antibody might be too high, or the antibody might stick to particular regions of the embryo and result in a high background. In these cases it might be necessary to pre-absorb the antibody against zebrafish acetone powder. This method can help considerably in reducing background staining, and some researchers include this step routinely in their staining protocol by absorbing all antibodies, primary and secondary, against acetone powder.

Protocol 6

Pre-absorption of antibodies

Reagents

- Blocking solution as in *Protocol 1*
- Acetone
- Filter paper

- Bovine serum albumin (BSA) (2 mg/ml) in PBST

Method

1 Collect a large number of zebrafish larvae and overanaesthetize (it is important to use larvae that do not express the protein that you intend to detect). Collect material in a minimal volume of embryo medium. Alternatively, use adult tissue that has been ground up in liquid nitrogen.

2 Add 4 volumes of ice-cold acetone, mix well and keep on ice for 60 min. Spin down at 10 000 rpm for 15 min. Discard supernatant, add 4 volumes of acetone and repeat centrifugation step.

Protocol 6 continued

3 Allow embryo/tissue powder to dry on filter paper. Transfer powder into pre-weighed tube and dissolve at 0.2% (w/v) in 2 mg/ml BSA in PBST. Store at 4 °C.

4 Add 2 μl antibody to 400 μl of this solution and mix gently overnight at 4 °C. Spin at 13 000 rpm for 20 min. Remove the required amount and dilute to the appropriate concentration in blocking solution of choice.

4 Mounting

Mounting is an essential part of analysing and documenting your specimen. If one considers the amount of time it usually takes to go through fixing your sample, preparing probes and antibodies and carrying out various incubations and staining procedures, the actual mounting takes very little time, but is often the point where least care is taken and most information is lost. There are different ways of mounting, depending on the kind of specimen and also on personal preference.

4.1 Viewing chambers for observing embryos

Viewing embryos using a compound microscope requires placing them between a glass slide and a cover slip (*Figure 4*). For constructing this simplest of viewing chambers, one small cover slip (22 × 22 mm) is glued on to each end of a glass slide, with one side of each cover slip being flush with the long side of the glass slide. A second pair of cover slips is then glued on top of the first set. Dechorionated embryos in E3 medium (*Protocol 4* of Chapter 1) are then placed into this viewing chamber, and a large cover slip put on top of them. With some practice, dechorionated embryos fit snugly into this set-up, and can be moved around by carefully adjusting the position of the large cover slip.

An alternative is to use kneading mass (plasticine) as the spacer material, placed at the four corners of the cover slip. While this allows for greater flexibility in adjusting the height of the viewing chamber, it requires careful placing of the cover slip to ensure that it is parallel to the glass slide.

4.2 Methyl cellulose mounting

In many labs a preferred way to look at living material is to mount in methyl cellulose (2–3% in E3 medium). Methyl cellulose is very viscous and takes some time to dissolve, but it is

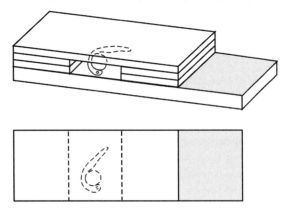

Figure 4 Viewing chamber.

optically clear and perfect for moving around even fragile blastula- and gastrula-stage embryos.

Protocol 7

Preparing and using methyl cellulose for mounting gastrula-stage embryos

Equipment and reagents

- Methyl cellulose at 2–3% in E3 medium
- Nylon loop or blunt-end glass rod
- Depression glass slide

Method

1 Prepare 3% methyl cellulose solution in E3 medium. Once dissolved, store at −20°C in small aliquots). When thawed, centrifuge briefly in a microfuge to remove all air bubbles.

2 Place two drops of methyl cellulose on to a glass depression slide (or in the middle of a viewing chamber as described above). Transfer embryo in minimum amount of E3 medium into methyl cellulose. Make sure the embryo is covered by methyl cellulose, as the E3 medium will be quickly taken up by the methyl cellulose and the embryo becomes hard to manipulate otherwise.

3 Using a nylon loop or a hair loop, orientate the embryo within the methyl cellulose.

4 If so desired, place a cover slip over the methyl cellulose using kneading mass (plasticine) as a support. Make sure that the embryo does not dry out, which is a danger when using methyl cellulose.

4.3 Araldite mounting

Araldite is a common embedding medium used for sectioning in electron and light microscopy. It is also very useful for obtaining beautiful flat-mounts from fixed and stained material between the tailbud and 15–18 somite stages.

Protocol 8

Flat-mounts of early to mid-somitogenesis stages in araldite

Equipment and reagents

- Araldite
- Dissection needle
- Depression glass slides

Method

1 Dehydrate the specimen in an increasing methanol series up to 100% methanol (30%, 50%, 70%, 90%, 96%, 100% and 100% methanol for 15 min each). With a dissection needle, remove as much of the yolk as possible without damaging the embryo proper.

2 Transfer embryo in very little methanol into a large drop of araldite on a glass depression slide. Gently move the embryo around to achieve some mixing of methanol and araldite. The embryo turns soft in araldite, and is no longer brittle.

3 Continue to remove yolk granules from the embryo with the dissection needle.

Protocol 8 continued

4 Using the needle, take the embryo up and place it into a small drop of fresh araldite on a glass slide. Orientate the embryo, and then carefully lower a cover slip on to the araldite. Practice to determine the amount of araldite required beforehand.

It is possible to make small final adjustments of the embryo's position by gently pushing the cover slip. In most cases, the embryo will not move easily and a bit of patience is required, as the araldite is very viscous. There is no need to bake the araldite, the preparation will harden over time.

4.4 Mounting in benzyl benzoate/benzyl alcohol

Compared to the previous method, benzyl benzoate/benzyl alcohol (BB/BA) has an advantage in that its optical refractory index is identical to that of fixed yolk, making the yolk virtually disappear in BB/BA preparations. This is often very desirable and makes up for the disadvantage of BB/BA, namely that it renders embryos extremely brittle, therefore making it difficult to orientate specimens without damaging them.

Protocol 9

Mounting in a 2:1 mixture of BB/BA

Equipment and reagents

- Benzyl benzoate and benzyl alcohol, mixed in a 2:1 ratio
- Viewing chamber as described in Section 4.1

Method

1 Embryos are dehydrated in an increasing methanol series and transferred to a 2:1 mixture of BB/BA. Allow the embryos to equilibrate for a few minutes.

2 Glue two coverslips, each at a distance of about 10–15 mm apart, on to a glass slide, to create a viewing chamber. Place two drops of BB/BA between the coverslips.

3 Carefully place the specimen on to the viewing chamber. Orientate the specimen and lower the coverslip on to the embryo. The BB/BA is rather liquid and, in more cases than not, the embryo will lose its desired orientation.

4 With a dissection needle, and by gently pushing the top coverslip, orient the embryo into its desired position. If possible, do so right next to the microscope that you intend to use to take pictures, as the orientation of the embryo is still sensitive to vibrations and to the glass slide being carried around.

5 After documentation or observation, and if you would like to salvage your specimen, place the slide and its contents into a flat dish containing methanol. Remove the cover slip (use extreme care), and transfer the embryo back to its original vial for further storage.

5 Preparing sections

In most cases, due to the small size of the zebrafish embryo and larvae, it is sufficient to carry out antibody staining or *in situ* hybridization in whole mounts and to examine the results without sectioning. However, there are cases where sectioning is necessary: First, there are tissues such as cartilage and bone that are so dense that, in the first place, staining might not work in whole mounts, due to penetration problems. Secondly, one might need to examine stained structures that are not accessible optically in whole mounts (for example, cell layers in the retina or proximal structures in a pigmented embryo). Thirdly, whenever fine resolution at the single-cell level is required, sections will, in the majority of cases, be superior to whole mounts.

The following protocols work well for most applications, but are by no means exhaustive. *Protocol 11* covers those cases where staining needs to be done on sectioned material.

The literature about preparation of material for electron microscopy and for preparing cryosections is extensive (5), and we have not attempted to cover this methodology here. The reason for this is that most labs that have access to an electron microscope facility also have access to trained and experienced personnel who can assist them in all aspects of tissue fixation and preparation. Also, there are monographs, such as the one from Bozzola and Russell (6), that cover this technology in great detail and that can be used for reference.

5.1 Plastic sections using Technovit®

Technovit® is one of the most commonly used embedding resins, the main advantages being speed of processing and ease of performing sections.

Protocol 10

Technovit® embedding and sectioning

Reagent

- Technovit® (Fluca)

Method

All incubations are done at room temperature and with gentle agitation.

1 Fix fish in 4% paraformaldehyde in 75 mM phosphate buffer, pH 7.4, for 30 min at room temperature and store at 4 °C overnight.

2 Dehydrate in 30%, 50%, 70%, 90%, 96%, 100% and 100% ethanol for 15 min each.

3 Incubate twice in 1 : 1 100% ethanol/Technovit® for 30 min.

4 Incubate in 100% Technovit® (2 × 60 min each).

5 Transfer fish into freshly made Technovit® containing starter (1 : 15) and orientate without delay in embedding moulds (solidifies rapidly depending on ambient temperature). Cover with plastic slips to avoid air contact and leave to polymerize overnight.

6 Section blocks at approximately 1–4 μm using a glass knife.

7 Transfer sections on to a drop of water on a glass slide; dry water off on a heated plate and stain sections with Toluidine Blue as described below.

5.2 Paraffin sections combined with antibody staining

In situ hybridization is an easy and widely applicable method for detecting the expression of genes of interest. The methods described above work robustly as long as the tissue or age of the specimen still allows the probe and the antibody, respectively, to penetrate the sample efficiently. For older samples (day 4 and older), this can be a limitation, and if, in addition, the tissue in question is very dense (e.g. bone and cartilage), other methods might be required in order to ensure that the *bona fide* expression can be detected.

Protocol 11

Preparing paraffin wax sections for subsequent antibody detection

Reagents

- Hyaluronidase (25 mg/ml in PBS)
- Avidin-peroxidase and DAB as in *Protocol 3*

Method

1 Fix material in 4% paraformaldehyde overnight at 4°C. Wash 2 × 5 min each in PBS.

2 Dehydrate through a series of 30%, 50%, 70%, 90%, 95%, and 100% ethanol. Transfer for 30 min each to ethanol/xylol, 100% xylol, paraffin/xylol, then to paraffin at 60°C. Change paraffin twice after 4 h. Orient embryos in embedding moulds and allow to solidify. Section using glass knives and mount sections on glass slides. Bake at 50°C.

3 Remove paraffin through 2 washes in xylol (5 min each) and 2 washes in ethanol. Rehydrate through washes in 70%, 50%, 30% ethanol (5 min each), then in PBS (2 × 2 min, 2 × 5 min). Digest with hyaluronidase for 30 min at 37°C.

4 Wash using PBST (2 × 2 min, 2 × 5 min).

5 Wash in methanol containing 1% peroxidase for 30 min at room temperature in order to block endogenous peroxidases.

6 Rinse with PBST (2 × 2 min, 2 × 5 min).

7 Incubate 30 min in 5% goat serum in PBST.

8 Incubate for 60 min in primary antibody in PBST containing 5% goat serum.

9 Rinse with PBST (2 × 2 min, 3 × 5 min).

10 Incubate in biotinylated secondary antibody in PBST for 60 min at room temperature.

11 Rinse with PBST (2 × 2 min, 3 × 5 min).

12 Incubate with avidin–peroxidase, diluted 1:200 in PBST for 30 min.

13 Rinse in PBST (2 × 2 min, 3 × 5 min).

14 Add DAB in PBST, final concentration 0.3 mg/ml. Incubate for 5 min.

15 Add H_2O_2, final concentration 0.003–0.03%.

16 Monitor staining reaction carefully, and stop by washing with PBST.

5.3 Embedding and sectioning using JB-4 resin

Protocol 12

Sectioning JB-4-embedded material

Equipment and reagents

- Ethanol
- JB-4
- Embedding moulds (plastic, gelatinous—do not use silicone)

Method

1 Anaesthetize fish in Mesab and kill using 50% ethanol.

2 Dehydrate through 70% and 100% ethanol (30 min to 4 hours, depending on size).

3 Transfer fish to 50% ethanol/50% JB-4, and incubate for 2–12 h.

Protocol 12 continued

4   Transfer fish to 100% JB-4 and incubate for a minimum of 12 h.

5   Wearing gloves, mix 15 parts JB-4 with 1 part 1 JB-4 Solution B. Embed fish in gelatinous capsules or plastic chambers (do not use silicon), then put into a nitrogen chamber and cool to 4 °C. Orient again after 30 min, then allow to solidify for a minimum of 12 h.

5.4 Toluidine Blue staining of semi-thin sections

This is one of the most routine procedures for staining sections. Most stains do not penetrate well in plastic sections (araldite, Epon, and mixtures thereof), but Toluidine Blue, Methylene Blue and basic fuchsin (pararosaniline) do, and the respective staining procedures are very quick and straightforward. If using a combination of Toluidine Blue and fuchsin, as described below, basophil structures (such as chromatin) and osmophilic structures (such as membranes) are stained bright blue (7, 8), while connective tissue stains red (7).

Protocol 13

Toluidine Blue staining of sections

Equipment and reagents

- 0.5% Toluidine Blue in a 1% sodium tetraborate solution. The solution needs to be filtered (5) and is then best stored as aliquots in centrifuge tubes. Centrifuge the solution just prior to use
- 1% Fuchsin (basic) in distilled water
- Glass slides

Method

1   Place sections (0.5 μm to 5 μm thickness) on a drop of 15% ethanol in distilled water on to a glass slide.

2   Stretch the section by placing the glass side on to a hotplate (60–80 °C). Let the section dry and stain with a drop of Toluidine Blue for 10–60 s.

3   Destain immediately under running tap water. The intensity of the staining depends on the thickness of the section, the staining time, the temperature during the staining step, the pH of the Toluidine Blue solution, and the hydrophilicity of the resin (epoxy resins stain somewhat weaker than methacryl resin, for example).

4   If required, overstained sections can be partially destained by treating the sections with solvents, such as ethanol at 70%.

5   Counterstain, if so desired, with fuchsin on a hotplate (60 °C) for 1–2 min. Rinse with tap water and dry.

6   If sections are to be kept in the long term, or if the sections are going to be examined using an oil-immersion lens, a permanent specimen can be obtained by overlaying the section with a drop of Epon or araldite that is kept for this purpose in small aliquots in Eppendorf tubes at −20 °C. Make sure to practise to establish the required amount of Epon/araldite to use on an empty slide. Carefully lower a large cover slip on to your section and monitor the even spreading of the Epon/araldite. Polymerize overnight in an oven (60–80 °C).

6 PTU treatment to prevent melanization of embryos

One of the main advantages of using zebrafish is the transparency of the embryos. The optical clarity of the embryos, which makes observing development at early stages so much easier than in other systems, starts to become restricted at around 32 hours post-fertilization, when melanization starts to get in the way of observing the segmented zebrafish embryo. One way around this is to use unpigmented embryos, e.g. mutants in the *albino* locus. Another, more practical, solution, which is applicable without crossing schemes involving *albino* mutant alleles, is the inhibition of melanization by exposing embryos to 1-phenyl-2-thiourea (PTU). While this method works very effectively when applied no later than 24 hours post–fertilization, it causes a mild retardation of the embryos/larvae and oedema often forms on days 4 and 5.

The embryos are placed in E3 medium containing PTU at a final concentration of 0.2 mM. Use gloves at all times and avoid contact with PTU-containing solutions as PTU is extremely toxic. Pronase may be used to remove the chorion.

7 Endothelial cell staining

At roughly 30 hours of development the heart starts to beat in the zebrafish embryo. A few hours later, one can see the first blood cells moving around and, slightly later, vessels form throughout the embryo. Endothelial cells line the lumen of the blood vessels. In the zebrafish larvae, they exhibit high levels of endogenous alkaline phosphatase, which makes them easily detectable by simple enzymatic staining (*Figure 5*, see also *Plate 4*).

The following protocol was used successfully in a large mutant screen, where thousands of embryos were routinely stained per week in parallel, and has proved to be extremely robust. The protocol can be applied for larvae between 72 and 120 hours of age.

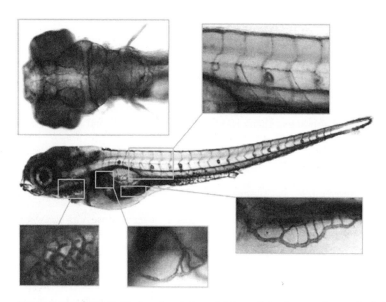

Figure 5 (see Plate 4) Alkaline phosphatase staining outlining the vasculature of a 4-day-old zebrafish larva. The larva was processed exactly as described in *Protocol 14*. Clockwise from upper left: head vessels, dorsal view; intersomitic vessels, lateral view; subintestinal vessel, lateral view; vessel of the pectoral fin, lateral view; gill vessels, lateral view.

Protocol 14

Alkaline phosphatase staining of endothelial cells

Reagents

- 4% paraformaldehyde in PBS/0.1% Tween-20 (freshly prepared, can be stored at −20 °C)
- Methanol
- Acetone at −20 °C
- 50 × PTU stock in H_2O: 0.01 M; final PTU concentration: 0.2 mM in E3

- NTMT buffer: 100 mM Tris pH 9.5, 50 mM $MgCl_2$ 100 mM NaCl, 0.1% Tween-20
- NBT: 75 mg/ml in 70% dimethylformamide
- BCIP: 50 mg/ml in dimethylformamide

Method

1 To prevent melanization of embryos, treat embryos with PTU, starting at 24 hours post-fertilization, by simply adding an appropriate amount of PTU stock to the embryo medium.

2 Fix embryos at 72–96 h with 4% paraformaldehyde in PBST for 30 min at room temperature.

3 Store in methanol, if so desired: transfer larvae into 50% methanol in PBST for 5 min, then into 100% methanol. Store at −20 °C.

4 Treat embryos with pre-cooled acetone for 30 min at −20 °C.

5 Rinse with PBST twice (5 min each) and equilibrate embryos with NTMT three times for 15 min at room temperature.

6 Start the staining reaction by incubating in 2.25 µl NBT, 1.75 µl BCIP per 1 ml NTMT. Staining usually takes about 15–30 min.

7 After staining is completed, wash three times in PBST.

8 Transfer embryos to 50% glycerol for 5 min, then to 87% glycerol. Larvae can be stored and documented in 87% glycerol. For long-term storage, keep specimen at −20 °C to inhibit bacterial growth.

Acknowledgements

We would like to acknowledge H. Habeck, T. Trowe, and R. Dahm, who have contributed protocols or have shared new developments with us in the preparation of this chapter. *Figure 1* was prepared by K.-H. Nill. *Figure 3* was contributed by Y. Kishimoto, *Figure 4* by E. Ober, and *Figure 5* by H. Habeck. We would also like to thank H. Schwarz, F. van Bebber, and R. Dahm for helpful comments on the manuscript.

References

1. Cox, K. H., DeLeon, D. V., Angerer, L. M., and Angerer, R. C. (1984) *Dev. Biol.*, **101**, 485.
2. *DIG user's guide*. Roche Diagnostics GmbH.
3. *In situ hybridization manual*. Roche Diagnostics GmbH.
4. Humason, G. L. (1979). *Animal tissue techniques*, (4th edn). W.H.Freeman and Company, San Francisco.
5. Glauert, M. G. (1975). *Fixation, dehydration and embedding of biological specimens*. North Holland Publishing Company, Amsterdam.
6. Bozzola, J. J. and Russell, L. D. (1999). *Electron microscopy*, (2nd edn). Jones and Bartlett Publishers.
7. Romeis, B. (1989). Urban & Schwarzenberg, München, (17th edn), p. 248.
8. Richardson, K. C., Jarret, L., and Finke, E. H. (1960) *Stain. Technol.* **35**, 313.
9. Schulte-Merker, S., Lee, K. J., McMahon, A. P., and Hammerschmidt, M. (1997). *Nature*, **387**, 862.
10. Schulte-Merker, S., Ho, R. K., Herrmann, B. G., and Nüsslein-Volhard, C. (1992). *Development*, **116**, 1021.
11. Ober, E. A. and Schulte-Merker, S. (1999). *Dev. Biol.*, **215**, 167.

Chapter 3

The morphology of larval and adult zebrafish

Thomas F. Schilling

Department of Developmental and Cell Biology, University of California, Irvine, CA 92697–2300, USA

1 Introduction

With its emergence as a popular model system in developmental genetic studies, the zebrafish, *Danio rerio* (Teleostei, Cyprinidae), needs accurate anatomical descriptions and guides to staging (1). Staging by morphology helps resolve the problem of asynchrony during development and provides better accuracy in developmental analyses. There is a staging series for the embryonic period in zebrafish (2), but not for larvae and adults. This chapter brings together information on larval and adult morphology, most of which has been published separately in experimental papers. Although not a detailed staging series or atlas, it is meant to serve as a brief guide to late stages of zebrafish development in the laboratory, and to summarize knowledge and publications on zebrafish anatomy.

2 Body proportions and surface features

2.1 Adult

Zebrafish are minnows native to streams and rivers of northern and central India (*Figure 1*) (1). Minnows belong to the superorder Ostariophysi, which contains three-quarters of the world's freshwater fishes and includes the Otophysi: the Cypriniformes (minnows and carps), the Characiformes (tetras), and the Siluriformes (electric fish and catfishes). Members of this group are characterized by specialized skeletal features, including a Weberian apparatus (bones and ligaments linking the swim bladder with the inner ear) as outlined by Fink and Fink (3). Like all minnows, zebrafish have a single dorsal fin and no adipose fin. Their fins do not contain true spines but instead have hardened rays. Minnows also lack teeth in the jaw, and instead they have pharyngeal 'jaws', with 1–3 tooth rows that grind food in the back of the throat. These teeth are usually fused to a modified pharyngeal bone of the most posterior gill arch. Although the phylogenetic relationships among the minnows are unclear, recent molecular phylogenies suggest that the *Danios*, including *Danio rerio* (formerly *Brachydanio rerio*), constitute a monophyletic group (4).

Like other fishes, zebrafish pass through larval and juvenile stages and then continue to grow as adults throughout life. In the laboratory, adults, defined here as older than 90 days post-fertilization (dpf), initially have an average total length of 2–3 cm when raised in ideal conditions, but continue to grow during a life span of 2–3 years and can reach sizes of 4–5 cm. General characteristics of zebrafish, including coloration and external morphology, have been described elsewhere (1, 5). Their bodies are shallow and elongate, with a short head, a pointed snout, and an upturned, slanted mouth (*Figure 1*). The upper jaw is highly protrusible,

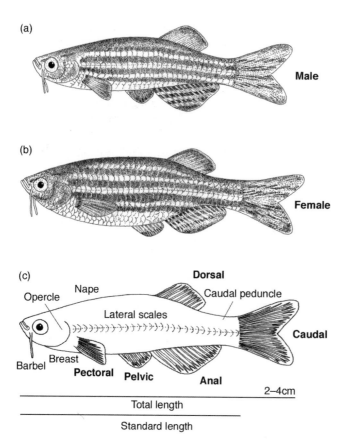

Figure 1 Camera lucida drawings of adult male and female zebrafish. As in most other figures, anterior is to the left and dorsal to the top. Pigmentation and surface features are omitted in (c); fin rays are outlined and major body landmarks and fins are labelled. Total and standard length measurements are indicated by bars at the bottom.

extending the gape of the mouth and aiding in suction feeding. On either side of the mouth are thin barbels (approximately 1.0 mm in length) that hang ventrally. Like other minnows, the single dorsal fin and paired pelvic fins are displaced posteriorly, providing rapid acceleration during swimming; the dorsal fin originates at the same anterior–posterior level as the origin of the anal fin. Males are generally more slender and darker in colour than females.

The most striking feature of the zebrafish is its pattern of longitudinal black and white stripes along the sides of its body and on the anal and caudal fins. Adults have five alternating blue-black stripes, containing two types of pigment cells, melanophores and iridiophores, and silvery-yellow stripes containing xanthophores and iridophores. From above, they appear dark olive or grey, males slightly more yellow/brown than females. They have lighter stripes along their sides, and dull yellow to white below, with the black lining of the peritoneum showing through. Breeding males are dark with a dark yellow tint to their undersides and fins. Of the total body length only 10–15% is head. The body is usually less than 1 cm in depth and even less wide, only 0.2–0.3 cm (*Figure 1c*). Zebrafish have a complete lateral line, running from the operculum to the tail fin. Specialized lateral scales (adults have 28–34) run along the midbody, within one of the silver-yellow stripes, and are associated with the lateral row of neuromasts. There are also species-specific numbers of rays in each of the fins (10 pectoral; 7 pelvic; 8 dorsal; 13 anal; 23 caudal), although these can vary slightly among individuals.

2.2 Larvae and juveniles

Larval life in the zebrafish officially begins at 72 hours post-fertilization (hpf), when the young fish are approximately 3.5 mm in length and nearly free-swimming (2). This is approximately 1 day after hatching and a few days before the switch from internal yolk reserves to external food sources. During the larval period, defined here as the following 27 days, zebrafish raised in the laboratory in ideal conditions grow an additional 7 mm, approximately 2 mm/week (*Tables 1, 2* and *Figures 2, 3*), but maturation is highly asynchronous, therefore in this chapter the stages are by body length (in millimetres).

Table 1 Periods of larval and adult development

Period	Day (length)	Description
Early larva	3 (3.5 mm)	Free swimming; startle response; upright orientation; optokinetic response
Mid-larva	14 (6 mm)	Swim bladder; food seeking; growth
Juvenile	30 (1 cm)	Adult fin and pigmentation patterns
Early adult	90 (2 cm)	Breeding
Lifespan	1000 (4–5 cm)	Death

Table 2 Stages of larval development

Stage	Day	Description
Early larva	3	TL = 3.5 mm; StL = NA; 6 teeth
		Open mouth protrudes anterior to eyes; iridophores in yolk stripe and half of eye; gill filament buds; cartilage in arches and pectoral girdle; operculum rudiment; cleithrum; motile intestine
	5	TL = 3.9 mm; StL = NA; 6 teeth
		Swim bladder inflates; active feeding; pronephric tubules
	7	TL = 4.5 mm; StL = NA; 8 teeth
		Tail fin narrows; cartilage in anterior vertebrae; 3 lines of trunk neuromasts
Mid-larva	14	TL = 6.2 mm; StL = NA; 10 teeth
		First hypural cartilage in tail fin; Rohon-Beard neurons completely replaced by dorsal root ganglia
	21	TL = 7.8 mm; StL = 7.0 mm; 10 teeth
		Ossified pharyngeal skeleton; radials in pectoral fins; dorsal and anal fin buds
Juvenile	30	TL = 10.0 mm; StL = 9.0 mm; 10 teeth
		Adult fin and pigmentation patterns; 7 hypural cartilages in tailfin; old mid-body lateral line shifts ventrally and new mid-body line forms; kidney haematopoietic; mesonephros replaces pronephros
	45	TL = 14.0 mm; StL = 12.5 mm; 12 teeth
Adult	90	TL = 18.0 mm; StL = 15.0 mm; teeth?

TL: total length; StL: standard length; NA: not applicable.

There are descriptions of the external morphology and skeletal development of larval stages, but no detailed staging series (5, 6). This chapter defines two periods, larval and juvenile, before and after 1 month (10 mm) of development. However, there is no dramatic metamorphosis between these periods. Rather, larval development ends and juvenile development begins when the adult pattern of fins and fin rays emerges, scale development has

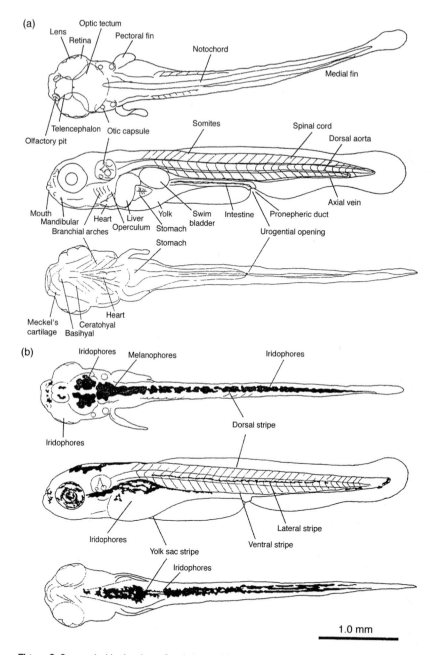

Figure 2 Camera lucida drawings of early larvae (4 mm), outlining major internal features (a) as well as larval pigmentation (b). In each series, larvae are shown in dorsal, lateral and ventral views, and major landmarks are labelled. Scale bar = 1.0 mm.

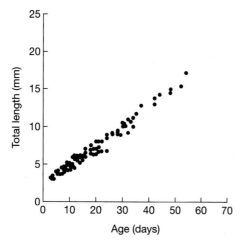

Figure 3 Rates of development for larvae and juveniles incubated at 28.5 °C. Total length in millimetres (mm) is indicated on the vertical axis, and age in days post-fertilization (dpf) on the horizontal axis. Notice in the figure that developmental rates are almost linear. Hence, in ideal conditions, the developmental rate may vary as a linear function of incubation temperature, as has been demonstrated for embryos (2).

begun, and ossification of the skull bones is complete. In zebrafish, as in other minnows, this transformation is gradual (7, 8).

Early larval zebrafish are long and narrow, with short, rounded heads that become more pointed as the mouth protrudes anteriorly (*Figure 2*). By 5 dpf (4 mm), larvae have well-developed vision and begin active feeding. A continuous, median fin surrounds the trunk and tail at this stage, with only a small interuption at the urogenital opening. Due to the transparency of zebrafish larvae in unpigmented regions, internal features are visible (*Figure 2a*). These can be seen even more clearly in pigmentation mutants, such as *golden, sparse, transparent* and *albino*, which develop normally in other respects, or in embryos which have been treated with 1-phenyl-2-thiourea (PTU) in order to inhibit melanogenesis (9) (see *Protocol 4*). There are well-developed sense organs, including paired olfactory pits, large eyes that move in synchrony, and otic capsules containing two highly refractile otoliths. When viewed dorsally, major lobes of the brain are visible, particularly the two large hemispheres of the optic tectum which forms the dorsal roof of the midbrain. The most prominent features in lateral views are the air-filled swim bladder and 30–32 muscle segments, or somites, in the trunk. The mouth opens into a segmented pharynx, which forms a series of gill openings between each of the branchial arches. The gut leads posteriorly to a small stomach, a large liver, and a relatively long intestine that empties at the urogenital opening. The intestine becomes motile at 72 hpf (3.5 mm). In ventral views, there are seven pharyngeal arches (mandibular, hyoid and five gill arches), lying between the mouth anteriorly and the pectoral fins, and jaw movements begin at 4–5 dpf (4 mm). The rapidly beating heart in its pericardial cavity lies just posterior to the arches, between the pectoral fins.

There are four early larval stripes of black pigment cells, or melanocytes (10) (*Figure 2b*):

- dorsal, a double row above the brain and spinal cord, which is expanded in the head region;
- lateral, along the horizontal myoseptum;
- ventral, a double row between the eyes and over the yolk sac, particularly around the swim bladder; and
- a yolk-sac stripe, extending in a diamond shape over the ventral surface of the yolk.

The number of melanophores varies within these early stripes, with, for example, up to 21 cells in the lateral band. Silver-white pigment cells, the iridiophores, are fewer in number and concentrated around and on the eyes, over the hindbrain and spinal cord, and along the top of the yolk. Yellow xanthophore pigmentation covers the entire dorsal half of the body at 72 hpf (3.5 mm) in regions devoid of melanophores. Xanthophores become granular at 5 dpf (4 mm) and are found throughout the body. At 8 dpf (5 mm), the lateral midbody stripes of melanocytes become more uniform in distribution as a secondary wave of melanophores develops, giving rise to a total of 30 cells on average along the lateral stripe.

The larval to juvenile transition is a gradual process, starting about 2 weeks post-fertilization (5 mm) (*Figure 4*). Three major changes occur: the development of the adult pattern of pigment stripes, the formation of the adult fins, and the ossification of the skeleton. Larvae retain the embryonic pattern of three major melanocyte stripes along the body—dorsal, median, and ventral; cells of the embryonic yolk-sac stripe intermingle with the ventral stripe. These stripes are then gradually added to, or replaced by, the adult pattern during the third and fourth weeks of development (6–9 mm), resulting in five stripes.

The various fins of the zebrafish develop at completely different stages from one another, replacing the single median fin of the embryo. Paired fins develop first, beginning during embryogenesis, and cartilages of the pectoral girdle differentiate by 72 hpf (3.5 mm) (6). At this stage the pectoral fins maintain a vertical orientation and the fin folds are supported by basal rays. These then gradually rotate into a more horizontal position and distal rays form within the fin folds by 8 mm. In contrast, median and pelvic fins only begin to develop at

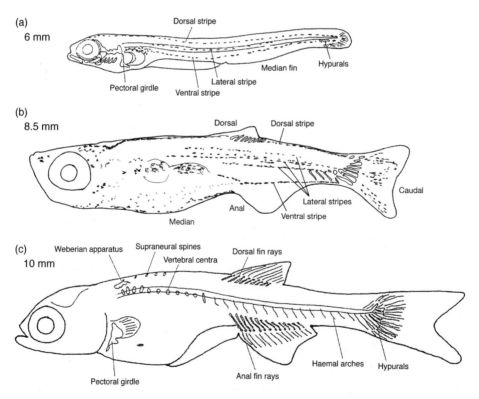

Figure 4 Camera lucida drawings of late larval (a, b) and early juvenile (c) stages, showing emergence of fins and adult pigmentation, from whole-mounted skeletal preparations (see *Protocol 1*).

8 mm. Dorsal, caudal, and anal fins replace the embryonic median fins, and these are first visible as small prominences when the embryonic fin begins to regress. In the tail, the end of the notochord (urostyle) first flexes upward and the caudal fin narrows (5 mm). Soon thereafter the first hypural cartilages can be detected by skeletal staining (6 mm) (*Figure 4a*). By 8.5 mm both dorsal and anal fins emerge, leaving only the most anterior portion of the ventral median fin from the embryo. Cartilage develops in the dorsal fin before it does in the anal fin, and in all fins by the beginning of the juvenile period (10 mm), as do the small ossified elements of the Weberian apparatus, dorsal to the first few vertebrae (*Figure 4c*).

Juvenile zebrafish have an adult pattern of five alternating black and white stripes. They also have scales, which develop starting along the lateral line near the caudal (tail) fin, and then spread anteriorly, as in many other teleosts. Once the full complement of scales is attained, the number remains fixed. Late larval and juvenile stages are hard to recognize, since they vary with the environments of different individuals. In other minnows, the numbers of fin rays, vertebrae, and scales have been shown to decrease with higher developmental temperatures (11). Our stages are based on laboratory fish raised at 28.5 °C and on measurements of maximal length, which should be more accurate than time in adjusting for differences in fish health. Useful measurements for staging are given here, including total body length, standard length, lateral scale number, fin spine numbers, and tooth numbers on pharyngeal jaws (*Figure 1* and *Table 2*).

3 Skeleton and musculature

3.1 Adult skeleton

Skull

Zebrafish, as ray-finned fishes, have many more bones in their skeletons than tetrapods; there are 74 ossifications in the skull alone (*Figure 5*). The anatomy of the cartilages, bones, and teeth of the skull, and their development, has been described in detail using differential staining of the skeleton with Alcian Blue and/or Alizarin Red in whole mounts (6, 11–13) (see *Protocol 1*). In adults, there are two types of cranial bones: endochondral, which make up most of the braincase and pharyngeal skeleton (45 bones), and dermal, which form later in development, primarily over the top of the brain and around the front of the head (29 bones). There are also several different types of cartilage, although most is cell-rich hyaline cartilage, which will not be dealt with in this chapter (14). Rows of teeth are attached to endochondral bone of the fifth branchial arch.

The braincase, or neurocranium, consists of four capsules of endochondral bone that surround the major subdivisions of the brain and sense organs: ethmoid, orbital, otic, and occipital (*Figure 5a–d*). The small bones of the ethmoid, such as the preethmoid and lateral ethmoid, surround the olfactory bulbs in a ring. Like other Cypriniform fishes, zebrafish have an additional bone in this region, the kinethmoid (3). They also lack the basisphenoid, which in most teleosts normally forms in the midline, between the eyes. Other sphenoid bones, together with the sclerotics, surround the orbit of the eye (*Figure 5c*). Three otic bones enclose the otic capsule. Posteriorly, the occipital bones of the basicranium surround the hindbrain and anterior spinal cord (*Figure 5d*).

The pharyngeal arch skeleton, or viscerocranium, contains seven segmentally reiterated sets of cartilages and bones that form a ladder-like array surrounding the pharynx, and here the bones are also endochondral (*Figure 5e*). Each arch may have up to five paired elements, including, from ventral to dorsal, the basi-, hypo-, cerato-, epi- and pharyngo-branchials, and these are each linked by cartilage at points of articulation. These elements are enlarged in the

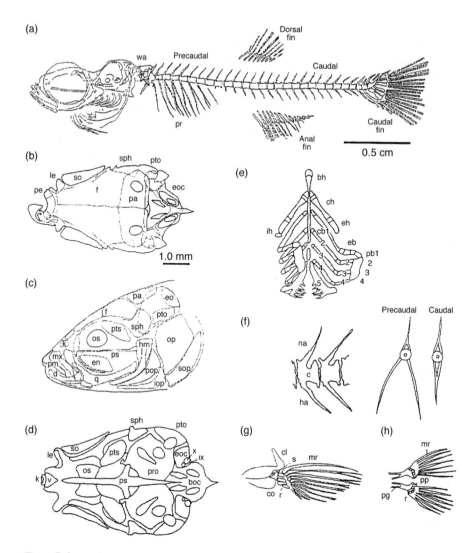

Figure 5 Camera lucida drawings of late juvenile and early adult skeletons, from whole-mounted skeletal preparations. (a) Lateral view of an entire juvenile (1.8 cm), highlighting vertebrae and median fin skeletons. pr, pleural ribs; wa, Weberian apparatus. (b) Dorsal view of juvenile skull (15 mm). eoc, epioccipital; f, frontal; le, lateral ethmoid; pa, parietal; pe, pre-ethmoid; pto, pterotic; so, supraorbital; sph, sphenotic. (c) Lateral view of early juvenile skull (10 mm). d, dentary; en, entopterygoid; eo, epioccipital; f, frontal; hm, hyomandibula; iop, interopercle; k, kinethmoid; mx, maxilla; op, opercle; os, orbitosphenoid; pa, parietal; pm, premaxilla; pop, preopercle; ps, parasphenoid; pto, pterotic; pts, pterosphenoid; q, quadrate; sop, subopercle; sph, sphenotic. (d) Ventral view of juvenile skull (12.5 mm). boc, basioccipital; eoc, epioccipital; k, kinethmoid; le, lateral ethmoid; os, orbitosphenoid; pro, prootic; ps, parasphenoid; pto, pterotic; pts, pterosphenoid; so, supraorbital; sph, sphenotic; v, vomer; IX, glossopharyngeal nerve; X, vagal nerve. (e) Dorsal view of early juvenile hyoid and branchial arches (10 mm). bh, basihyal; cb, ceratobranchials; ch, ceratohyal; eb, epibranchials; eh, epihyal; ih, interhyal; pb, pharyngobranchials. (f) Lateral and frontal views of juvenile vertebrae. c, centrum; ha, haemal arch; na, neural arch. (g) Ventral view of adult pectoral skeleton. cl, cleithrum; co, coracoid; mr, median rays; r, radials; s, scapula. (h) Ventral view of adult pelvic skeleton. mr, median rays; pg, pelvic girdle; pp, pelvic process; r, radials. Scale bars: A = 0.5 cm; B–H = 1.0 mm. From refs 6 and 16.

two anterior arches, the mandibular and hyoid, which form the jaw and its suspension, while the five smaller posterior arches support the gills. Like other Ostariophysans, the pharyngeal jaws (the fifth gill arch) contain a much larger ceratobranchial than other gill arches, and this is attached to several rows of teeth.

The dermatocranium contains 29 bones that form the outer covering of the skull, surrounding the anterior braincase and the first two pharyngeal arches (*Figure 5c*). There are no dermal bones in the occipital skull or in any of the branchial, gill-bearing arches. Only the mandibular and hyoid arches contain some dermal bone, while the gills are covered by a dermally ossified operculum. The olfactory organs and capsule are surrounded by the median vomer and paired nasal bones, as well as the median ethmoid and kinethmoid (*Figure 5d*). There are also dermal sphenoid and sclerotic bones, in addition to their endochondral counterparts, that surround the orbits. Further posteriorly the large frontal bones cover much of the dorsal braincase, a median parasphenoid bone lies between the eyes, and there are five infraorbitals and supraorbitals.

Vertebral column and fins

Zebrafish have relatively few vertebrae 30–32, as compared with many other fish, which are highly similar to one another along the anterior–posterior axis. Each is composed of a centrum encircling the notochord, as well as dorsal neural arches extending above it, around the spinal cord and ventral haemal arches positioned below (3) (*Figure 5a, f*). In teleosts the vertebral centrum strongly constricts the notochord. Each vertebra contains up to eight components: centrum, neural and haemal arches, pleural and epipleural ribs, as well as basiventral and basidorsal cartilages to support the ribs, and possibly a supraneural spine. The dorsal neural spine houses a neural arch and the spinal cord; the ventral haemal arch surrounds the dorsal aorta. Despite their relative uniformity, there are two major classes of vertebrae along the anterior–posterior axis, precaudal in the trunk and caudal in the tail, as well as specialized vertebrae at the anterior and posterior ends of the column (*Figure 5f*). Precaudal vertebrae back to the level of the urogenital opening have pleural ribs, with the anteriormost eight vertebrae supporting the longest, while caudal vertebrae lack ribs and have only haemal arches. In addition to pleural ribs, which form at the junction of the myoseptum with the coelomic wall, there are also dorsal, epipleural ribs in the horizontal myoseptum in some segments. The Weberian apparatus, which connects the swim bladder to the ear, consists of a series of modified neural and supraneural spines of the anterior vertebrae. In the first two vertebrae, the axis and atlas, the centra are expanded and the neural and haemal arches flattened where they connect to the occipital skull.

The paired fins of virtually all freshwater fishes are supported by skeletal bars radiating from the body at the girdle, with integumentary fin rays occupying the distal, vane-like part of the fin and primary fin rays hidden within the body wall musculature (6, 15, 16). In zebrafish there are two paired fins: pectoral and pelvic; and three unpaired fins: dorsal, anal, and caudal (*Figure 5g, h*) The pectoral girdle and fins lie between the head and the trunk. The pectoral girdle is divided into three endochondral bones (coracoid, mesocoracoid, scapula) which support the radials, and four dermal bones (cleithrum, clavicle, postcleithrum, supracleithrum, posttemporal), that connect to the occipital skull, although only a subset are visible in ventral views (*Figure 5g*). There are two rows of radials, proximal and distal. Distally, characteristic numbers of exoskeletal lepidotrichia (fin rays) develop in the pectoral fin folds. These rays are subdivided along their proximodistal axis, and may be branched or unbranched. The pelvic girdle is much simpler (*Figure 5h*). It has no dermal elements, and only one pair of skeletal elements, puboischia, which remain unfused. These articulate with two or three endoskeletal radials.

In the tail of the zebrafish the tail-fin rays are symmetrical along the dorsoventral axis but attach to a series of asymmetrical tail vertebrae that provide their support (*Figures 5a, 7*). Vertebrae in the tail curve dorsally, and their ventral processes, known as hypurals in the most posterior segments and haemal arches anteriorly, extend into the base of the tail. Finally, scales in zebrafish, like other teleosts, are cycloid and composed of an outer bony surface and inner collagenous fibrillar base.

Protocol 1

Differential staining of cartilage and bone in whole larvae and adults by Alcian Blue and Alizarin Red[a]

Reagents

- 0.4% Tricaine® stock (3-amino benzoic acidethylester) (Sigma, USA): 400 mg : 97.9 ml distilled water (dH$_2$O) : 2.1 ml 1 M Tris (pH = 9.0); working solution 0.08%
- Ethanol series (70%, 80%, 96%)
- Acetone
- 0.3% Alcoholic Alcian Blue (0.3 g Alcian Blue, 70 ml absolute ethanol, 30 ml acetic acid)
- 0.1% Alcoholic Alizarin Red S (0.1 g Alizarin Red, 100 ml 96% ethanol)
- 1.0% Aqueous potassium hydroxide (10 g KOH, 1 litre dH$_2$O)
- Glycerine series (20%, 50%, 80%, 100% in 1% KOH)

Method 1: Young larvae

1 Anaesthetize fish in 0.08% Tricaine®.
2 Fix in 3.7% formaldehyde/phosphate-buffered saline (PBS) for 24/72 hours.
3 Rinse in PBS.
4 Stain cartilage in 0.1 mg/ml Alcian Blue in ethanol/acetic acid (4:1).
5 Rehydrate through ethanol series (90%, 50%, 30% in water).
6 Bleach in 1% H$_2$O$_2$, 1% NH$_3$ solution after Alcian Blue staining (optional).
7 Digest overnight in 50 mg/ml trypsin in 30% sodium tetraborate in water.
8 Stain bone with 0.4 ml of Alizarin Red S solution in 10 ml 0.5% KOH.
9 Destain in 1% KOH/glycerine series and store in glycerol.

Method 2: Older larvae or adults

1 Anaesthetize fish in 0.08% tricaine.
2 Fix in 80% ethanol.
3 Remove skin and viscera as soon as firm.
4 Transfer to 96% ethanol for 24 hours.
5 Place in acetone for 48 hours to remove fat.
6 Rinse in 96% ethanol for 1–2 hours.
7 Stain for 2–6 hours, depending on size, in 20 ml of fresh, filtered Alcian Blue solution at 37 °C.
8 Rinse in 96% ethanol for 1–2 hours.
9 Stain for 2–6 hours in 20 ml of fresh Alizarin Red solution at 37 °C.
10 Wash in running tap water for 1–2 hours.
11 Clear in 1.0% KOH until the skeleton has become clearly visible (12–36 hours).

12 Clear for 1 week in graded series of glycerine/KOH.

13 Store in 100% glycerine.

ᵃ Modified from the methods given in refs 6 and 9.

3.2 Larval and juvenile skeletal development

During larval development:

(i)   an early cartilaginous skeleton is largely replaced by a bony one;

(ii) a dermal skull develops superficially

The larval anatomy of the skull and the paired fins has been described in detail (*Figure 6*) (6, 9, 15, 16) as has the larval development of the pharyngeal muscles (9, 17).

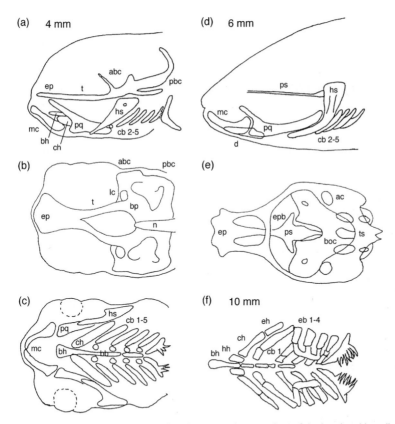

Figure 6 Camera lucida drawings of whole-mounted preparations of the larval and juvenile skull. (a, d) Lateral views and (b, e) ventral views of the neurocranium. abc, anterior basicranial commissure; ac, auditory capsule; bh, basihyal; boc, basioccipital; bp, basal plate; cb, ceratobranchials; ch, ceratohyal; d, dentary; ep, ethmoid plate; epb, epiphysial bar; hs, hyosymplectic; lc, lateral commissure; mc, Meckel's cartilage; n, notochord pbc, posterior basicranial commissure; pq, palatoquadrate; ps, parasphenoid; t, trabeculae; ts, tectum synoticum. (c, f) Ventral views of the viscerocranium. bb, basibranchials; bh, basihyal; cb, ceratobranchials; ch, ceratohyal; eb, epibranchials; eh, epihyal; hh, hypohyal; hs, hyosymplectic; mc, Meckel's cartilage; pq, palatoquadrate.

Skull

The cartilage of the neurocranium and viscerocranium develops by 3.5 mm, and is then partially replaced in the following few weeks by endochondral bone. In contrast, dermal bones that form the cranial vault during the same period do not have cartilage precursors. Early larvae (3–4 mm) have a rudimentary ethmoid and neurocranial base plate underlying the brain, and all seven sets of cartilages of the viscerocranium which form a ladder-like array in ventral views, including Meckel's cartilage (mandibular), the ceratohyal (hyoid), and five ceratobranchials (*Figure 6A, C*) (6). No bone has formed at this stage. Cartilage develops in a ventral to dorsal sequence within each arch segment, and the segments form in an anterior to posterior sequence. Thus Meckel's cartilage and the ceratohyal develop first, in the two most anterior arches at the end of embryogenesis, and elongate as thick bars towards the ventral midline at 3.5 mm. Subsequently, dorsal cartilages develop in these arches and ventral elements form in the more posterior gill arches. The major exception to the dorsal–ventral sequence are the ventral midline cartilages, the basihyal, basibranchials, and hypobranchials, which form at 3.6 mm and subsequently fuse. The last cartilage to chondrify in the early larval skull is the lateral commissure, at 3.8 mm (*Figure 6b*).

At 4 mm the skeleton is functional, and larvae begin feeding (*Figure 6a–c*) (6). There are no additional chondrifications at this stage, but the first bone appears, which ossifies around the pharyngeal jaws (ceratobranchial 5). There are also three early dermal bones: the opercle, the median parasphenoid, and the branchiostegal rays. Ossification begins slowly, and after 1 week of development only one bone of the occipital skull has ossified. However, by 2 weeks of age (approximately 6 mm), most of the bones of the mandibular and hyoid skeletons ossify, as well as the occipital skull (*Figure 6e*). During the third week of development (8 mm), the branchial arches ossify (*Figure 6f*). Last are the bones of the dermatocranium, including those of the ethmoid region, sclerotics, infraorbitals, basi, and hypobranchials. The sequence of skull ossification is complete by approximately 60 dpf (17 mm).

At 4 mm, larvae have 3–4 pharyngeal teeth on each side. Development of the pharyngeal dentition in larvae up to 6 mm has been described, during which time several rows of additional teeth form in a stereotyped pattern (12).

Vertebral column and fins

During larval development in the zebrafish trunk and tail:

(i)   vertebrae form in an anterior to posterior sequence; and

(ii) paired and median fin skeletons differentiate.

The vertebral column develops later than the skull. Cartilage of the most anterior vertebrae first appear around 7–8 dpf (5 mm), although this has not been described in detail (see *Figure 5*). Where dorsal, ventral, and horizontal myosepta intersect, four paired cartilage condensations appear in a segmental pattern of rings around the notochord. These are the arcualia, the primordia of the vertebral centra. Later, pleural ribs form in the peritoneal membrane and attach to the developing centra of every precaudal vertebra except the first two. A Weberian apparatus above the anterior vertebrae is visible by 3 weeks (8 mm).

The appendages emerge at widely different stages. Paired fins develop first, in the embryo, and cartilages of the pectoral girdle are visible at 3.5 mm (*Figure 7*) (6, 15, 16). Originally in a vertical orientation, these gradually rotate into a near-horizontal position and fin rays (lepidotrichia) develop within the fin folds. The early pectoral girdle contains the scapulocoracoid, and anteriorly the dermal cleithrum, and these support the larval appendage (*Figure 7a*). Out in the pectoral fin, a cartilaginous disc forms, and actinotrichia extend into the forming fin fold. By 5 dpf (4 mm), the scapulocoracoid attaches to the cleithrum and, apart from growth, little changes over the next 2 weeks of development. The pectoral fin then undergoes a transition during the third week (*Figure 7b–d*). The larval endoskeletal disc in the pectoral fin becomes subdivided into four stripes of cartilage, prefiguring the pattern of the four proximal radials at the 8 mm stage and distinct radials form by the beginning of the juvenile period (10 mm). Additional elements also arise in the pectoral endoskeleton, the mesocoracoid, and the dermal skeleton, the supracleithrum, and the postcleithrum (not shown).

Median fins and paired, pelvic fins start developing much later. Cartilage differentiates in the dorsal fin before it does so in the anal. All fins contain cartilage and some bone by juvenile onset at 10 mm (see *Figure 4*). A notable difference between pectoral and pelvic fins of

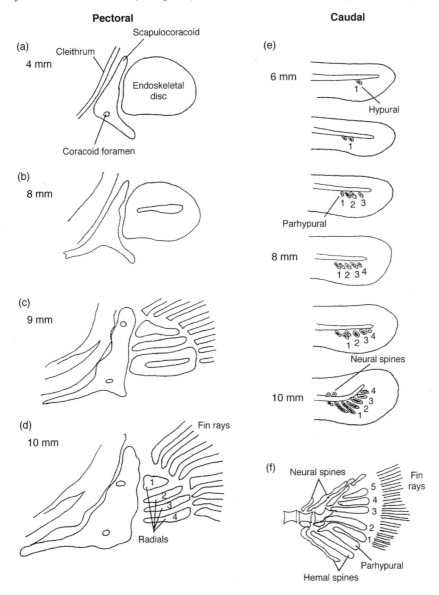

Figure 7 Camera lucida drawings of larval and juvenile pectoral and tail fins, from whole-mounted preparations. (a–d) Dorsolateral views of larval (a–c) and juvenile (d) pectoral girdle and fin. (e) Stages of tail-fin vertebral modifications. (f) Higher magnification of an adult tail-fin skeleton. From refs 15 and 16.

the zebrafish is that the latter leads directly to the formation of the adult fin structure, without any larval appendage. Development of the tail-fin skeleton occurs between 2 and 4 weeks of development, and can be a useful staging tool in late larvae (*Figure 7e, f*). Hypural supports in the tail vertebrae develop in sequence, such that the first one develops at 14 dpf (6 mm) and another every 2 days from then on, to form a total of 7 at 9–10 mm. Pelvic fin buds appear during the third week and the skeleton is complete by juvenile onset. Chondrogenic condensations of the radials in the pelvic fins form after the condensation of the pelvic girdle has become visible; either two or three distal radials develop.

Scales in most fishes begin to develop during the late larval or early juvenile stages, and Sire and colleagues (18) have observed similar patterns in zebrafish. Scales arise as bony plates in the dermis. Osteoblasts lay down layers of concentric circles of bone, starting in the region of the lateral line. These grow by accretion as more bone is added along their periphery, expanding their diameters. This can be a useful means of staging individuals in wild fish populations, but probably not in the laboratory, where daily and annual cycles are less pronounced.

Protocol 2

Trichrome stain[a]

Equipment and reagents

- Rotary microtome for paraffin sectioning
- Coplin jars for slide staining
- Bouin's fixative: 75 ml saturated aqueous picric acid, 25 ml formalin, 5 ml glacial acetic acid
- Lugol solution: 1.0 g iodine crystals, 2.0 g potassium iodide in 100 ml dH_2O
- Sodium thiosulphate solution, 5.0 g in 100 ml dH_2O
- Mallory I: 1.0 g acid fuchsin in 100 ml dH_2O
- Phosphomolybdic acid: 1% in dH_2O
- Mallory II: 0.5 g Aniline Blue, 2.0 g Orange G in 100 ml dH_2O
- Ethanol series (70%, 100%)

Method

1. Anaesthetize fish in 0.08% Tricaine® until movements cease.
2. Fix in Bouin's for 24–48 hours.
3. Wash extensively in 70% ethanol.
4. Dehydrate through two changes of absolute ethanol.
5. Section in paraffin (see Protocol 5).
6. Soak in saturated aqueous $HgCl_2$, plus 5% acetic acid: 10 min.
7. Deparaffinize and hydrate slides to water: remove $HgCl_2$.
8. Wash, treat with Lugol's and sodium thiosulphate, wash and rinse in dH_2O.
9. Stain in Mallory I: 15 seconds.
10. Rinse in dH_2O to differentiate reds: 10 s or more
11. Treat with phosphomolybdic acid: 1–5 min.
12. Rinse in dH_2O.
13. Stain in Mallory II: 2 min.
14. Rinse in dH_2O.
15. Differentiate in Aniline Blue in 90% alcohol.
16. Dehydrate, clear in xylene and mount in Permount®.

[a] Modified from the methods in refs 19 and 20.

3.3 Adult musculature

In contrast to the skeleton, no study has carefully described the anatomy of the musculature of the adult zebrafish. In this chapter I extrapolate from numerous studies of muscles in other teleosts (21, 22) as well as my own observations of larval cranial muscles (*Figure 8*) (9). Most muscles are skeletal and striated, including all of the body wall muscles used in swimming, and all of the pharyngeal muscles used in feeding and respiration. Smooth muscles line the gut, the blood vessels, and some glands and visceral organs, while cardiac muscles line the single atrium and ventricle of the heart.

Cranial muscles

In the head, there are six pairs of skeletal muscles that move the eyes, and at least 30 pairs that expand or contract the pharyngeal cavity (*Figure 8a–f*). The extrinsic eye muscles include two pairs of obliques and four pairs of recti, which, as in all vertebrates, are innervated by the oculomotor (III), trochlear (IV) and abducens (VI) nerves. Muscles of the pharyngeal arches form a ladder-like array when viewed ventrally, similar to the pharygeal skeleton (*Figure 8e*). These are subdivided into dorsal and ventral muscle groups in each of the seven segments, and contract or expand the pharyngeal cavity, or in the case of the fifth branchial arch, process food (*Figure 8c*). There are five main functional muscle groups: (i) dorsal and (ii) ventral constrictors, that surround and compress the pharynx; (iii) levators, that elevate the gills for pharynx compression, (iv) adductors, that bend each arch by drawing dorsal and ventral halves together; and (v) interarcuals, that bend the dorsal ends of the arches posteriorly. Muscles within any one of these groups form a series of segmental homologues with similar shapes in each arch. Thus muscles in the cheek, such as the levator arcus palatini of the mandibular arch, have homologues in the hyoid arch, the levator hyomandibulae (*Figure 8b*). Likewise, there is a segmental series of ventral constrictor muscles that includes the inter-mandibularis anterior (mandibular, arch 1), interhyal (hyoid, arch 2), and transverse ventrals (branchials, arches 3–7), each perpendicular to the ventral midline (*Figure 8c, e*) (9). Mandibular muscles are innervated by the trigeminal (V), hyoid muscles by the facial (VII), first branchial muscles by the glossopharyngeal (IX), and all other branchial muscles by the vagal nerve (X). The largest ventral muscles are the hypobranchials, which include the long sternohyal muscles that extend from the coracoid to the hyoid arch, and these are innervated by spinal motor nerves (*Figure 8f*). Zebrafish have no true tongue, rather a slight pad or projection of the basihyal bone or cartilage, and thus no true tongue muscles.

Smooth muscle lines the walls of the digestive tract. Here there are both longitudinal and circular muscles that work in opposition to one another. Smooth muscles are also associated with the swim bladder, as well as reproductive and excretory ducts, and some other organs and glands (see below). Cardiac muscle is found only in the heart, and is thickest in the walls of the ventricle.

Trunk and tail

The axial musculature in adult zebrafish is subdivided into approximately 32 somitic segments along the anterior–posterior axis. Each myotome is also subdivided dorsoventrally by the horizontal connective tissue myoseptum (*Figure 8j, k*). Epaxial muscles form dorsal to this septum, hypaxial muscles form ventrally, and these major compartments are easily visible morphologically in living animals. Along the mediolateral axis within a myotome there are two different types of muscle fibre with differing contractile characteristics: white and red (as seen in transverse sections) (*Figure 8k*). Fast-twitch, white fibres make up the majority and lie medially, while the much smaller population of slow-twitch, red fibres, which are rich in mitochondria, lie in a lateral, superficial sheet under the skin (23, 24). Muscle fibres in teleosts are limited to one segment; their myotomes do not fuse. The muscle fibres have many different alignments, depending on their dorsoventral positions. In the median fins protractors erect the dorsal and anal fins and retractors depress them (*Figure 8a*). Lateral inclinators bend the soft rays of these fins. In paired fins, single dorsal adductors pull the fin dorsally and caudally, opposed by single ventral abductors.

3.4 Larval and juvenile muscle development

In early zebrafish larvae: (i) the jaw becomes motile, and pharyngeal muscles form a ladder-like array of segmental homologues, (ii) the fish become free-swimming, and slow- and fast-twitch muscle fibres proliferate in the somites, and (iii) the gut becomes motile, and this reflects the development and innervation of smooth muscle. The spatial and temporal

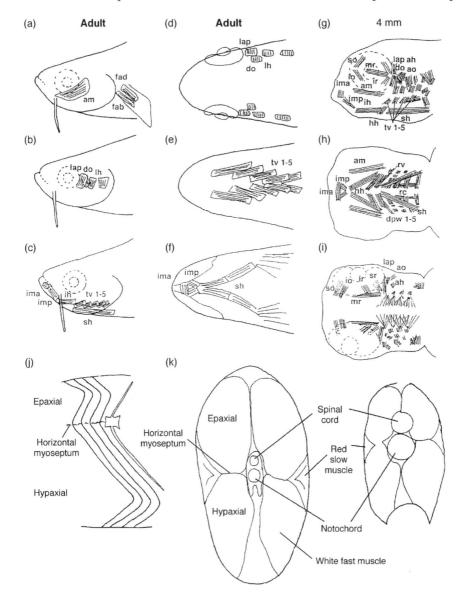

patterns of muscle development, which can be made visible in whole mounted larvae with polarized light or immunohistochemical detection of muscle myosins, have been described for the first 5 days of development (*Figure 8g–i*) (9). Development of cardiac muscle has been described (25; reviewed in 26), but smooth muscle has not. Less is known about later larval and juvenile changes in muscle pattern.

Cranial muscles

In the head, extrinsic eye muscles develop first, followed by skeletal muscles in the pharyngeal arches, and both the eyes and the jaw are motile by early larval stages. At 4.0 mm (*Figure 8g, i*) larvae have all six pairs of eye muscles (27). The pharyngeal muscles, like their adjacent cartilage, develop in an anterior to posterior sequence, and within an arch, myogenesis occurs in a ventral to dorsal sequence (9). All five of the main muscle groups found in the adult arches are already present in the larva in the mandibular (arch 1) and hyoid (arch 2) arches at 3 dpf (3.5 mm), including two dorsal and two ventral constrictors, as well as one adductor. In contrast, only ventral transverse muscles have developed in the gill arches at this stage. Dorsal muscles of these arches do not develop until 4 dpf (4 mm). Eventually all of the gill arches have a similar set of 2–4 muscle pairs, including branchial levators and transverse ventrals. Between the segmental gill muscles there are rectus ventralis muscles spanning branchial arches 1–3 (*Figure 8h*). Such intersegmental muscles are specialized features of the musculature of minnows (22). Like the pharyngeal cartilages, there are also enlarged branchial arch muscles on the fifth gill arch, the pharyngeal jaws which are specialized for feeding. Finally, ventral to the branchial muscles the sternohyals, already very long at this stage, stretch from the pectoral girdle to the hyoid arch.

Trunk and tail

Dorsoventral subdivisions of the somites into epaxial/hypaxial muscles, and mediolateral subdivision into slow/fast territories, are largely complete by early larval stages. At 72 hpf (3.5 mm), slow red muscle fibres are segregated to the periphery and can be distinguished from more medial fast-twitch, white fibres immunohistochemically (28). Both fibre types are polyneuronally innervated. By 7 dpf (4.5 mm), differences in muscle fibre physiology appear, including differences in ATPase reactivity, and accumulation of mitochondria, and these characteristics become more prominent at later stages (*Figure 8k*) (23). At 3.5 mm the ratio of muscle fibres to neurons is 10:1. After this, the muscle continues to proliferate while no more motoneurons are generated, up until a ratio of 45:1 is reached in animals aged 4–5 months.

Figure 8 Camera lucida drawings of adult and larval musculature. Adult (a–f, j and first part of k), 4 mm (g–i), and 6 mm (second part of k). Whole-mounted preparations (a–j), or transverse plastic sections stained by the trichrome method (k; see *Protocols* 2 and 5). (a–c) Lateral views of adult head. am, adductor mandibulae; do, dilator operculi; fab, fin abductor; fad, fin adductor; ih, interhyal; ima, intermandibularis anterior; imp, intermandibularis posterior; lap, levator arcus palatini; sh, sternohyal; tv, transverse ventrals. (d–f) Ventral views of adult head. do, dilator operculi; ih, interhyal; ima, intermandibularis anterior; imp, intermandibularis posterior; lap, levator arcus palatini; sh, sternohyal; tv, transverse ventrals. (g) Lateral view of early larval head. ah, adductor hyomandibulae; am, adductor mandibulae; ao, adductor operculi; do, dilator operculi; hh, hyohyal; ih, interhyal; ima, intermandibularis anterior; imp, intermandibularis posterior; io, inferior oblique; ir, inferior rectus; lap, levator arcus palatini; mr, medial rectus; sh, sternohyal; so, superior oblique; tv, transverse ventrals. (h, i) Ventral views of early larval head. ah, adductor hyomandibulae; am, adductor mandibulae; ao, adductor operculi; dpw, dorsal pharyngeal wall; hh, hyohyal; ima, intermandibularis anterior; imp, intermandibularis posterior; io, inferior oblique; ir, inferior rectus; lap, levator arcus palatini; mr, medial rectus; rc, rectus communis; rv, rectus ventralis; sh, sternohyal; so, superior oblique; sr, superior rectus. (j) Lateral view of a series of three adult somites and one vertebra. (k) Transverse sections through adult and larval trunks, showing compartments and distribution of muscle fibre types. From refs 9 and 22.

4 Nervous system

4.1 Adult central nervous system

As visualized in dissected specimens, the zebrafish brain, as in other teleosts, is dominated by the large optic tecta of the dorsal midbrain that receive primary visual input (*Figure 9*) (29). There are also specialized facial and vagal lobes of the dorsal hindbrain, which are specific to cyprinids, and which receive gustatory input from the huge numbers of taste buds in these fish. When viewed dorsally, the lobes of the telencephalon are small and oblong in comparison to the tectum (*Figure 9b*). They attach anteriorly to the olfactory bulbs, which lie closely apposed to the brain. The diencephalon is covered by the rostral optic tectum and out of view. From lateral or ventral views the optic nerves, hypothalamus, and pituitary of the diencephalon are visible externally (*Figure 9a*). The hypothalamus consists of one medial and paired lateral lobes. The cerebellum is subdivided into a large anterior corpus and a smaller, posterior crista. Cranial nerves (V, VII, VIII, IX, X) emerge from the lateral rhombencephalon and the medulla, which grades posteriorly into the spinal cord.

The internal anatomy of the adult zebrafish CNS has been described in detail using serial sections stained with the Bodian method to co-label neuronal cell bodies and their axons (*Figure 9*) (29) (see *Protocol 3*). In ray-finned fishes the telencephalon receives a huge olfactory input (*Figure 9c, d*). Olfactory fibres enter the periphery of the olfactory bulbs and synapse in a more internal glomerular layer. Cells further within the bulb, which receive the primary olfactory input, then give rise to medial and lateral olfactory tracts that project posteriorly. Unlike other vertebrates, the telencephalon in ray-finned fishes also appears to be 'everted' dorsally, such that the ependymal lining of the telencephalic ventricle appears at the surface (*Figure 9d*). For this reason, in part, it remains difficult to assign homologies between fish and mammalian forebrain areas. Behind the olfactory bulbs, the telencephalon is subdivided into dorsal and ventral areas, much of which receive primary olfactory fibres. Ventrally there are several nuclei that have migrated away from the ventricular layer to surround the anterior commissure (*Figure 9d*). Posterior to this commissure the lateral and medial forebrain bundles coalesce and descend into the diencephalon.

The zebrafish diencephalon is subdivided into five regions from dorsal to ventral, including the epithalamus, dorsal and ventral thalamus, posterior tuberculum, and hypothalamus (*Figure 9e*). Anteriorly the preoptic and suprachiasmatic nuclei lie next to the entry point of the optic nerves. Dorsally the photoreceptive epiphysis or pineal organ extends above the habenula. Further ventrally, the thalamic nuclei lie close to the ventricular layer. In teleosts the posterior tuberculum, lying beneath the thalamus, is particularly large and includes both posterior tuberal and paraventricular nuclei, as well as a collection of preglomerular nuclei that have migrated well away from the ventricular layer (*Figure 9f*). These include some cells that receive tertiary gustatory input. The diencephalon also includes the pretectum, which is primarily a visual area. The largest region of the diencephalon is the ventral hypothalamus. This includes separate inferior lobes in its dorsal regions and thick periventricular nuclei ventrally which are attached caudally to the pituitary (*Figure 9e, f*). There is a separate pituitary, with distinct neurohypophysis and adenohypophysis (*Figure 9a*).

The mesencephalon is classically subdivided dorsoventrally into the tectum and tegmentum and, in zebrafish, between these lies a lateral sensory torus semicircularis (not shown), as well as a medial torus longitudinalis (*Figure 9f*). The large optic tectum is the most complex layered structure in the brain and receives a topographic map of retinal projections, as well as many other sensory inputs. The torus semicircularis lies just lateral to the third ventricles and mediates auditory and mechanoreceptive senses. The more ventrally located tegmentum contains motor neurons of the oculomotor (III) and trochlear (IV) nerves that innervate eye muscles, as well as interneurons that project caudally and form the medial longitudinal fascicle that extends into the spinal cord. Trochlear motor axons traverse the dorsal mesencephalon, crossing in the cerebellum and exiting between the torus semi-circularis and the rhombencephalon.

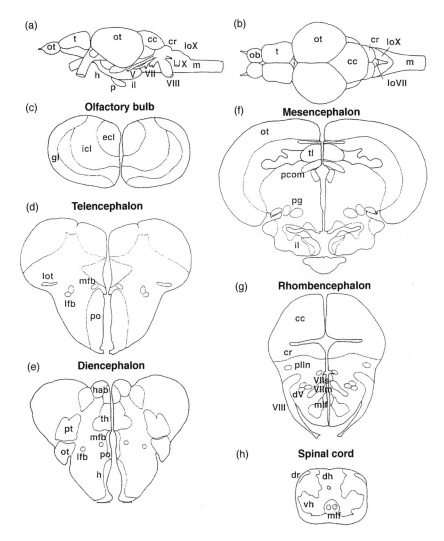

Figure 9 Camera lucida drawings of the adult zebrafish brain, as a whole mount (a, b) and in a series of transverse paraffin sections stained by the Bodian method to reveal neurofilaments (c–h; see *Protocols 3 and 5*). Representative sections are shown for each major subdivision along the anteroposterior axis. Major tracts and cell nuclei are outlined. cc, corpus cerebelli; cr, crista cerebelli; dr, dorsal root; dh, dorsal horn; ecl, external cellular layer; gl, glomerular layer; h, hypothalamus; hab, habenula; icl, internal cellular layer; il, inferior lobe; lfb, lateral forebrain bundle; lot, lateral olfactory tract; loVII, facial lobe; loX, vagal lobe; m, metencephalon; mfb, medial forebrain bundle; mlf, medial longitudinal fascicle; ob, olfactory bulb; ot, optic tectum; p, pituitary; pcom, posterior commissure; pg, preglomerular nucleus; plln, posterior lateral line; po, preoptic area; pt, pretectum; t, telencephalon; th, thalamus; tl, torus longitudinalis; V, trigeminal nerve; vh, ventral horn; VII, facial nerve; VIII, acousticolateral nerve; X, vagal nerve. From ref. 29.

THOMAS F. SCHILLING

The rhombencephalon, or hindbrain, consists anteriorly of the cerebellum and posteriorly of the medulla oblongata (*Figure 9a, b, g*), which contains cranial motor and sensory nuclei, as well as reticular interneurons that project to the spinal cord and mediate swimming and escape, among many other behaviours (30). In all teleosts there is a dorsal corpus cerebelli while, deeper within the cerebellum, cells are organized into a medial valvula cerebelli and a vestibulolateralis lobe (not shown). The rhombomeres of the hindbrain medulla contain a segmental series of sensory terminal fields and motor neurons that form the cranial nerves (*Figure 9g*; trigeminal, V; abducens, VI; facial, VII; acousticolateral, VIII; glossopharyngeal, IX; and vagal, X). The sensory component of the facial nerve, carrying gustatory information, terminates in a rostral medullary structure known as the facial lobe, which in zebrafish appears from the surface as a single lobe in the midline. Gustatory sensory components of the vagal nerve terminate in paired vagal lobes further posteriorly in the rhombencephalon, and in zebrafish these form elaborate, layered structures (*Figure 9b*).

In the spinal cord axons descend in a ventral medial longitudinal fascicle, and ascend in a dorsal root (*Figure 9h*). There are primary and secondary motorneurons located in the ventral horns of the grey mater of the spinal cord, which project to skeletal muscles, and axons of incoming axons of the dorsal roots terminate in the dorsal horns. Individual muscle fibres are polyneuronally innervated in adult zebrafish by a single primary and one or more secondary motoneurons; the ratio of muscle fibres to motor neurons is nearly 50:1 (24, 31). Large motoneurons that innervate white muscle lie medially, close to the central canal, while smaller cells innervating slow red fibres lie further laterally.

Protocol 3

Bodian Silver/Cresyl Violet method to visualize neurofilaments

Equipment and reagents

- Rotary microtome for paraffin sectioning (see *Protocol 5*)
- 2-litre glass beaker
- Coplin jars for staining sections
- Copper shot
- 0.4% Tricaine® stock (3-amino benzoic acid ethylester) (Sigma): 400 mg : 97.9 ml dH_2O : 2.1 ml 1 M Tris (pH = 9.0); working solution 0.08%
- AFA: 90 ml 80% ethanol, 5 ml 37% formalin, 5 ml glacial acetic acid
- Ethanol series: 30%, 50%, 70%, 95%, 100% EtOH
- Poly-L-lysine (MW 70 000–150 000; Sigma)
- Protargol® solution, prepared by adding 1 g of Protargol® to 100 ml dH_2O, by sprinkling on the surface of water in a large beaker. DO NOT STIR
- Reducing solution (HF) prepared by adding 1 g of hydroquinone and 5 ml of formalin to 100 ml of dH_2O
- Gold chloride: prepared by adding 1 g gold chloride to 100 ml dH_2O
- Oxalic acid: 2 g oxalic acid in 100 ml dH_2O
- 5% Sodium thiosulphate
- Cresyl Violet counterstain, prepared by first making a stock of 0.2 g Cresyl Violet (Merck) in 150 ml dH_2O, then adding 6–12 ml of this solution to 100 ml of buffer solution (94 ml of 0.1 M acetic acid + 6 ml of 0.1 M sodium acetate).
- Entellan® (Merck)

Method[a]

1. Anaesthetize larvae or juveniles in Tricaine®, by dissolving 4.2 ml stock in 100 ml fish water.
2. Fix in cold AFA (4°C) for at least 1 month.
3. Dehydrate through an ethanol series with at least two changes of 100% EtOH.

Protocol 3 continued

 4 Embed in paraffin and section at approx. 8 μm (see *Protocol 5*).

 5 Mount on poly-L-lysine-coated slides.

 6 Deparaffinize and rehydrate through ethanol series.

 7 Impregnate in Protargol® solution at 37°C for 12–24 hours in a Coplin jar containing 6 g of copper shot.

 8 Wash in distilled water, several changes.

 9 Reduce for 15 min in HF solution.

 10 Rinse in 6 changes of dH_2O: 1 minute total.

 11 Develop in 2% aqueous oxalic acid, while monitoring under microscope, until background appears grey and fibres are sharply defined: approximately 3 min.

 12 Wash in 6 changes dH_2O: 1 min each.

 13 Fix in 5% aqueous sodium thiosulphate: 5 min.

 14 Wash in running water: 5–10 min.

 15 Rinse in dH_2O.

 16 Counterstain with Nissl-stain Cresyl Violet.

 17 Treat with acidified water: 5 min.

 18 Dehydrate through ethanol series and coverslip with Entellan®.

 [a] Method after ref. 32.

4.2 Larval and juvenile neural development

There is a large gap between our understanding of embryonic brain development in zebrafish and adult brain morphology (*Figure 10*) (33). Best studied are the early larval stages of forebrain and hindbrain development, including development of retinal projections to the diencephalon and tectum (34), and the hindbrain reticulospinal system (35, 36). During larval development:

(i) secondary neurons join the primary neurons formed in the embryo;

(ii) major sensory tracts and maps are established; and

(iii) cells migrate from the ventricular zone to establish nuclei.

By 5 days, all the major components of the brain are present, although a limited number of cells have migrated away from the ventricular zone (33). Between 3.5 and 4 mm, the ventral kink in the brain at the mesencephalon gradually straightens as the jaw extends underneath. During this period, the main tracts of the early embryo persist, including the nucleus and the tract of the postoptic commissure, as well as the posterior commissure (*Figure 11A*) (37).

The larval telencephalon already shows signs of an everted organization at 5 dpf (4 mm), with the ventricular region lying externally, but relatively few cells have migrated away from the ventricle. There are well-developed olfactory bulbs, and an olfactory map is established by this stage (*Figure 10b, c*) (38). The olfactory bulbs give way posteriorly to a telencephalon subdivided into a dorsal pallium and ventral subpallium in front of the anterior commissure. There are also large lateral forebrain bundles of axons from telencephalic cells projecting into the diencephalon.

In the larval diencephalon, all of the major adult subdivisions, including epithalamic, thalamic, posterior tubercle, pretectal and hypothalamic areas, are present at 5 dpf (4 mm) (*Figure 10e*). There is also a well-defined preoptic area. Optic axons enter and cross the midline

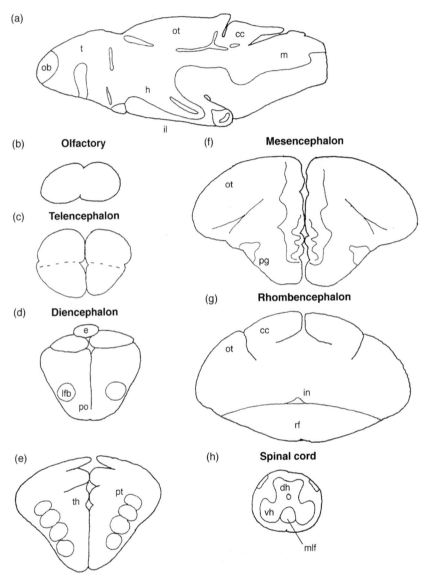

Figure 10 Camera lucida drawings of the larval (4 mm) zebrafish brain in parasagittal (a) or transverse (b–h) paraffin sections stained by the Bodian method (see *Protocols 3 and 5*). Representative sections are shown for each of the major subdivisions along the antero-posterior axis. Major tracts and cell nuclei are outlined. cc, corpus cerebelli; dh, dorsal horn; h, hypothalamus; il, inferior lobe; in, interpeduncular nucleus; lfb, lateral forebrain bundle; m, metencephalon; mlf, medial longitudinal fascicle; ob, olfactory bulb; ot, optic tectum; pg, preglomerular nucleus; po, preoptic area; pt, pretectum; rf, reticular formation; t, telencephalon; th, thalamus; vh, ventral horn. From ref. 33.

ventrally at the optic chiasm, and project to nine foci within the neuropil of the pretectum, as shown by anterograde labelling with horseradish peroxidase (HRP) (see *Protocol 4*), although the cell bodies of neurons that will constitute these nuclei have not yet migrated away from the ventricular zone where they are born (33, 34). At the ventricular layer, dorsal thalamic cells overlie an intermediate layer of cells of the posterior tuberculum. In contrast, cells of the

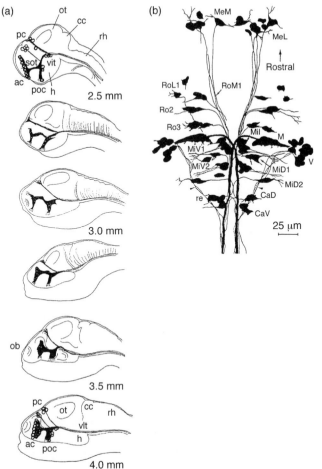

Figure 11 Camera lucida drawings of larval brains stained to reveal neuronal projections immunohistochemically (a) or by backfilling axons with hoseradish peroxidase (HRP) (see *Protocol 4*), from whole-mounted preparations (b). (a) Lateral views of late embryos (2.5 mm) and early larvae stained with an anti-acetylated tubulin antibody, revealing the major tracts (black). ac, anterior crista; cc, corpus cerebelli; h, hypothalamus; ob, olfactory bulb; ot, optic tectum; pc, posterior commissure; poc, postoptic commissure; rh, rhombencephalon; sot, supraoptic tract; vlt, ventrolateral tract. (b) Dorsal view of the early larval (4 mm) hindbrain in which the reticulospinal neurons have been backfilled from the spinal cord with HRP. There are lateral and medial clusters within the mesencephalon (MeL and MeM), as well as clusters in each of the rostral (Ro1–3), middle (Mi1–2) and caudal (Ca) segments of the rhombencephalon. M, Mauthner; V, vestibular; From ref. 36.

preglomerular complex, part of the posterior tuberculum, have already migrated away from the ventricle by this stage (*Figure 10e*). The pituitary has also formed at the ventralmost extremity of the diencephalon.

In the mesencephalon a retinotectal map is established during late embryonic and early larval stages, and the tectum begins to take on a layered, cortical organization at 5 dpf (4 mm) (*Figure 10f*). The tectum overlies the developing torus semicircularis, lying in a far lateral position. There is a large ventral tegmentum in the larval mesencephalon, with relatively few cellular landmarks other than the nucleus of the medial longitudinal fascicle anteriorly and a ventral interpeduncular nucleus posteriorly (*Figure 10f*).

THOMAS F. SCHILLING

The larval rhombencephalon contains an early ladder-like array of segmental neurons of the reticular formation, branchiomotor neurons, and commissural interneurons, which have been revealed using retrograde labelling with HRP or fluorescent dextrans, as well as immunohistochemically (35, 36) (*Figure 11b*; see *Protocol 4*). By 5 dpf (4 mm) there are 50 hindbrain reticulospinal neurons on each side, including the large Mauthner cells, that project to the spinal cord and mediate the escape reflex. These cluster in seven bilateral, segmental groups and can be classified based on their projections along the medial longitudinal fascicle (mlf; MiD, MiV) or along the lateral longitudinal fascicle (llf; RoL; *Figure 11b*). For example, the Mauthner neuron is of the MiD class. There is a prominent cerebellum overlying the dorsal hindbrain in early larvae, but no evidence of the gustatory facial and vagal lobes over more posterior regions of the medulla oblongata at this early stage.

The larva initially retains the embryonic pattern of primary and secondary neurons in the spinal cord. While spinal neurons are segmentally organized during embryonic development, they appear less so at later stages, and in early larvae (3.5 mm) secondary motoneurons form a continuous column throughout the ventral spinal cord. The number of these motoneurons progressively increases until 30 dpf (10 mm), at which time the ratio of muscle fibres to neurons is 10:1. After this, muscle cells continue to proliferate while no more motoneurons are generated, increasing this ratio to approximately 45:1 in adult animals 4–5 months of age (23). In addition to motoneurons there are at least four classes of early interneurons with ascending projections in the dorsolateral fascicle, or descending projections within the ventrolateral fascicle (39, 40). There are also large dorsal Rohon–Beard neurons, the somatosensory cells of the embryo and early larva, which have central axons running along the dorsolateral fascicle of the spinal cord, and peripheral axons in the skin. Subsequently, however, the Rohon–Beard neurons die off between 4 and 6 mm (39, 41). Thus there is a switch from embryonic to a more adult pattern of neurons during early larval development.

Protocol 4

Horseradish peroxidase or dextran fills of neuronal projections in older fish

Reagents
- 1-phenyl-2-thiourea (PTU), to inhibit melanogenesis, 0.003% in 10% Hank's saline
- Horseradish peroxidase (HRP, Sigma Type VI), prepared by adding 0.1 ml distilled water to 2000 units of HRP (approx. 7% solution)
- Tetramethylrhodamine isothiocyanate (TRITC)–dextran or biotinylated dextran (10 000 MW; Molecular Probes), prepared in 0.1 M KCl in a 3% solution.

Method[a]
Lesion:

1 Anaesthetize in 0.003% Tricaine®.
2 Embed fish in 1.2% agar on a glass slide in 10% Hank's saline at 37°C.
3 Remove a small amount of agar over the region to be labelled.
4 Dry the hole with tissue paper and apply tracer (HRP or dextran) to the agar hole.
5 Lesion the area of interest with broken micropipette.
6 Allow 5 min recovery in the agar, and then release and raise overnight.

Injection:

1 Anaesthetize in 0.003% Tricaine®.

2 Embed fish in 1.2% agar on a depression slide in 10% Hank's saline at 37°C.

3 Remove small amount of agar over the region to be labelled and immerse in Hank's.

4 Place under dissecting microscope and use a micromanipulator and pressure inject through a micropipette into the area of interest.

5 Allow 5 min recovery and then release and raise.

Development:

1 Anaesthetize fish in 0.08% Tricaine®.

2 Remove body regions outside the area of interest to provide better penetration,

e.g. for the brain, isolate heads for fixation, for spinal cord fills remove heads and just fix tails.

3 Fix in 4% paraformaldehyde in 0.1 M phosphate buffer for 1–3 hours at room temperature or, in the case of fluorescent dextrans, visualize in living preparations.

4 Embed in gelatin and cut 10 μm sections on a cryostat (see *Protocol 5*).

[a] Method after that described in ref. 40.

4.3 Adult peripheral nervous system and sense organs

Less is known about the anatomy of the peripheral nervous system in adult zebrafish, although there are descriptive studies of most of the primary sensory systems. As in other vertebrates, primary sensory neurons in the trunk and tail are enclosed in dorsal root ganglia (DRG), with peripheral axons that bifurcate into dorsal and ventral territories in each body segment (*Figure 12a*). Somatosensory neurons of the DRG take over the sensory roles of the Rohon–Beard cells, which are lost in young larvae (41). Cranial sensory neurons are found in ganglia of the cranial nerves I–X. The most anterior of these, the trigeminal (V), mediates sensory input from several branches above and below the eye, as well as along the mandible. A similar pattern of branches around the eye and jaws is seen in the innervation of the anterior lateral line. The sensory neurons of other cranial nerves (VII, IX, X) predominantly innervate taste buds of the branchial arches, and lie in epibranchial ganglia that form above each arch. One exception to this is the posterior lateral line which innervates four lines of neuromasts along the trunk and tail. There are also peripheral autonomic ganglia, sympathetic or parasympathetic, which innervate smooth muscles of various organs and glands, as well as pigment cells in the skin, and enteric neurons that innervate the gut.

Sense organs

Olfactory

The olfactory organs are paired sacs located just anterior to the eyes and above the mouth. In zebrafish, these open to the exterior via external nares (*Figure 12b*) (42, 43). The olfactory organ in fish is composed of many different cell types, including basal cells, ciliated and microvillous receptor cells, support cells, and crypt cells. Non-sensory areas also contain goblet, ciliated non-sensory, and epidermal cells. The olfactory nerve (cranial nerve I) carries olfactory information to the olfactory bulb. A typical olfactory bulb has four laminae:

(i) olfactory nerve; (ii) glomerular; (iii) mixed mitral cell/plexiform; and (iv) granule cell layers.

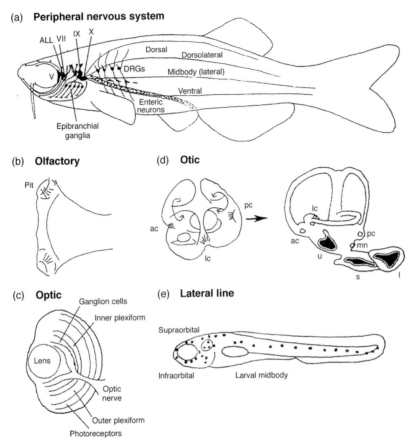

(a) Peripheral nervous system

(b) Olfactory

(d) Otic

(c) Optic

(e) Lateral line

Figure 12 Schematic drawing of the adult (a) peripheral nervous system, and camera lucida drawings of larval sense organs (b–e). Whole-mounted preparations (b, d, e) and from plastic horizontal sections stained with Methylene Blue (c; see *Protocol 5*). (a) Lateral view of an adult (2.0 cm), based on information from other species and from larval stages, illustrating the major components of the peripheral nervous system and their approximate locations. ALL, anterior lateral line; DRGs, dorsal root ganglia. (b) Dorsal view of early larval (4 mm) olfactory pits. (c) Dorsal view of early larval (4 mm) eye, from plastic horizontal section, showing major cellular layers. (d) Lateral views of larval (4 mm) and adult (2.0 cm) otic capsules, showing three semicircular canals and hair-cell patches. ac, anterior crista; l, lagena; lc, lateral commissure; mn, macular node; pc, posterior crista; s, sacculus; u, utricle. (e) Lateral view of early larva (4 mm), stained to reveal the pattern of neuromasts on the anterior (infra- and supraorbital) and posterior (midbody) lateral lines. From refs 45, 47.

On average there are 80 glomeruli per bulb, and these all exhibit bilateral symmetry. In fact, many glomeruli appear to be identifiable from animal to animal, suggesting that there is a spatial map of olfactory information in the bulb (42, 43).

Optic

The zebrafish retina forms a similar set of neuronal laminae as found in other vertebrates (*Figure 12c*) (44). However, teleost fish are unusual in that the retinal margin continues to proliferate and produce neurons throughout life. The photoreceptor layer of the retina contains five types of photoreceptors: rods, short single cones, long single cones, and short and long double cones. These synapse with bipolar and amacrine cells in the outer plexiform layer, with projections on to the inner plexiform layer, and an innermost layer of ganglion cells which project through the optic nerve to pretectal and tectal targets.

Ganglion cells of the retina project to the contralateral optic tectum in a topographic array, forming a reversed map of the visual world, with nasal axons projecting to posterior tectum and temporal axons projecting anteriorly. These two types of axons occupy different regions of the optic nerve and tract. There are also multiple projection sites of retinal ganglion cell axons within the pretectum of the diencephalon (34).

Otic

The ear serves two functions, as an auditory system to detect sound waves, and as a vestibular system to maintain balance. There is a membranous labyrinth in the inner ear, and three semicircular canals arranged orthogonally to each other (*Figure 12d*) (45). These canals contain otoliths, which are calcareous structures of species-specific shapes and sizes. Fish do not possess a cochlea, instead the macular organs of the sacculus, lagena, and utricle detect sound waves. In Otophysan species the gas bladder is used to enhance auditory stimulation (46). The mechanosensory hair cells detect vestibular and auditory stimuli and are arranged in thickened patches of sensory epithelium known as maculae or cristae. Hair cells stain with fluorescent dyes, notably DiAsp, that label their actin-rich cytoskeleton, and this simple method can be used to quickly determine their pattern (45). In the adult there are an estimated 13 000 hair cells in all of the maculae together. It is possible that, as in some other teleosts, hair cells and neurons of the auditory and vestibular systems in zebrafish are generated throughout life. The auditory nerve (cranial nerve VIII), carries auditory information from the hair-cell patches of the inner ear to auditory centres within the hindbrain.

Lateral line

Like other aquatic vertebrates, zebrafish have a pressure-sensitive lateral line system consisting of a series of hair-cell organs (neuromasts) on the head and stretching in lines along the body axis (*Figure 12a*). Neuromasts are surface structures, easily identified by the circular arrangement of epithelial cells around a pit, slightly raised from the surface and containing hair cells. These are also useful staging tools, since the pattern of neuromasts, like hair cells in the ear, can be determined quickly by staining with fluorescent cytoskeletal markers such as DiAsp (47). In the head, the anterodorsal branches of the anterior lateral line (ALL) run above (supraorbital) and below (infraorbital) the orbit and ear (otic), and the anteroventral branches run along the jaw (mandibular) and operculum. In the trunk and tail, neuromasts of the posterior lateral line (PLL) form four rows:

(i)   dorsal, running immediately under the dorsal fin;

(ii)   dorsolateral, halfway between the dorsal line and the lateral scales;

(iii) midbody, running along the lateral scales; and

(iv) a ventral line.

Adult zebrafish have well over 100 neuromasts along the midbody line (48) (see *Figure 1*). These lie in dorsoventrally oriented rows, or 'stitches', at the base of each scale, and are organized around one primary neuromast that develops initially on each scale (see below). Cells innervating the neuromasts of the ALL lie in a ganglion just anterior to the otic capsule, while the ganglion of the PLL is located just posterior to this capsule (*Figure 12a*).

Taste buds

Taste receptors in fishes, like their olfactory receptors, can respond to chemical stimuli from distant sources. The gustatory system is highly elaborated in cyprinids such as zebrafish.

Taste buds are densely packed throughout the oropharyngeal epithelium, and also occur in the external skin, particularly around the mouth and on the barbels. There are also solitary chemosensory cells in many places throughout these epithelia. The elaborate system of taste buds is innervated by neurons in epibranchial ganglia of the VII, IX, and X nerves, which project to stratified lobes in the dorsal hindbrain. The facial nerve innervates the extraoral surface, including lips, barbels, and snout, and has a recurrent branch innervating stray taste buds on the flank. The glossopharyngeal nerve (cranial nerve IX) innervates oral taste buds, while the vagal nerve (X) innervates pharyngeal receptors, including the enlarged, muscular palatal organ of cyprinids.

4.4 Larval and juvenile development of the PNS and sense organs

Peripheral nervous system

During larval development of the PNS:

(i)   DRG neurons replace the Rohon–Beard cells as the primary sensory system;

(ii) cranial ganglia expand and a well-developed gustatory and lateral line system emerges;

(iii) autonomic ganglia appear; and

(iv) enteric neurons innervate the gut.

By 72 hpf (3.5 mm) most of the basic components of the PNS are distinguishable (49–51). HRP backfills reveal that at 72–96 hpf (3.5–4 mm) there is only one bilateral pair of DRG cells per segment that has undergone axogenesis.

Sense organs

Olfactory

The olfactory organ develops from a subepidermal layer to form all of the major cell types by 4 dpf (4 mm)—basal cells, ciliated and microvillous receptor cells, support cells, ciliated non-sensory cells—as well as several others (*Figure 12b*). Even at this early larval stage, olfactory axons already project into the forebrain (38). The olfactory epidermis then separates to form a pit, which is a rosette of cells with olfactory lamellae. However, while early olfactory differentiation proceeds rapidly, the olfactory organs and bulbs then grow gradually to assume their adult proportions (52).

Optic

By 72 hpf (3.5 mm) zebrafish larvae have an optokinetic response, which by 4 dpf (4 mm) nearly equals that of the adult (27). By this stage nearly all retinal neurons are postmitotic and the major neuronal laminae are distinguishable morphologically (*Figure 12c*). Eye development in the embryo and early larva has been described in detail (53–58). Neurons are generated within a short window of time just prior to the end of embryogenesis, except for the retinal margin which continues to grow and produce neurons throughout life. Of the photoreceptors, long and short cones and double cones become evident morphologically at 4 dpf (4 mm). Rods develop later, but by 12 dpf (5.5 mm) all of the photoreceptor classes can be distinguished (58). Rows of cones are separated by rods in the photoreceptor array.

The first optic axons reach the optic tectum late in embryogenesis (reviewed in 59), and by 72 hpf (3.5 mm) have also arborized in nine other distinct regions within the diencephalon (34). These projections are maintained during the first week of development and correspond to the pretectal arborization fields found in the adult, although the cell bodies of the recipient nuclei have not yet migrated peripherally within the brain (33). A retinotopic map is established on the optic tectum in which axons of ganglion cells in the nasal retina grow to targets in the posterior tectum, while temporal axons grow anteriorly. At 72 hpf (3.5 mm), nasal axons have arrived at the tectum.

Otic

An acoustic startle response and an ability to orient relative to gravity can be observed at 4 dpf (4 mm) (60), indicating that auditory neuronal connections have been established. The larval otic vesicles are subdivided by three semicircular canals: anterior, posterior, and horizontal canals (*Figure 12d*). These canals contain the calcareous otoliths, which in many species grow by daily accretion of calcium carbonate crystals and therefore can be used for highly accurate staging (61). Ear development has been described in detail (45–47). At 72 hpf (3.5 mm) the otic vesicle is well developed, and contains two ventral sensory patches of hair cells, or maculae, each overlain by an otolith. There are 10–20 hair cells in each macula. This number grows to 80/macula by 7 dpf (5 mm). Hair cells also develop in three cristae, sensory patches for the semicircular canals. Thus by 1 week of development all of the key components have developed. The ventral anterior macula of the early larva becomes the utricle, the ventral posterior patch develops into both saccule and the lagena, and a third otolith develops by 10 dpf (5.5 mm). There is no endolymphatic duct.

Lateral line

The pattern of neuromasts in the lateral line changes dramatically during development, and can be used as a tool to stage both larval and juvenile stages, although this chapter does not attempt to cover all of the stages here (*Figure 12a, e*). Most work has focused on the development of the posterior lateral line in the early larva (47, 48, 62, 63), although there have been some recent observations at older stages (Ledent and Ghysen, personal communication). The primary lateral line of the embryo consists of six neuromasts along the horizontal myoseptum, the first lying at approximately somite 7, and the others spaced 4–6 somites apart along the body. Secondary neuromasts form during the first week of larval development and this pattern is more variable (*Figure 12e*). On average, 4–5 neuromasts appear between somites 7–15, and 2–3 in the tail. The lateral line then undergoes a dramatic shift, leaving the myoseptum and moving ventrally except at its anterior and posterior ends. At early juvenile stages (12 mm) there is one neuromast for each somite along this now ventral line, and 3–4 neuromasts appear along the myoseptum forming a new midbody line. By 2 months all four lines of neuromasts found in adults are visible, 6–8 in each of the dorsal and dorsolateral lines, a new lateral, midbody line, and 30 or so in the ventral line. All of these lines form stitches. Thus the line of stitches of neuromasts that develop along the lateral scales of the adult are not derived from the primary embryonic line, but from a new line that develops in the juvenile.

Taste buds

While few data are available for the development of taste buds (64), more is known about the formation of secondary epidermal solitary chemosensory cells (SCCs) (65). SCCs penetrate the epithelial surface at 72 hpf (3.5 mm), prior to the appearance of taste buds at 5 dpf. The number of SCCs then increases dramatically to cover the entire body surface, while taste bud pores remain confined to the oral epithelia and barbels.

5 Cardiovascular, digestive, and reproductive organs

5.1 Adult

In comparison with other organ systems, attention has only recently turned to the development of the haematopoietic system, the visceral organs, and the gonads in zebrafish (*Figure 13*). This chapter therefore extrapolates in some cases from general descriptions for teleosts as found in Harder (66). In adults, the heart and its pericardial cavity are separated from the peritoneum which houses the other organs (*Figure 13a*). Posterior to the pharynx, the gut forms a short oesophagus and stomach, followed by a long, looping intestine. The liver and pancreas surround the anterior region of the coelom. The swim bladder is located further dorsally and is subdivided into anterior and posterior chambers. The kidney runs along the dorsal roof of the coelom. In females, ovaries extend throughout most of the length of the cavity. Digestive and urogenital ducts open at a common point between pelvic and anal fins.

Vascular system, heart, kidney, and haemopoietic tissues

Teleosts have a relatively simple vascular system with a two-chambered heart that sends blood directly through the gills to be oxygenated and into the head before passing posteriorly

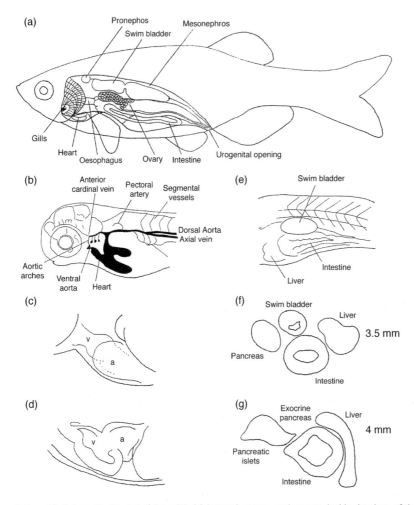

Figure 13 Schematic drawing of the adult (a) internal organs, and camera lucida drawings of the vascular system (b–d) and visceral organs (e–g). Whole-mounted preparations (b–e) and from plastic transverse sections stained with Methylene Blue (f, g). (a) Lateral view of an adult (2.0 cm), based on dissected specimens, illustrating the major internal organs and their approximate locations. (b) Lateral view of an embryo (2.5 mm), showing heart and major vessels. (c) Lateral view of the early larval (3.5 mm) heart. a, atrium; v, ventricle. (d) Lateral view of a late larval (5 mm) heart. a, atrium; v, ventricle. (e) Lateral view of visceral organs in the early larva (3.5 mm). (f) Transverse section showing visceral organs in an early larva (3.5 mm) and (g) slightly later larva (4 mm). From refs 66 and 71.

through the rest of the body (*Figure 13b–d*) (reviewed in ref. 26). The heart is located posterior and ventral to the gill arches. A sinus venosus brings blood into the single atrium and into the ventricle, which then pumps it out through the trunk of the bulbus arteriosus. The heart is separated by a transverse septum, which serves as a pathway for vessels carrying blood to the heart. A ventral aorta in the midline below the pharynx carries blood away from the heart and into the carotid arteries. Internal carotids then transport blood forward into the head. The aortic arches carry blood from the ventral aorta to dorsal vessels. In the gills, blood flows through lamellae, narrow closely spaced ridges to maximize surface area, that form on both sides of the gill chamber. From the gills, blood then travels into the dorsal aortas, paired vessels in the head region that carry blood posteriorly, uniting at the level of the liver and extending into the trunk and tail. Other vessels include the coeliac and mesenteric arteries to visceral organs, parietals supplying the body wall, genitals and kidneys, and hepatic portal veins that empty into the sinus venosus (not shown). There is a renal portal system, drained by subcardinal veins. Blood returning in the anterior and posterior cardinal veins meet at the common cardinals, which empty into the sinus venosus and the heart.

The adult kidney is composed of mesonephric tubules, the pronephros having regressed and condensed to form the haematopoietic tissue of the inter-renal gland (*Figure 13a*) (67). In a 6-month-old adult the pronephros has completely degenerated and the mesonephri have fused. Within the kidney, which extends nearly the full length of abdominal cavity, there are internal glomeruli, as well as proximal and distal convoluted tubules. The teleost kidney also serves haematopoietic, reproductive (epididymis), and urinary functions. The glomeruli form a small, tubular capillary network supplied by afferent renal veins branching from renal portal veins. Rather than true adrenal glands, chromaffin and cortical tissue form in suprarenal tissue in small aggregations next to the dorsal aorta, near the sympathetic ganglia. Chromaffin tissue, which produces adrenaline and noradrenaline, is associated with the postcardinal veins. Also suprarenal or inter-renal cells are present which produce corticosteroids.

Viscera, glands

The digestive system and glands in zebrafish are similar to those of other vertebrates, although in many of the fish's organs the cells are more dispersed (66). The pleuroperitoneal cavity, or coelom, containing the digestive organs, is separate from the pericardial cavity (*Figure 13a*). Teleosts have a short oesophagus which is continuous with the stomach, and this ends in a pyloric sphincter separating it from the intestine. The liver is not lobular, and is connected to the gut by intrahepatic bile ducts and a gall bladder for the storage of bile. The pancreas is composed of at least one prominent islet connected to the gut by a pancreatic duct, and accessory islets and exocrine cells are scattered along the intestine. As is typical of minnows, the zebrafish intestine has several loops. The intestinal epithelium contains absorptive, endocrine, and goblet cells, like the mammalian small intestine, but lacks crypts. Enteric nerves innervate circular and longitudinal smooth muscles that surround the gut. There is no real cloaca, but intestinal and urogenital ducts open at the same location.

In the gill region, the thyroid gland consists of isolated follicles scattered along the ventral aorta and its branches in the gill arches, and forms a midventral evagination of the pharyngeal wall (not shown). In teleosts, the thymus is thought to be derived from four pairs of pharyngeal pouches. The ultimobranchial gland, a calcium-regulating organ in teleosts incorporated into the thyroid in mammals, occupies the posterior pharynx adjacent to the pericardium (68). The swim bladder, also an outpocket of the gut, forms an air-filled sac in the anterior coelom. This plays a specialized function in buoyancy and in hearing by transmitting vibrations to the inner ear through the bony Weberian apparatus.

Gonads

The gonads of teleosts are paired structures, both in the male and female, developing along the splanchnic coelom (66, 69). Short gonoducts expel gametes to the exterior. Zebrafish achieve external fertilization using their pelvic and anal fins. The testes lie along the swim bladder and contain a mass of elongated, branching tubules which have secondary and tertiary branches, where spermatogenesis follows the normal vertebrate pattern. Sperm leave the testes through coiled, mesonephric kidney tubules. The ovary is a hollow, bilobed sac ventral to the swim bladder (*Figure 13a*). It is suspended in the body cavity by a vascularized mesovarium, that opens into an oviduct which is not a modified kidney tubule but a continuation of the ovary. At maturation the oocyte becomes free of the follicle during ovulation, and water is taken up, causing the egg to swell. The ovary occupies an extensive part of the coelom and can occupy 10–20% of the weight of a female zebrafish when gravid. In many teleosts, environmental factors, such as water temperature, influence sex determination, but these factors are unclear in zebrafish.

Protocol 5

Sectioning and histological stains of larval or adult tissue

Equipment

- Cryostat for frozen sections
- Rotary microtome for paraffin sections

Frozen sections

1 Fix tissue and wash several times in PBS prior to embedding.
2 Place in moulds and remove excess PBS.
3 Fill mould with molten agarose–sucrose at slightly less than 50 °C.
4 Orient the specimen as agarose solidifies.
5 Remove block from mould and trim to a pyramidal shape, with the specimen near the narrow top cutting surface. Try to have a block face of at least 5 mm, with plenty of agar surrounding the specimen in case of tearing during sectioning.
6 Sink block in 30% sucrose at 4 °C overnight.
7 Cool cryostat to −30 °C.
8 Freeze a platform of OCT or similar cryo-embedding compound on to specimen holders.
9 Drain specimen block on filter paper and mount on to OCT platform with some additional fresh OCT. Then freeze on to holder by immersing its base in liquid nitrogen.
10 Place in cryostat and cut 7–10 μm sections, collect on TESPA-coated microscope slides and air dry for several hours.

Plastic sections

1 Embed embryos in JB4 resin (Agar Scientific) or araldite.
2 Cut 7–10 μm sections using a glass or tungsten knife on an electron microscopy (EM) microtome.
3 Collect sections individually on drops of water on an uncoated glass microscope slide and dry on a hot plate at 70 °C.
4 Sections may be counterstained with Toluidine Blue and coverslipped.

Selected histological stains

- Cresyl Violet (see *Protocol 3*)—for Nissl substance in nerve cells

- Methylene Blue/azure II—for general cellular staining
- Haematoxylin and eosin—for general cellular staining
- Periodic acid-Schiff staining—for intestinal glycogen
- Wright-Giemsa—for blood cells
- Diaminofluorene—for haemoglobin peroxidase

5.2 Larval and juvenile organogenesis

Heart, blood, kidney, vasculature, and haematopoiesis

Teleosts such as zebrafish, that hatch as small larvae, depend on cutaneous respiration across the body surface. Early larvae then switch to gill respiration and haemoglobin appears in erythrocytes. During larval development of haematopoietic tissues:

(i) zebrafish switch to adult/definitive haematopoiesis in the spleen and kidney, rather than the bone marrow;

(ii) the pronephros gives way to a mesonephros.

At 72 hpf (3.5 mm) there are five aortic arches and a simple circulation through several cephalic arteries, a dorsal aorta, segmental vessels in each somite, and blood returns to the heart through the axial vein (*Figure 13b*). An atlas of vascular anatomy in larvae between 3 and 7 dpf, based on microangiography, is now available (70). Embryonic erythrocytes travelling through these vessels are flattened and approximately 5.6 μm in diameter, while by 5 dpf circulating erythrocytes are larger and resemble those of adults (7×3 μm). Endocardial cells at these stages appear to bud off basiphilic cells resembling precursors of granulocytes and lymphocytes. There are glomeruli in the pronephros at 72 hpf (3.5 mm), and tubules form by 5 dpf (4 mm) (71, 72). During the second week of development, haematopoiesis from the endocardium ceases.

By 4 weeks of development (10 mm) the kidney becomes haematopoietic (67). Pronephric tubules degenerate and the mesonephros takes over the role of waste eradication. Granulocytes, lymphocytes, and erythrocytes are generated within the inter-renal tissue, particularly around the posterior cardinal veins. In a 6-month-old adult the pronephros has completely degenerated or transformed into the haematopoietic 'head kidney', and the mesonephri have fused.

Viscera

Teleost larvae depend on the yolk for nutrition into early larval stages, until they begin feeding. Likewise, digestive organs do not develop until larval stages. The major events include onset of gut motility and differentiation of secretory cells of the liver and pancreas. Development of the digestive organs has been described for early larvae (71). By 72 hpf (3.5 mm), representatives of most cell types have differentiated (*Figure 13e–g*). The pharynx, oesophagus, and intestine are all distinguishable, and the intestine undergoes peristalsis. By 4 dpf (4 mm) exocrine cells of the pancreas differentiate and the intestinal cells form a microvillar brush border (*Figure 13g*). Liver cells and bile are visible by 3 dpf (3.5 mm) and hepatic blood flow by 4 dpf (4 mm).

Little is known about the development of the glands associated with the pharynx. One exception is the thymus, which has been identified in embryos by its expression of *recombination activating genes* (*rag*) which identify the primordium of the developing thymus in late embryogenesis, and this enlarges at 4 dpf (4 mm) (73, 74). The thymus achieves its mature histological organization, including lymphocytes, at the beginning of juvenile period. In the pronephros, haematopoiesis begins at 2 weeks (6 mm). Mature lymphocytes are also distinguishable here, in the tissue surrounding the renal tubules.

Gonads

Gonadogenesis in zebrafish has been described (not shown) (75). The exact stage at which the gender is determined in many species of fish is controversial, and completely unclear in zebrafish. Minnows may go through a prematurational sex change, differentiating but not maturing first as females, with some individuals later changing to males. The gonads are paired and develop from a single epithelial cortex, in close association with the nephric system. Primordial germ cells, specified in the early embryo, colonize the posterior part of the nephric ridge well before 72 hpf (3.5 mm) (76). Development after this stage is unclear in zebrafish. In males, testes are divided into compartments called acini. In these the spermatogonia pass through the spermatid stage. The oogonia become surrounded by a single layer of follicular cells. Developing eggs are then supplied with yolk by the follicular cells during vitellogenesis.

Acknowledgements

I am very grateful to many people for contributing to this work, including Paolo Sordino, Sam Cooke, Alain Ghysen, Nicolas Gompel, Mario Wullimann, and Steve Wilson for unpublished data. A special thanks to Ruth BreMiller, who has perfected many of the standard histological techniques now used in zebrafish. I am also very grateful to the late Nigel Holder for his support of this project.

References

1. Hamilton-Buchanan, F. (1822). *An account of the fishes found in the River Ganges and its branches.* Archibald Constable, Edinburgh.
2. Kimmel, C. B., Ballard, W. W., Kimmel, S. R., Ullmann, B. and Schilling, T. F. (1995). *Dev. Dyn.,* **203**, 253–310.
3. Fink, S. V. and Fink, W. L. (1981). *Zool. J. Linn. Soc.,* **72**, 297–353.
4. Meyer, A., Ritchie, P. A. and Witte, K.-E. (1995). *Phil. Trans. R. Soc. Lond. B Biol. Sci.,* **349**, 103–11.
5. Kirschbaum, F. (1975). *Roux's Arch. Dev. Biol.,* **177**, 129–52.
6. Cubbage, C. C. and Mabee, P. M. (1996). *J. Morphol.,* **229**, 121–60.
7. Kendall, A. W., Ahlstrom, E. H., and Moser, H. G. (1984). In *Ontogeny and systematics of fishes.* Am. Soc. Ichthyol. Herpetol, Spec. Publ., Vol. 1, pp. 11–22.
8. Brown, D. D. (1997). *Proc. Natl Acad. Sci,.* **94**, 13011–16.
9. Schilling, T. F. and Kimmel, C. B. (1997). *Development,* **124**, 2945–60.
10. Milos, N. and Dingle, A. D. (1978). *J. Exp. Zool.,* **205**, 205–16.
11. Lindsey, C. C. (1988). *Fish Physiol.,* **11**, 197–274.
12. Huysseune, A., Van der Heyden, C. and Sire, J. Y. (1998). *Anat. Embryol.,* **198**, 289–305.
13. Pashine, R. G. and Marathe, V. B. (1973). *J. Univ. Bombay,* **42**, 53–62.
14. Benjamin, M. (1990). *J. Anat.,* **169**, 153–72.
15. Sordino P., Van der Hoeven, F., and Deboule, D. (1995). *Nature,* **375**, 678–81.
16. Grandel, H. and Schulte-Merker, S. (1998). *Mech. Dev.,* **79**, 99–120.
17. Miyake, T., McEachran, J. D., and Hall, B. K. (1992). *J. Morphol.,* **212**, 213–56.
18. Sire, J. Y., Allizard, F., Babiar, O., Bourguignon, J., and Quilhac, A. (1997). *J. Anat.,* **190**, 545–61.
19. Pantin, C. F. A. (1946). *Notes of microscopical technique for zoologists.* Cambridge University Press, Cambridge.
20. Humphrey, C. and Pittman, F. (1974). *Stain Technol.,* **49**, 9–14.
21. Edgeworth, F. H. (1935). *The cranial muscles of vertebrates.* Cambridge University Press, Cambridge.

22. Winterbottom, R. (1974). *J. Acad. Nat. Sci. Phil.*, **125**, 225–317.
23. Van Ramsdonck, W., Pool, C. W., and te Kronnie, G. (1978). *Anat. Embryol.*, **153**, 137–55.
24. Van Ramsdonck, W., Mos, W., Smit-Onel, M. J., Van der Laarse, W. J., and Fehres, R. (1983). *Anat. Embryol.*, **164**, 63–74.
25. Hu, N., Sedmera, D., Yost, H. J., and Clark, E. B. (2000). *Anat. Rec.*, **260**, 148–57.
26. Fishman, M. C. and Chien, K. R. (1997). *Development*, **124**, 2099–117.
27. Easter, S. S. Jr and Nicola, C. N. (1996). *Dev. Biol.*, **180**, 646–63.
28. Devoto, S. H., Melancon, E., Eisen, J. S., and Westerfield, M. (1996). *Development*, **122**, 3371–80.
29. Wullimann, M. F., Rupp, B., and Reichert, H. (1996). *Neuroanatomy of the zebrafish brain*. Birkhauser Verlag, Berlin.
30. Lee, R. K. and Eaton, R. C. (1991). *J. Comp. Neurol.*, **304**, 34–52.
31. Westerfield, M., McMurray, J. V., and Eisen, J. S. (1986). *J. Neurosci.*, **6**, 2267–77.
32. Romeis, B. (1948). *Mikroskopische Technik*. Leibniz Verlag, Munchen.
33. Wullimann, M. F. and Puelles, L. (1999). *Anat. Embryol.*, **199**, 329–48.
34. Burrill, J. D. and Easter, S. S. (1994). *J. Comp. Neurol.*, **346**, 583–600.
35. Trevarrow, B., Marks, D. L., and Kimmel, C. B. (1990). *Neuron*, **4**, 669–79.
36. Metcalfe, W. K., Mendelson, B., and Kimmel, C. B. (1986). *J. Comp. Neurol.*, **251**, 147–59.
37. Cooke, S. (1999). Undergraduate thesis, University College London.
38. Dynes, J. L. and Ngai, J. (1998) *Neuron*, **20**, 1081–91.
39. Myers, P. Z. (1985). *J. Comp. Neurol.*, **236**, 555–61.
40. Bernhardt, R. R., Chitnis, A. B., Lindamer, L., and Kuwada, J. Y. (1990). *J. Comp. Neurol.*, **302**, 603–16.
41. Williams, J. A., Barrios, A., Gatchalian, C., Rubin, L., Wilson, S. W., and Holder, N. (2000). *Dev. Biol.*, **226**, 220–30.
42. Baier, H. and Korsching, S. (1994). *J. Neurosci.*, **14**, 219–30.
43. Byrd, C. A. and Brunjes, P. C. (1995). *J. Comp. Neurol.*, **358**, 247–59.
44. Marcus, R. C., Delaney, C. L., and Easter, S. S. Jr (1999). *Vis. Neurosci.*, **16**, 417–24.
45. Haddon, C. and Lewis, J. (1996). *J. Comp. Neurol.*, **365**, 113–28.
46. Fay, R. R. and Popper, A. N. (1974). *J. Exp. Biol.*, **62**, 379–87.
47. Whitfield, T. T., Granato, M., van Eeden, F. J., Schach, U., Brand, M., Furutani-Seiki, M., Haffter, P., Hammerschmidt, M., Heisenberg, C. P., Jiang, Y. J., Kane, D. A., Kelsh, R. N., Mullins, M. C., Odenthal, J. and Nusslein-Volhard, C. (1996). *Development*, **123**, 241–54.
48. Metcalf, W. K. (1983). Ph.D. thesis, University of Oregon.
49. Raible, D. W. Wood, A., Hodsdon, W., Henion, P. D., Weston, J. A., and Eisen, J. S. (1992). *Dev. Dyn.*, **195**, 29–42.
50. Schilling, T. F. and Kimmel, C. B. (1994). *Development*, **120**, 483–94.
51. Metcalf, W., Kimmel C. B., and Schabtach, E. (1985). *J. Comp. Neurol.*, **233**, 377–89.
52. Hansen, A. and Zeiske, E. (1993). *J. Comp. Neurol.*, **333**, 289–300.
53. Nawrocki, L. (1985). Ph.D. dissertation, University of Oregon, Eugene, OR.
54. Larison, K. D. and BreMiller, R. (1990). *Development*, **109**, 567–76.
55. Schmitt, E. A. and Dowling, J. E. (1994). *J. Comp. Neurol.*, **344**, 515–36.
56. Li, Z., Joseph, N. M., and Easter, S. S. Jr (2000). *Dev. Dyn.*, **218**, 175–88.
57. Malicki, J., Neuhauss, S. C. F., Schier, A. F., Solnica-Krezel, L., Stemple, D. L., Stainier, D. Y. R., Abdelilah, S., Zwartkruis, F., Rangini, Z., and Driever, W. (1996). *Development*, **123**, 263–73.
58. Branchek, T. and BreMiller, R. (1984). *J. Comp. Neurol.*, **224**, 107–15.
59. Karlstrom, R. O., Trowe, T., and Bonhoeffer, F. (1997). *Trends Neurosci.*, **20**, 3–8.
60. Kimmel, C. B. (1972). *Dev. Biol.*, **27**, 272–5.
61. Panella, G. (1971). *Science*, **173**, 1124–7.
62. Raible D. W. and Kruse, G. J. (2000). *J. Comp. Neurol.*, **421**, 189–98.
63. Alexandre, D. and Ghysen, A. (1999). *Proc. Natl Acad. Sci. USA*, **96**, 7558–62.
64. Hatta, K., BreMiller, R., Westerfield, M., and Kimmel, C. B. (1991). *Development*, **112**, 821–32.
65. Kotrschal, K., Krautgartner, W. D., and Hansen, A. (1997). *Chem. Senses*, **22**, 111–18.
66. Harder, W. (1975). *The anatomy of fishes*. Eschweizerbartsche Verlagshuchhandlung, Stuttgart.
67. Al-Adhami, M. A. and Kunz, Y. W. (1977). *Dev. Growth Diff.*, **19**, 171–9.

68. Yamane, S. (1978). *Bull. Fac. Fish. Hokkaido Univ.*, **29**, 213–22.

69. Bauer, M. P. and Goetz, F. W. (2001). *Biol. Reprod.*, **64**, 548–54.

70. Isogai, S., Horiguchi, M., and Weinstein, B.M. (2001). *Dev. Biol.*, **230**, 278–301.

71. Pack, M., Solnica-Krezel, L., Malicki, J., Neuhauss, S. C. F., Schier, A. F., Stemple, D. L., Driever, W., and Fishman, M. C. (1996). *Development*, **123**, 321–8.

72. Drummond, I. A., Majumdar, A., Hentschel, H., Elger, M., Solnica-Krezel, L., Schier, A. F., Neuhauss, S. C., Stemple, D. L., Zwartkruis, F., Rangini, Z., Driever, W., and Fishman, M. C. (1998). *Development*, **125**, 4655–67.

73. Willett, C. E. , Zapata A. G., Hopkins N., and Steiner, L. A. (1997). *Dev. Biol.*, **182**, 331–41.

74. Willett, C. E., Cortes, A., Zuasti, A., and Zapata, A. G. (1999). *Dev. Dyn.*, **214**, 323–36.

75. Frolander, H. T. (1950). Master's thesis, Brown University.

76. Braat, A. K., Zandbergen, T., van de Water, S., Goos, H. J., and Zivkovic, D. (1999). *Dev. Dyn.*, **216**, 153–67.

Chapter 4

Cell labelling and transplantation techniques

Donald A. Kane
Department of Biology, University of Rochester, 213 Hutchison Hall, Rochester, NY 14627, USA

Yasuyuki Kishimoto
Japan Science and Technology Corporation, 14 Yoshida-kawaramachi, Sakyo-ku Kyoto 606–8305, Japan

1 Introduction

What with their genetics, zebrafish have become the *Drosophila* of the vertebrates. But these little fish have many other advantages. As the developmental biologists of the mid-twentieth century knew (1), zebrafish are superb for observational studies of development. Their eggs are fertilized externally, so that it is a simple matter to have the embryos on a compound microscope a few moments after fertilization. There, the transparent embryo reveals the early cell cleavages of the blastula, the cell movements of the gastrula, or, later, the forming internal organs of the larva. *In situ* hybridization and antibody reactions are done as whole mounts, for sectioning can be done later, if at all. And if lineage tracer is injected into cells, as we describe in this chapter, the cells can be recorded for days, and sometimes weeks, afterwards.

The technique of moving about pieces of tissue from embryo to embryo has been an essential part of developmental biology for nearly a century. By placing developing cells into ectopic positions, we test the potential of the transplanted cells to develop into fates appropriate for the new position. These techniques, described in this chapter, are essential for connecting the developmental biology of the zebrafish to that of more traditional experimental vertebrates, such as *Xenopus* and chick, and are also necessary for the analysis of gene action in the zebrafish mutants.

In this chapter we first describe some of the general methods for injecting lineage tracers in to the embryo. These techniques can be used for creating labelled embryos (for later experiments) or for determining cell fate. Next we describe methods for moving cells from one embryo to another, methods that are used to test cell fate or cell autonomy. Finally, we describe methods for the analysis of these experiments.

Many of the methods we describe are imagined as difficult stuff, performed by those possessing calm temperaments and quiet hands. In fact, we have introduced some of these methods into undergraduate laboratory sessions, where students, after dechorionating embryos with watchmaker forceps, label cells with hand-held pipettes controlled by breath pressure, and later move cells about with small pieces of cactus spines and broken glass. Most who attempt these techniques a few times will have them working. In fact, the ultimate challenge is not performing the manipulations themselves, but analysing and interpreting the huge amount of data often produced from seemingly small and simple experiments.

2 Cell labelling techniques

Fate maps define what cells at some early stage later become. These maps are a starting point for the description of the normal development of the embryo. The labelling techniques described in this section are used to construct these maps, and are the methods used for the now classic Kimmel, Warga, and Schilling fate map of the zebrafish blastula (2). Still, much fate mapping remains in progress, either of finer regions or later stages (3–7). Much of this new work is motivated by the expression patterns of new cloned genes in zebrafish, where we wonder what fates lie in these expression patterns; other work is motivated by mutants that change expression patterns, where we wonder if fates are changed as well. Thus, in both cases, to properly answer the question, we must label cells and ask what they become.

Fluorescent and biotinylated dextran conjugates are the molecules of choice for labelling blastomeres of zebrafish embryos (*Table 1*). Having a high molecular weight, good water solubility and low toxicity, dextrans are hydrophilic polysaccharides that are stable in the intracellular space. Because dextran conjugates are membrane impermeable, they can be kept in clonal progeny of labelled blastomeres for many days, and, for some early differentiating cells, several weeks.

With a bright-red fluorescence, the photostable tetramethylrhodamine dextran is the preferred dye for repeated observation with epi-illumination. While the fluorescein dextrans are often used as a second colour, the photostability of the fluorescein conjugates is not as high as that of tetramethylrhodamine dextran and prolonged exposure causes substantial photodamage to the cells, making the dye less desirable for repeated observations by fluorescent microscopy. A rhodamine derivative, RhodoB, is now available that fluoresces with the

Table 1 Common vital dyes and terminology

Rhodamine (tetramethyl-rhodamine)	Vital fluorescent dye; red colour when excited with green light; good for light intensifiers with sensitivities biased toward infrared; excellent resistance to photobleaching and phototoxicity
RhodoB	Vital fluorescent dye; green colour when excited with blue light; good resistance to photobleaching and phototoxicity
Texas Red	Vital fluorescent dye; red colour when excited with green light; good for light intensifiers with sensitivities biased toward infrared; excellent resistance to photobleaching and phototoxicity
Cascade Blue	Vital fluorescent dye; blue colour when excited with UV light; poor for light intensifiers with sensitivities biased toward infrared; excellent resistance to photobleaching and UV but is toxic over intervals used for time lapse
Fluorescein (fluorescein isothiocyanide)	Vital fluorescent dye; green colour when excited with blue light; good for light intensifiers with sensitivities biased toward infrared; poor resistance to photobleaching; moderate phototoxicity
DMNB-caged fluorescein	Vital fluorescent dye; little initial fluorescence; after 'activation' with UV light, the dye gives a green colour when excited with blue light
DiO, DiI	Lipophilic dyes for marking cell membranes
Biotin	Added to lineage cocktail for later fixation, amplification with avidin-biotin and visualization with horseradish peroxidase colour reaction
Dextran	Sugars covalently bound to tracers to increase molecular weight
10 000 Da	Normal weight for lineage analysis
70 000 Da	High weight for lineage analysis of marginal blastomeres at 16- to 64-cell stage
2 000 000 Da	Ultra high weight for lineage analysis of 2- to 16-cell stages
Anionic, cationic	Charge on dextrans to aid in electrophoresing tracers into cells. Neutral dyes are fine for simple pressure injection

fluorescein filter set, and this new dye may be a partial replacement for fluorescein. Another alternative colour is Texas Red, which can be seen on the rhodamine filter set. Cascade Blue, which uses the UV DAPI/Hoechst filter set, seems an obvious third colour; however, its use as a vital dye is somewhat limited because of its UV excitation wavelength, which is cytotoxic, and its blue emission, which is invisible to many intensified cameras. Note that most of these dyes come in lysine-fixable versions, allowing covalent linkage to surrounding tissue during fixation.

Due to their low diffusion, newer fluorescent and biotinylated dextrans of super high molecular weight (500 000 and 2 000 000 Da) are particularly useful for fate-mapping blastomeres of the early stage embryo. The lower molecular-weight dextrans, under 50 000 Da MW, leak from blastomeres during early stages, when the cells are cytoplasmically bridged to the yolk, and then leak from the yolk and label other blastomeres. Another advantage of the large MW dyes is that they are excluded from the nuclei, allowing counterstaining of nuclei to visualize nuclear labels. However, clogging needles and low solubility somewhat limit the convenience of the large molecular-weight dyes.

For particularly difficult injection locations, the so-called DMNB-caged fluorescein (Molecular Probes, Eugene, Oregon, USA) dextrans can be used to label portions of the embryo. These dyes are injected at early developmental stages, and then later the target cells, which may be small or buried in tissue, are photo-activated with UV, uncaging the fluorescent molecule. Some of these dyes are fixable, e.g. the uncaged fluorescein can be late visualized with the normal antibody against fluorescein.

2.1 Labelling whole embryos with lineage tracer dyes

For transplantation techniques, cells from donor embryos must be marked to distinguish them from host cells. In classic work, this labelling was achieved by using donors possessing different genetic characteristics, such as different cytological characteristics (8) or lack of pigment (9). These days, whole embryos are labelled by injecting fluorescent and biotinylated dextrans into the yolk prior to the 16-cell stage (*Figure 1*). When the blastomeres of the early embryo are cytoplasmically continuous with the yolk, low molecular weight tracer dyes mix with cytoplasm amongst the yolk platelets and rapidly move into blastomeres carried by the early cytoplasmic streaming of the embryo. Clutches containing mutant embryos can be subdivided into a number of groups of embryos, and each group labelled with a different colour, leaving one group unlabelled for later use as hosts. Embryos, later to be used for transplantation, must be dechorionated, either manually with watchmaker forceps or enzymatically with 0.5 mg/ml of proteinase K.

Protocols 1 and *2* are appropriate for large-scale injections, where hundreds of embryos can be labelled in a single sitting. With little modification, this method is also appropriate for injection of DNA and RNA constructs, or for the injection of pharmaceutical agents that do not readily enter the embryo, such as ethanol, α-amanitin, or cytochalasin B. Alternative steps are outlined for small-scale injections, where tens of embryos are injected, for which we prefer to dechorionate the embryos manually and then inject them in an agar-bottomed Petri dish.

With the appropriate lineage tracers, this method can be extended to a crude sort of lineage technique. If a single blastomere is injected with dyes that have molecular weights in the 70 000 αDa size range, or, in the case of the 2- to 4-cell stage, the 2 000 000 Da size range, the dye only very slowly diffuses into the adjacent cells, effectively labelling the progeny of the targeted cell. For general lineage work, these clones are too large to analyse because of the large region of fluorescent label which occludes underlying cells, and because these early clones contribute to the majority of the tissues of the embryo. However, the method is very useful as a teaching aid (10) and, in certain experiments, it is useful in the marking of co-injected DNA or RNA constructs (11).

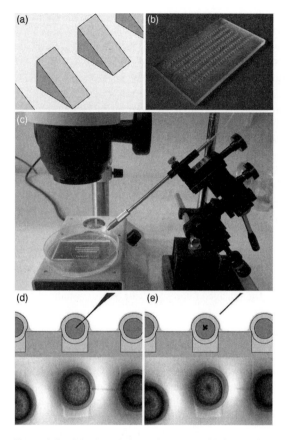

Figure 1 Dye injections during early cleavage. (a) Shape of teeth on plastic mould. Each tooth is approximately 0.8 mm wide. (b) Entire plastic mould. (c) The set-up of the dissecting microscope, agar plate with embryos, and pipette held by a micromanipulator. (d and e) Schematic and photographs of embryos in an injection plate. The embryos, protected by their chorions, are held in the slots by surface tension. In (e) the injection volume is a ball of dye about 1/10 to 1/6 the diameter of the yolk.

Protocol 1

Agarose plates for holding embryos

Equipment and reagents

- 60 or 100 mm plastic Petri dishes
- 1.5% agarose solution (in E2 medium) (see note on agar media in *Table 2*)
- Plastic mould, glass 1–1.5 mm capillary pipettes, or glass depression slides

Method

1 Pour liquefied 1.5% agarose into a plastic Petri dish to a depth of about 5 mm.

2a Float a plastic mould into the agarose, taking care to avoid trapping air bubbles below the mould.

Protocol 1 continued

2b Alternatively: float several capillaries (1–1.5 mm diameter) in the liquid agar.

2c Alternatively: tip a thick glass microscope slide (such as a depression slide) from the bottom of the dish to the edge, to form a trough in the liquid agar.

3 When the agarose is completely solidified, gently lift the mould or capillaries, or slide away from the agarose using coarse forceps or a sturdy needle.

Table 2 Common transplantation/lineage analysis solutions

E2	15.0 mM NaCl, 0.5 mM KCl, 1.0 mM $CaCl_2$, 1.0 mM $MgSO_4$, 0.15 mM KH_2PO_4, 0.050 mM Na_2HPO_4, 0.70 mM $NaHCO_3$ Make as combined 20× stock, except for sodium bicarbonate, which is made as a 200× or 500× stock. (For most uses, sodium bicarbonate can be omitted) pH of final stock should be 7.0–7.5
Danieau's	58 mM NaCl, 0.7 mM KCl, 0.4 mM $MgSO_4$, 0.6 mM $Ca(NO_3)_2$, 5.0 mM HEPES pH 7.1–7.3 Make each component as a separate 50× solution
Streptomycin/penicillin	50 μg/ml streptomycin–50 U/ml penicillin Purchase and store as frozen 1000× solution. Use media within 24 h of preparation
Methylcellulose	Make as 3% in either water or E2 medium Dissolve by mixing over several hours while partially freezing, clarify by refrigeration overnight; dispense in 1 or 5 ml syringes; refrigerate
MESAB (also called Tricaine® or MS322)	4 mg/ml ethyl -m-aminobenzoate methanesulphonate, 1% Na_2HPO_4, pH 7.0–7.5; refrigerate For use, dilute 1:20 to 1:100; solution will darken on exposure to light
Petroleum jelly (Vaseline®)	Melt in boiling water, pour into the back of 5 ml syringes, and put the plungers back in; dispense through a cut-off 18-gauge needle
Agar/ agarose media	1.5% agar for lining plastic Petri dishes; 1.5% agarose for moulds; 0.1% agarose for immobilizing embryos for time-lapse photography Made in Danieau's, E2, or water The yolk cell membrane will stick to dried agar surfaces; this can be corrected by hydrating the plates overnight. Check agarose brands before use, the yolk cell ticks to many that are normally used for electrophoresis; this can be corrected by cutting 10 parts agarose:1 part agar
Watchmaker's forceps (No. 5 Dumont)	Use three in normal work. The sharpest, just out of the package with 20 μm tips, is the tool for dechorionation and fine manipulations, such as removing blastoderms, etc. Slightly duller, with 50 μm tips, is a tool for holding the chorion or embryo. The dullest, with 100 μm or more tips, is a good tool for general embryo work, such as sorting and cleaning embryos or orienting embryos in methylose. Even straight out of the package, touch up forceps with a fine stone under a dissecting microscope; all forceps, fine to coarse, must make first contact at the very end of the tongs
Nylon 'hair' loops	Use lowest test weight fishing line, about 100–150 μm diameter, to fashion small 1–2 mm loops at the end of a glass capillary; fix to a glass capillary with glue or wax
Tungsten needles	Sharpen a tungsten wire by electrophoresis in a solution of 0.5 M NaCl. The needle is attached to the positive pole of a DC source; graphite or a pencil lead is used as a cathode; fix the sharped needle to a glass capillary with glue or wax
Cactus needles	Pick the finest needles from the top of the cactus plant; harden the needles by heating on an incandescent light bulb; fix to a glass capillary with glue or wax. For continued good results, water the cactus plant once or twice a year

Protocol 2

Early cleavage injections

Equipment and reagents

- Dissecting microscope, equipped with transmitted light
- Micropipettes. Make micropipettes from 1.0 or 1.2 mm thin-walled capillary glass filaments (e.g. TW120F-4 from World Precision Instruments) The capillary is pulled by an electrode puller (e.g. Flaming Brown) to produce a fine needle on the tip. This needle should be thick enough to penetrate the chorion without breaking, but thin and long enough (about 10 μm thickness and 1 mm length) so the embryos are not dragged when the needle is withdrawn from the chorion. In the alternative method (not penetrating the chorion) the needle can be thinner

- Micromanipulator. One movement should be set so that the micropipette moves parallel to the injection needle
- Microinjector. Most pressure injectors will work; we prefer models that are compact and manually controlled. A simple and adequate set-up is that by ASI with a pressure gauge, a switch (for pressure on/off), and mechanical pressure regulator. The unit should be able to attain 50 psi
- Agarose plates (*Protocol 1*)
- Fine forceps

Method

1a Place 1- to 16-cell stage embryos with chorions in the holes or slots in the agarose plate. If holes, each hole can hold one embryo. As much as possible, remove extra medium on the agarose plate with a pipette. Embryos will be held firmly in the holes by the surface tension of the medium.

1b Alternatively: dechorionate the embryos and place in an agar mould, or on a plain agar plate. In this case do not remove extra media.

2 Fill the micropipette with the dye solution. The capillary will draw the solution to the tip of the pipette. Special thin-gel loader tips (e.g. Eppendorf Micro loader) are sometimes useful for this step. This step can be done the evening before the injections and the needles stored in a humid chamber; extra time helps the tips fill better and allows air bubbles at the tip of the pipette to escape.

3 Set the micropipette in the needle holder connected to the pressure injector.

4 Break the tip of the micropipette with forceps or by touching lightly against a piece of coverslip.

5 Place the tip in a droplet of mineral oil, and work the injector to make a small droplet of dye solution. Adjust output pressure and injection time, so that the size of the droplet is between 50 and 60 μm (roughly 1/10 the diameter of the yolk cell).

6 Place the agarose plate with embryos on the stage. Move the micropipette tip to penetrate the chorion and place the tip at the centre of the yolk by using the micromanipulator and/or by movement of the agarose plate.

7 Inject the dye solution by pressure, and withdraw the micropipette from the embryo.

8 Repeat steps 6–7 for each embryo.

9 Add E2 medium on the agarose plate. Collect and transfer the embryos from the agarose plate to Petri dishes containing E2 medium.

10 Incubate the injected embryos away from direct light.

2.2 Labelling individual blastomeres after 256-cell stage

Cells labelled before the late blastula or gastrula stage tend to give rise to many unrelated tissue types. This is due in part to the short cell cycle times in the early blastula that produce some 20 to 50 cells, and in part to the spreading into many different positions on the fate map driven by the radial intercalation movements of early epiboly. Labelling shortly after the 1000-cell stage does not eliminate these problems—epiboly is yet to begin—but it helps to manage the problem by limiting labelling to a smaller number of cells that tend to remain together. The methods described here have been used to create the blastula fate map. For later fate maps, this method has been extended into the gastrula and segmentation periods (7, 12). These later maps have less variability because the movements of local regions of the embryos become more predictable, and cell mixing is less of a problem.

To label cells after the 1000-cell stage, we resort to the technical methods of the cell physiologist, using electronics to enter the cell and even to inject dye. The general method, outlined in *Protocol 3*, is to place the tip of the pipette on the cell membrane of the cell to be injected, depolarize the cell membrane (which creates a very small hole through which the pipette enters), and then use a combination of pressure and current to push dye into the cell.

Cells labelled in this manner often give rise to two related clones, one bright and one dim (see the clone in *Figure 8*). The brightly labelled clone is descended from the injected cell; the weakly labelled clone is descended from the brightly labelled cell's sister, which was labelled by leakage of dye across the midbody remaining from the mitotic division preceding the injection. Because cell divisions at the early blastula cell stages tend to give rise to widely separated siblings, in this case the founder cells of each clone, these subclones can sometimes be mapped as independent clones. To avoid such dual clones, inject cells that have rounded up for mitosis; these cells have usually lost their midbody remnant.

Protocol 3

Late cleavage injections

Equipment and reagents

- Microscope: upright design equipped with DIC optics, epifluorescence, 10× and 20× dry objective lenses and 40× or 63× water immersion objective lenses
- Micromanipulator: with fixed-stage microscopes, any of the heavy mechanical micromanipulators (such as those manufactured by Leitz) are acceptable. If the microscope has a movable stage, then a light micromanipulator (such as those manufactured by Narishige) is necessary. The micromanipulators should be set with the z-direction square to the horizontal (aligned with the movement direction of the objective) so that movement along this axis causes no apparent x and y movement when watching the pipette through the microscope. In the case of the stage-mounted manipulators, it is very useful to have the

z-direction controlled hydraulically, such as in the Narishige MO155. In all cases, besides x, y, and z, a fourth hydraulically controlled movement aligned along the shaft of the pipette is very useful, allowing the simple retraction of the injection pipette out of the cell after injections
- Microinjector as in *Protocol 2*
- Electronics. The electronics set-up is typical for cell physiology; expertise and advice on these techniques is often found in local neuroscience departments. The central apparatus consists of an intercellular amplifier that can detect and control the voltage and current at the tip of the injection pipette. An essential accessory is an electronics oscilloscope to visualize the output of the intercellular amplifier. The equipment should be able to measure current

Protocol 3 continued

at levels of 2 mA. Also useful are the so-called 'cap comp' electronics, and a stimulator to control the amplifier to deliver a 10 ms voltage spike of about 10 V, both used for clearing clogged pipettes or entering recalcitrant cells. Completing the circuit is, on the pipette side, the pipette itself (back filled with 1 M KCl) in a pipette holder with a silver bridge electrode holder (such as a World Precision Instruments MEH900S), and, on the bath side, a silver chloride agar bridge, made by inserting a silver wire into a 10 cm long plastic capillary filled with 1.5% agar in 1 M KCl

• Glass pipettes. Preparation of proper glass pipettes for injection is crucial for this procedure. Again, a neuroscience department often has the personnel that can help the inexperienced. Use 1.2 mm thin-walled borosilicate capillary glass filaments (e.g. TW120F-4 from World Precision Instruments). Using a electrode puller, such as the Flaming

Brown P80/PC electrode puller, this glass is pulled with a steep shank to produce a moderately stiff needle with a little whip. For late-blastula labels (at the 1000- through 4000-cell stages), the resistance of the pulled needle should be over 300 MΩ. Immediately before use, the tips of these needles are broken to a resistance of about 120 MΩ on a glass ball. For gastrula labels, pull thick-walled borosilicate capillary glass that has a hole that results in a resistance of about 100–300 MΩ, trying to make a very steep shank, producing a stiff needle with very little whip. Gastrula needles will not be broken before use. Using the capillary of the pipette, the pipettes can be back filled several hours prior to the experiment (or the night before), and stored at 100% humidity. In general, the fine glass tips of the capillaries dull on storage; thus, the glass should be pulled within 48 h of the experiment

Method

1 Dechorionate embryos.

2 Break pipette against glass ball (if injecting at the blastula stage).

3 Place embryos in a depression slide in E2 medium.

4 Focus microscope on the nucleus of the cell to be labelled.

5 Bring focus on the lens up about 1–2 mm.

6 Position pipette in light path and lower towards embryo, chasing the z-position of the needle with the plane of focus of the microscope.

7 Place the pipette above the cell and then lower the needle against the cell, dimpling it.

8 Depolarize the cell membrane using a voltage pulse.

9 Verify that the pipette is in the cell (based on a change in voltage).

10 Inject using pressure or electrophoresis.

11 Back the needle out of the cell using the axial drive.

2.3 Labelling groups of cells by photo-activation

One method that has been gaining popularity in recent times is the marking of cells by photo-activation of a dye that was initially injected into the entire embryo (13, 14). This method has the advantage that cells that are difficult to access with an injection needle can be labelled, and embryos are not traumatized during the sensitive gastrula stages by an injection needle. The labelling of the initial clone is best accomplished with a laser focused through the objective lenses of a compound microscope; an adequate replacement is to activate using a small pinhole in the aperture of the UV epi-illumination, as we outline in *Protocol 4*.

Typically, the labelled cells are not true clones, in that initially more than one cell is activated (or 'labelled'). Also, after repeated illumination, the activated cells and the unactivated cells immediately surrounding them seem to slowly assume the same level of fluorescence. Nevertheless, the method is ideal for intermediate fate maps, where lineage restrictions are not being tested and where vector movements of tissue layers are being analysed.

One useful variation of this method is to co-inject a rhodamine-labelled dye with the photo-activatable dye during late cleavage, and, at a later time point, activate the photo-activatable dye in a portion of the clone. This method produces a small subclone of cells which is framed by the movements of parent clone. This variation gives rise to a more manageable number of progeny, akin to a late blastula lineage label.

Protocol 4

DMNB-caged fluorescein photo-activation

Equipment

- Microscope. Compound microscope equipped with epifluorescent illumination, 20 × dry (0.5 NA), 40 × water (0.7–0.9 NA), and 63 × water (0.8–1.0 NA). DIC optics and epifluorescence for rhodamine, fluorescein, and DAPI/Hoechst. There must be an aperture at the virtual image of the epifluorescence path that reduces the UV beam to a narrow beam. If there is no aperture, introduce a slider, containing foil with a pinhole aperture, at the appropriate position. It is also helpful to have an appropriate combination of neutral density filters and an aperture at the virtual focus of the epifluorescence path to control UV intensity. An alternative method for photo-activation, not discussed here, is to activate the cells using a focused beam from a laser

Method

1 Inject embryos during early cleavage with caged dye.

2 Incubate embryos to desired stage. Protect from light.

3 Using the highest practical power/NA objective, focus in the embryo the precise location to be labelled. Use the fluorescein filter with the aperture closed to a pinhole to establish the area to be labelled.

4 Turn off the epifluorescence and slide the DAPI/Hoechst filter set in.

5 Briefly illuminate the specimen, observing the change in colour visually.

2.4 Labelling groups of cells by lipophilic membrane dyes

An alternative cell labelling technique is to label the cell membrane of cells. This is typically done with the fluorescent dyes DiI and DiO, which can be visualized with the rhodamine and fluorescein filter sets, respectively (15). These dyes, if injected or 'stabbed' with a tungsten needle into the embryo, will incorporate into the bilipid cell membrane of nearby cells. As for photo-activation, the labelled cells are not clones, in that more than one cell is typically labelled. However, this is a particularly nice method for labelling cells or axons that migrate from the labelled site, and the method has been used successfully to label migrating neural crest cells and the migrating growth cones and axons of the motoneurons of the spinal cord.

3 Transplantation techniques

A central question in many developmental biology problems is knowing when and where cells begin to enter an irreversibly committed state. To test such ideas, we describe here methods for moving cells from one place in an embryo to another place in another embryo, testing the state of commitment of cells (16).

The transplantation technique is also essential in distinguishing the autonomous and non-autonomous effects of gene activity (17). Here the technique is used to identify which cells express the gene of interest and what effects these cells have on other cells in the embryo that are not expressing the gene. Hence, the transplantation is not from location to location but from one genotype to another.

In both of the above cases, we are looking for differences in the fate of the transplanted cells. For most normal work, the use of double transplantations, where the behaviour and fate of one group of 'experimental' cells is compared with a 'reference' control transplant, reveals slight differences in movement or fate between the groups. When looking for cell autonomous effects of mutations, the double transplant method also has the advantage that the number of hosts containing mutant cells nearly doubles, which doubles the efficiency of documentation on subsequent recording sessions.

Finally, transplantations can, with care, be used for fate mapping. This method was the method of choice for classic work on amphibian and chick fate maps. At late stages, moving small labelled cells is easier than injecting them. While these late transplantations must be homotopic, and control experiments with actual injections would be necessary, transplantations could be used to form the bulk data set.

3.1 Precision transplantation of small groups of cells

Transplantation of single cells is a superb test of cellular determination. Large groups of cells and large pieces of tissues may contain many of the intracellular components that may have initially conferred fate on to the tissue. A single cell has a far better chance of losing such intracellular baggage. Actually, in the zebrafish, even somewhat larger transplantations, in the range of 10–25 cells, are also an acceptable test, as the introduced group of donor cells normally spreads and mixes into the surrounding host tissue, and the intracellular material is certainly diluted. Even in the case of single-cell transplantations, some extracellular material is also drawn into the pipette. This material may be important for the immediate survival of the blastomeres: single blastomeres often begin to lyse when the transplantations are done in a leisurely fashion, perhaps because of diffusion of the extracellular mêlée.

Protocol 5 and *Figure 2* outline the construction of the transplantation pipette. Basically, the pipette is crudely broken to a spear tip of the correct diameter, and then the end is smoothed and sharpened on a microforge. The 'perfect size' for the pipette is such that the blastomeres just fill the bore. Our simplistic chipping method for fashioning the spear can be replaced by grinding on a needle beveller, such as supplied by Narishige. The fashioning of a spear tip with a sharp whisker allows easy entry through the enveloping layer of the blastoderm. Without the whisker, the pipette tends to push the embryo out of the holding preparation. However, if an agar mould is used to hold the embryo, then a large-bore broken and chipped pipette can be used without microforging.

Small groups of cells can be moved between labelled donors and unlabelled hosts (*Protocol 6* and *Figure 3*, see also *Plate 5*). The embryos are immobilized in methyl cellulose during the transplantation operations. Typically, for the most accurate placement of cells, these operations are performed under DIC illumination using a 10× to 20× objective. With a properly fashioned transplantation pipette, the pipette enters the embryo using a fine glass whisker to penetrate the enveloping layer, and then, using suction, deep blastomeres are gently drawn into the pipette. Then, after leaving the donor and entering the host embryo, the end of the pipette can be positioned in the appropriate place of the embryo and the cells expelled.

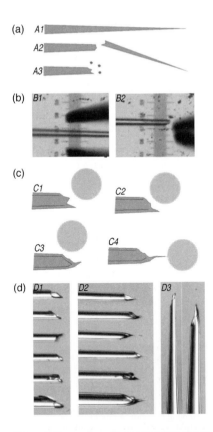

Figure 2 Preparation of transplantation pipettes. (a) and (b) Breaking THE pipette. (A1) selection of correct thickness away from the steep shank of the pipette; (A2) fracture of pipette with razor blade or fine forceps; (A3) forming of pipette end to approximately a 45° angle by chipping. (B1) Use of a stage micrometer to select size and, (B2) fracture of pipette with a forceps. Remove glass debris from the pipette with compressed air. (c) and (d) Microforging of pipette tip. Schematic of microforge operations: (C1) circle indicates filament cross-section. Inspection of the tip to assure that no glass shards remain inside; (C2) warming of tip to round edges; (C3) heating from side to bend tip over centre line of pipette; (C4) finishing of pipette by touching end to filament and pulling out a whisker of glass. (d) Before (D1) and after (D2) examples of finished pipettes. While on the microforge, the pipettes are measured (20, 25, 30, 40, 50, 60, or 80 μm) and subjectively graded (a, b, or c). With a split edge, the fifth pipette from the top is a 'c' grade. (D3) Low-power view (left) and a high-power view (right) of an overtapered pipette; this pipette will be difficult to control because as the oil–water meniscus moves along the conical barrel, the changes in diameter of the meniscus amplify microinjector movements. Also, the blastomeres tend to settle in the pipette tip and become fouled by oil.

The transplantations are easiest to perform between 30% to 50% epiboly, a period when the blastomeres are quite loose, and, also at this time, the geometry of the embryo allows easy placement of the blastomeres. These conditions are very important when targetting cells to the mesodermal and endodermal tissues, because cells must be placed very close to the blastoderm margin. What with the ideal period for doing the transplantations so short, it is most efficient to use two stations, one at a dissecting microscope for mounting the embryos and one at a compound microscope for performing the transplantations.

With care, experiments can be done before and after the optimal times. Before 30% epiboly, the blastomeres are very fragile and the cell lethality is high (the cells have the handling

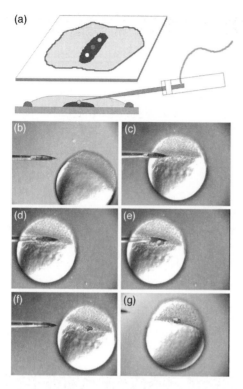

Figure 3 (see Plate 5) Transplantation technique. (a) Schematic of holding preparation. On a 25 × 25 mm coverslip, embryos are held in a strip of methyl cellulose overlaid with working medium. A ring of Vasoline® traps the droplet on the coverslip. The transplantation pipette approaches at about a 10° angle from the horizontal. (b) Pick-up of donor blastomeres. In this case, not all the collected blastomeres were labelled, stressing the need to check donor blastomeres for actual label. (c) The pipette is positioned to enter host. Normally this is the last movement of the pipette. (d) Using the stage control, the host embryo is moved on to the pipette, being careful to not pierce the yolk cell. To check if the pipette is in proper region of the embryo, a few blastomeres of the host can be drawn back into the pipette. (e) Expelling donor blastomeres. By moving the stage back and forth as the blastomeres are expelled, they can be mixed. (f) Movement of the host embryo off the pipette using the stage control. (g) Example of a transplantation in an early blastula, at the 2K-cell stage.

characteristics of water balloons). Also, because of the rapid cell cycle in the early blastula, the blastomeres are quickly decreasing in size throughout the transplantation period, necessitating frequent changes to smaller transfer pipettes.

In later transplantations, i.e. after 50% epiboly, the blastomeres begin to become sticky and difficult to draw into the transfer pipette. Also, as the cell layers in the gastrula become thinner than the outer diameter of the transfer pipette, it is easy to injure the yolk cell. However, after this 'difficult' period, transplantations become easier as the embryo acquires axial thickness in the mid segmentation stages, and transplantation of neural crest progenitors is quite possible.

While more time consuming than that for large-scale transplantations, this method is preferred for initial training in the cell transplantation technique. Seeing the tip of the pipette, the embryo, and the cells and their nuclei helps to quickly correct common mistakes in manufacture of the transfer pipette and treatment of the embryos.

Protocol 5

Construction of transfer pipettes

Equipment

- Dissection microscope
- Pipette puller
- Microforge
- Thin-walled glass pipettes

- Razor blades or fine forceps
- Stage micrometer or calibrated eyepiece ocular grid

Method

1 Pull pipette from thin-walled glass tubing (outer diameter (OD) 1.0–1.5 mm) with no capillary. The shape of the pipette should have a long taper, narrowing from 10 µm to 100 µm over about 5 mm.

2 Using a razor blade, break the glass to the appropriate diameter (OD of about 40 µm for late blastula work).

3 Working under a dissecting microscope, chip back glass until the tip is spear shaped. At this point the glass can be used for crude transplantation experiments.

4 Using a microforge, melt the glass at the edges of the opening.

5 Using a microforge, fashion a fine glass whisker on the tip of the spear by lightly touching the pipette to the microforge filament and pulling away.

Protocol 6

Single-cell transplantation

Equipment and reagents

- Room. The room that the microscope is placed in should be of uniform temperature and free of drafts. Temperature shifts will cause fluctuations in the movement of the oil in the microinjector. In a drafty room, it is sometimes helpful to construct a box around the sides and back of the microscope
- Microscope. Upright compound microscope equipped with DIC optics and a 10× lens. A left-handed mechanical stage allows the micromanipulator to be placed on the right side. Note that the mechanical stage should be smooth and have no backlash. Epifluorescent illumination is recommended
- Micromanipulator. If not using a fixed-stage microscope, the micromanipulator must move with the stage. A fixed-stage microscope and a table-mounted micromanipulator are recommended but expensive. A light manipulator, such as the Narishige MN151, attached to the stage

mount of a normal compound microscope, is quite acceptable and inexpensive. Set the angle of the micromanipulator to about 15° to perpendicular

- Microinjector. The cells at the tip of the pipette are controlled hydraulically using a hydraulic microinjector, such as the Narishige IM-5B. An inexpensive injector can be made with a Hamilton 1702TRLT Threaded Syringe attached to about 60 cm of stiff tubing ending in a World Precision Instruments MPH4 or MPH3 pipette holder, taking care to epoxy the tubing to the syringe and pipette holders. Carefully inspect the entire apparatus and remove any small air bubbles
- 3% methyl cellulose
- Coverslips, microscope slides, forceps
- Vasoline®
- E2 or Danieau's. All media must contain antibiotics

Protocol 6 continued

Method

1 Dechorionate embryos.

2 Inject embryos with lineage tracer before the 16-cell stage.

3 Prepare 22 × 22 mm coverslips with Vasoline® rings.

4 Wet adjust transfer pipette in droplet on coverslip. Set up so that the pipette can be backed out of the droplet using the *x*-axis drive, allowing enough room to move the coverslip in and out. Adjust the oil meniscus to the tip of the pipette.

5 For each set of embryos:

   • Place a 4 ×10 mm swath of 3% methyl cellulose in the centre of the coverslip.

   • Place a droplet of medium (E2 or Danieau's) over the methyl cellulose.

   • Position embryo in methyl cellulose, arranging donor and host embryos in a line perpendicular to the transfer pipette.

6 Advance the transfer pipette into the droplet, positioning the tip of the needle over the donor embryo.

7 Lower the pipette to dimple the surface of the embryo and advance the needle, sliding it into the embryo, taking care not to nick the yolk cell.

8 Back oil meniscus into pipette, pulling cells into pipette.

9 Slide pipette out of embryo.

10 Repeat steps 7–9 for additional donors as desired.

11 Repeat steps 7–9 for host, only expelling cells in step 8 by advancing the oil meniscus.

12 Transfer coverslips to 50 mm Petri dishes filled with E2 medium. The coverslips can be floated on the top of the medium until the embryos have completed epiboly and later transferred to the bottom of the Petri dish and the coverslip removed.

3.2 Large-scale transplantations of early blastomeres

When the difficulty of placing donor cells into a particular host tissue increases, the desire for greater numbers of experiments increases. The fine-scale transplantation technique normally yields about 15–30 hosts, and using two donors per host only doubles this number. By using dissecting microscopes, simple micromanipulators, and agar plates, hundreds of experiments can be done in a single sitting. Furthermore, many workers tend to favour the convenience of using their own dissecting microscopes for transplantation experiments rather than competing for the laboratory compound microscope. Thus, the large-scale method outlined in *Protocol 7* has become quite popular.

As in the fine-transplantation technique, cells are moved with a hydraulically controlled pipette between labelled donors and unlabelled hosts. However, during the transplantation operations the embryos are immobilized in various types of crafted agarose surfaces and the operations are typically performed under 50× magnification on a dissecting microscope. Because the field of view is larger and the depth of focus is greater, the pipette can be moved quickly from one embryo to the next.

These transplantations are easiest to perform in the late blastula, a period when the large-diameter transfer pipette can be kept away from the large yolk cell. Although at this time the geometry of the embryo does not allow easy placement of the blastomeres, the scatter of the blastomeres during epiboly tends to distribute the donor cells to many tissues of the hosts.

The misplacement and death of embryos can be rather high in this experiment. 'Jumping' of embryos from one hole to another can be controlled with a layer of methyl cellulose. Misplacement errors can be minimized using an alternative experimental design where only wild-type embryos are labelled and cells are transferred into host embryos produced from mutant clutches. Thus, all unlabelled embryos in a Petri dish are hosts. For the survival of the embryos, the use of antibiotics at all stages after the operation is essential. To prevent cross-infection from dying embryos, embryos should be segregated into individual Petri dishes or wells by the early segmentation stages.

Protocol 7

Large-scale transplantation

Equipment and reagents

- Room, micropipettes, and microinjector as in *Protocol 6*
- Dissecting microscope, equipped with transmitted light, and powers from 20× to 50×
- Micromanipulator, such as the Narishige MN151. Set the angle of the
- micromanipulator to about 45° to perpendicular
- 3% methyl cellulose
- Agarose plates (*Protocol 1*)
- Watchmaker forceps

Method

1. Collect, dechorionate, and label embryos as in *Protocol 6*.
2. Prepare the agarose plates by placing a bed of methyl cellulose in the troughs or holes, afterwards scraping off the excess from the agar surface. Overlay the plate with 2–5 mm of medium.
3. Transfer donor and host embryos to the troughs of the agarose plate, so donor and host embryos are placed side by side.
4. Place the agarose plate under the dissecting microscope and move the agarose plate and the micropipette so that the tip of the pipette is touching the one labelled donor embryo.
5. Insert the micropipette into the one labelled donor embryo, and draw cells into the pipette. Withdraw the pipette from the donor embryo.
6. If using a second donor embryo, repeat step 5 with the next embryo.
7. Place the tip of the micropipette near the surface of a host embryo.
8. Insert the micropipette and transplant the donor cells into the host embryo.
9. Withdraw the pipette from the host embryo.
10. If possible, allow the embryos to remain undisturbed until epiboly is completed. Then transfer the embryos to small agar dishes or microtitre plates, one embryo to a dish or well.

3.3 Transplantation of pieces of embryos

In certain experiments it is desirable to transfer or remove a portion of the embryo, including the intracellular environment. In this case, the cells must stay together, forming a small community. One method to accomplish this is to replace the deep-cell domain with a large group of transplanted deep cells, creating a chimera in the embryo in which a portion is mutant and a portion is host. This can be accomplished by aspirating half to three-quarters of the cells out of the host embryos and then transplanting half to three-quarters of the cells of a donor blastoderm into the host embryo, using one donor for each experiment. Large groups of cells moved in this manner mix only at the border between the donor and the host cells. However, in such an experiment, intracellular factors are dispersed.

An alternative method is to transplant pieces of embryos. Using cactus or tungsten needles, the blastoderm can be dissected off a 128-cell embryo at 2 h post-fertilization and placed on the yolk cell of another embryo from which the blastoderm had been previously removed. Using a small piece of agar, or a hole in an agar surface, the donor blastoderm can then be held in place until it has healed in. In these experiments there is a ring of cells along the margin of the blastoderm that are derived from the yolk cell before the formation of the yolk syncytial layer, but the majority of the blastoderm is from the donor.

When zebrafish blastoderms are removed from the yolk cell after the 32-cell stage and placed in Danieau's medium (18), the blastoderms continue to develop, generating stereotypic morphogenetic movements, as shown in *Figure 4* (see also *Plate 6*). Later, although histogenesis is difficult to interpret, similar to experiments in *Fundulus* (19), markers of cell specification aid in the analysis of the embryoids (20). One interesting variation of this procedure is to heal two blastoderms together, ventral face to ventral face, and examine the interaction of cells at the interface. An example of this is shown in *Figure 4e–j* (see also *Plate 6*).

4 Procedures for observing labelled cells

Data analysis is a challenging aspect of clonal and cell transplantation experiments. A single labelled cell at the blastula stage can divide into a 24-hour clone containing some 20–60 cells.

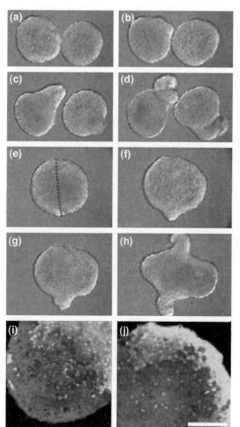

Figure 4 (see Plate 6) Development of isolated blastoderms. At the 128-cell stage, blastoderms were dissected off yolk cells using cactus needles, allowed to develop in Danieau's medium, and time-lapse recorded in 0.1% agarose. (a–d) A pair of embryoids dissected at 2 h. By the beginning of the experiment, the blastoderm, which after dissection resembles a hemisphere, has fused at the vegetal margin; the little 'tail' grows from the vegetal side. Recordings are at 2.5 h (a), 4 h (b), 7 h (c), and 10 h (d). (e–h) A pair of blastoderms dissected at 2 h and fused vegetal side to vegetal side. The dotted line on the embryoid indicates the border between the two blastoderms. Recordings are at 3 h (e), 4 h (f), 7 h (g), and 10 h (h). (i and j) Examples of the cell mixing that occurs in embryoids between the fused blastoderms. In this experiment, blastoderms were labelled with rhodamine or fluorescein, and dissected and fused between 2 and 3 h. The embryoids were fixed at 8 h, and the images were produced using a Confocal microscope at 40×. Scale bar: 300 μm for a–h, and 100 μm for i and j.(i and j were kindly contributed by R. M. Warga.)

Each of these cells assumes a unique location, and a unique cellular morphology, amongst more than 150 different possible histotypes.

4.1 Procedures for observing live material

For many experiments, intermediate observations of the labelled material are desired. In some cases, the original coordinates for fate mapping must be collected in the late blastula and early gastrula. In other cases, intermediate observations document cell trajectories during gastrulation (*Figure 5*, see also *Plate 7*); and in other cases yet, observations of intermediate morphologies are necessary for the documentation of late differentiating cells as they assume their final morphologies (*Figures 6* and *7*, see also *Plates 8* and *9*).

For these intermediate observations, we image cells visualized with fluorescence and recorded with low-light cameras, using either silicon-intensified target cameras or light intensifiers attached to normal cameras. These devices amplify light with quantum efficiency, capturing about 50% of the light photons emitted by the sample to produce a usable image. Other cameras, such as cooled CCD cameras will capture low-light images, but only with long scan times or with higher intensity light—lysing the labelled cells in the process. Note that, having been developed by the military to see heat-emitting mammals in the infrared spectrum, most of the intensified cameras work best on the red side of the visible spectrum.

Protocol 8 outlines many different methods for mounting embryos for observations. Of them, two of the procedures are commonly used during clonal analysis. In the first, we record the starting position of cells relative to the dorsal side of the embryo at about 6 hours development. This requires two views, one from the animal pole and one from a side view. The animal-pole view must be square and record the distance between the shield and the labelled

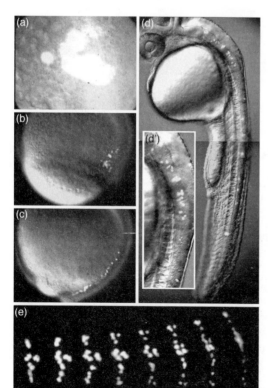

Figure 5 (see Plate 7) Behaviour of donor blastomeres after transplantation. (a) Transplantation of marginal (red) and animal-pole (green) blastomeres to host margin. At 70% epiboly (b) and tailbud (c), marginal donors have involuted and are moving past animal-pole blastomeres. (d) These marginal donors later form muscle at 24 hours. Animal-pole blastomeres make ectodermal derivatives. (e) From a different experiment; spreading and intercalation of transplanted blastomeres as they move into the shield.

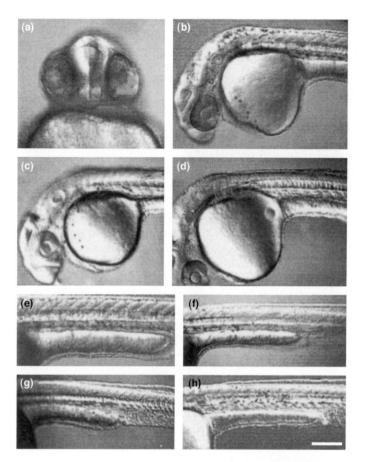

Figure 6 (see Plate 8) Examples of histology of transplantations double labelled with rhodamine (red) and fluorescein (green). (a) Nasal view where one group of transplanted blastomeres formed cells in both neural retinas with a few cells in the telencephalon; the co-transplanted blastomeres formed cells in the telencephalon, the neural retina and the lens. (b) Side view of head, with muscle and hatching gland and notochord. (c) Side view of head, with blood (on yolk sac), endothelium (in head), fin mesenchyme, pronephric duct, and muscle and muscle and mesenchyme. (d) Side view of head, with CNS cells, including floorplate and muscle. (e) Side view of trunk, a thin row of floor plate and, ventralwards, larger notochord cells. (f) Side view of trunk, with pronephric duct and a single endothelium cell on the yolk extension and muscle. (g) Side view of trunk, with muscle and mesenchyme. (h) Side view of trunk, with muscle and notochord sheath and cells of the spinal cord. Clones for a, b, c, and d are sketched in *Figure 9*. Scale bar: 200 μm. (Figure contributed by R. M. Warga and D.A.K.)

cells. The side view must record the number of tiers of cells between the margin of the blastoderm and the labelled cells. Both these views are typically mounted with methyl cellulose and recorded with the 0.30 NA/10 × power objective of a compound microscope. Working with embryos at this stage in methyl cellulose requires practice and patience. For finding the clones and orienting the embryos, we have found the new epilumination dissecting microscopes to be very helpful.

The second procedure is the quick scan of 30-hour fish. For this we use bridged coverslips. The best are made with two pairs of No. 1.5 22 × 22 coverslips, one pair glued on each side of a 24 × 60 coverslip. The glass surface of the coverslip spacers slides better and does not

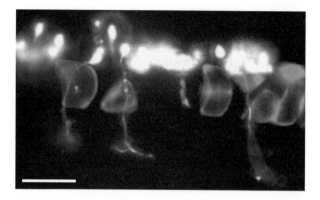

Figure 7 (see Plate 9) High-power view of donor cells in the ventral spinal cord (green) and in the notochord (red). Motoneurons are extending axons into adjacent somites. Scale bar: 50 μm. (Figure contributed by R. M. Warga.)

deteriorate, allowing one bridged coverslip to be used for an entire session. The fish are anaesthetized by transferring them to a solution containing MS322 for 1 or 2 minutes. Then, using a Pasteur pipette, the fish are transferred in between coverslips, where they lie nicely on their sides. This side view is very useful for marking the location of labelled material. If there are cells on the 'wrong' side of the embryo, the coverslip can be flipped and looked at from the other side. For the majority of the cells in the embryo, the location of the progeny of the labelled cells will remain unchanged. Hence this is a good starting point for observations that might continue through the larval period.

For microscopy, most brands of microscope work, but many lenses do not. The dry 16× or 20× lenses are adequate for scanning the embryo. But for cellular morphology, the 40× or 63× water immersion lenses, with NAs of 0.8–1.0, are essential. The dry version of the 40× and the glycerol/oil versions of the 63× lenses do not have the working distance to get deep into the preparation. More importantly, perhaps because of the of thickness of the sample, the water lenses seem to process superior optics compared to their counterparts. Note here that we are referring to live observations; all lenses work equally well on fixed tissue.

Protocol 8

Observing living fluorescent cells

Equipment and reagents

- Microscope. Compound microscope equipped with epifluorescent illumination, 5×, 10×, 20×, 40× dry, and 40× W (0.7–0.9 NA) or 63× W (0.8–1.0 NA). DIC optics and epifluorescence for rhodamine and fluorescein. It is also helpful to have an appropriate combination of neutral density filters and an aperture at the virtual focus of the epifluorescence path, to control UV intensity

- Cameras. Any camera that will capture faint UV images. For intermediate recordings, low light intensity and short duration are necessary (especially at the fluorescein wavelengths), therefore intensified cameras are usually the better choice. However, if the specimens are fixed or about to be fixed, cooled CCD cameras are excellent. Also, if the cells are somewhat overlabelled, we have had success with black and white high-resolution cameras from DAGE-MTI. The images are typically digitized through a video capture board on to a computer, and saved digitally

Protocol 8 continued

- Dissecting microscope
- 65 × 25 mm coverslips with 400 μm spacers for 24-h embryos or 600 μm spacers for younger embryos. The best 400 μm spacers are two 1.5 coverslips
- Boiled 0.1% agarose. Note, this medium appears as a so-called 'semi' solid. However, it

is not liquid; not only do the particles of unmelted agar harm embryos, but the agarose will not hold the embryo on cooling. Thus, be sure that the medium has actually been boiled

- Brass or steel block, for quickly cooling agar preparations to room temperature

Method

1a For oriented views of embryos, mount with methyl cellulose. Lay a bed of 3% methyl cellulose, 2 × 10 × 0.5 mm high, on the coverslip in the middle of a Vasoline® ring; overlay the methyl cellulose with 0.5 ml medium, using the ring of Vasoline® to prevent spreading of the medium. Working under a dissecting microscope, partially sink the embryos into the surface of the methyl cellulose, manipulating the embryos using dull forceps or nylon 'hair' loops. This preparation can be observed directly with low-power lenses or coverslipped for higher-power views.

1b For quick side views of 24-h to 72-h embryos, anaesthetize embryos in a solution of 0.01 mg/ml MS322, and slip in-between coverslips.

1c For immobilizing embryos for repeated observations (or time-lapse recordings), mount the embryos between coverslips in 0.1% agarose. Ring the agarose with a wall of Vasoline® to prevent evaporation.

1d For immobilizing embryos in oriented views, mount in 0.1% agarose in small 0.1 ml droplets. Working under a dissecting microscope, as the agarose solidifies, manipulate each embryo using dull forceps or nylon 'hair' loops. If more embryos are to be added to the preparation, add each to a separate droplet, orienting each embryo before adding the next. Ring the preparation with a wall of Vasoline® to prevent evaporation. Then overlay the entire preparation with 0.1% agarose and cover with a coverslip. Wait at least 5 min before moving the preparation, to allow the agarose to completely solidify.

2 Observations using UV epiflorescence must use low illumination and be short, especially when using the fluorescein filter set. When scanning the embryo, use the 10× objective; low numerical aperture lenses are less damaging to labelled material. At high power, do not use the UV to find cells and do not 'study' the illuminated embryo. Learn to use the intensified camera to observe the embryo and not the eyepieces. (The intensified camera can see dimmer cells at lower levels than human eyes.)

3 Rescue embryos from the preparations by immersing in a 150 mm Petri dish with 10–15 ml of medium. Carefully separate the coverslips under the surface of the medium, using a dissecting microscope to watch the embryos . For preparations made as in 1a, the embryos can be swept from in-between the coverslips with a stream of medium from a Pasteur pipette.

4.2 Procedures for preparing fixed material

In many cases, fixed material is desired, either for a permanent record or because there is inadequate time for immediate analysis. Sometimes particular views require sectioning or cutting away part of the animal. It is possible to visualize the fluorescently labelled cells in fixed material. However, conventional staining is easier to see in a complex tissue, with DIC optics rather than fluorescence. Also, being more sensitive than fluorescence, biotin staining reveals dimly labelled cells.

After following the embryos using fluorescence, a common preparation has rhodamine/biotin in one group of cells, and fluorescein in another group of cells. Typically at the end of the experiment, the fish are anaesthetized with MS322, and fixed with 4% paraformaldehyde for 2–3 hours at room temperature or overnight at 4°C. If the embryos are younger than tailbud stage, the fixation step should be done on an agar surface to prevent damage to the yolk cell.

Our basic protocol for the double staining of fixed material (*Protocol 9*) can be edited to include many variations. The simplest is to stain only the biotin of one clone; in this case, the cobalt could be eliminated, leaving the brown DAB (3,3'-diamino-benzidine) staining. (This variation is an excellent protocol for use in an advanced undergraduate laboratory.) Sometimes, an antibody reaction is necessary to reveal some useful marker; if the final amplification is planned to be done with avidin/biotin, then the biotin-labelled cells will come up in the final staining reaction. Alternatively, the antibody reaction could be done first with cobalt (or one of the many other colour reactions available) and biotin used for the counterstaining of the clone. *Figure 8* shows several examples of these stains.

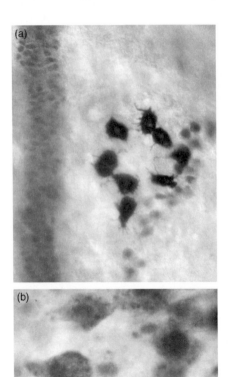

Figure 8 (see Plate 10) Examples of double-labelled material, using high-power views of endodermal progenitors lateral to the notochord. (a) Notochord progenitors labelled with antibody to Ntl and visualized with biotin/avidin. The biotin/avidin stain also detected a clone of endodermal progenitors that was labelled in the blastula at the 2K-cell stage with biotin and rhodamine. Note the light and the dark portions of the clone, caused by leakage of label across the midbody into the sister of the injected blastomere. (b) Endodermal progenitors double labelled with biotin lineage tracer (large purple-blue cytoplasmic staining of blue peroxidase substrate reaction from Vecta stain) and antibody to Forkhead2 (brown nuclear staining of DAB). Note that the nuclei can be detected faintly in the blue intensified cells. Scale bar: 50 μm for a and 30 μm for b. (Figure contributed by R. M. Warga.)

The mounting procedure is a dehydration into Permont® (*Protocol 10*). As this preparation ages, we have had good experience with the preservation of the DAB and DAB-coca labelled material, as well as many other stains. Note that many dyes fade over several days or weeks if the preparations are stored in solutions of benzyl benzoate and benzyl alcohol.

Protocol 9

Procedure for whole-mount staining of biotinylated and fluorescein dextran labelled cells

Equipment and reagents

- Ice-cold acetone
- Vectastain® Elite or Standard ABC kit (Vector Laboratory)
- PBST (phosphate-buffered saline plus Tween) must be made with Triton X-100 instead of Tween 20 if doing the $CoCl_2$ intensification
- DAB/$CoCl_2$ presoak solution: add 25 μl DAB stock solution (40 mg/ml in water) and 10 μl $CoCl_2$ stock solution (60 mg/ml in water) to 2 ml PBST. Final concentration of DAB is 0.5 mg/ml and final concentration of $CoCl_2$ is 0.3 mg/ml. Add the $CoCl_2$ second to prevent precipitation. This solution must be prepared fresh
- DAB/$CoCl_2$ staining solution: add 0.3% H_2O_2 1:1000 to the DAB/$CoCl_2$ presoak solution.

- This solution must be prepared fresh
- DAB presoak solution: dilute DAB stock solution (40 mg/ml in water) 25:2000 in PBST. Final concentration of DAB is 0.5 mg/ml. This solution must be prepared fresh
- DAB staining solution: add 0.3% H_2O_2 1:1000 to the DAB presoak solution. This solution must be prepared fresh
- Blocking solution: 2% normal goat serum (NGS), in PBST with 1% dimethyl sulphoxide (DMSO)
- Anti-fluorescein Fab-peroxidase (Boehringer Mannheim)
- Quenching solution: 0.5 ml 3% H_2O_2 added to 2.5 ml absolute methanol

Method

Permeabilization of embryos:

1 Transfer fixed embryos into distilled water for 2 min (for gastrulae) to 5 min (for 24-h embryos).

2 Remove water (leaving embryos covered).

3 Add ice-cold acetone for 3 min (for gastrulae) to 7 min (for 24-h embryos).

4 Remove acetone (leaving embryos covered).

5 Watching the embryos under a dissecting microscope in a spot plate, add water. As embryos begin to swell, gently add back PBST to stop swelling.

6 For embryos older than 2 days: transfer to quenching solution for 30 min. Afterwards, wash 3–5 times with PBST.

7 If an antibody staining reaction for endogenous markers is to be done, insert method here.

Staining of biotin-labelled cells:

8 Preparation of pre-complexed AB solution: dilute solutions A and B from the Elite ABC kit 1:200 in 1% DMSO in PBST. (For the standard kit, 1:100). Mix this solution for 20 min before using (to pre-complex the avidin–biotin horseradish peroxidase (HRP)).

9 Incubate embryos with pre-complexed AB solution for 1–2 h.

10 Wash with PBST for 15 min, then 30 min, and then 1 h.

11 Incubate for 15 min in DAB/$CoCl_2$ presoak solution.

Protocol 9 continued

12 Watching under a dissecting microscope, develop the embryos with DAB/CoCl$_2$ staining solution (to a blue-black colour)

13 Stop the reaction by washing 3–5 times with PBST.

Staining of fluorescein-labelled cells:

14 Transfer embryos to blocking solution for 1–2h at room temperature or overnight at 4°C.

15 Incubate with in anti-fluorescein (dilute 1:1500 in block) for 2–4h at room temperature or overnight at 4°C.

16 Wash with PBST for 15min, then 30min, and then 1h.

17 Incubate for 15min in DAB presoak solution.

18 Watching under a dissecting microscope, stain the embryos with DAB staining solution (to a brown colour).

19 Stop the reaction by washing 3–5 times with PBST.

Protocol 10

Mounting procedure for stained embryos

Equipment and reagents

- Ethanol series: 30%, 50%, 70%, 80%, 90%, 100%, 100%
- 2:1 benzyl benzoate and benzyl alcohol
- Permount®
- Bridged coverslips: 60 × 25mm coverslips with two No. 2 thickness 22 × 22mm on each

end, glued together with Permount®. Make several days beforehand
- 40 × 24mm coverslips

Method

1 Wash embryos in distilled water.

2 Dehydrate the embryos by washing through the ethanol series, 5min at each step.

3 Transfer the embryos to a 2:1 mixture of benzyl benzoate and benzyl alcohol, and wait until the embryos sink (about 15min).

4 Transfer the embryos into a small (about 10–20 μl) droplet of the benzyl benzoate and benzyl alcohol solution on a bridged cover slip. Carefully remove excess solution with a tissue.

5 Overlay with Permount®, orient the embryos, and then cover with a 40 × 24mm coverslip.

6 Slide and wiggle coverslip around to adjust orientation of embryo.

7 The preparation is permanent, and lasts several years.

4.3 Procedures for analysing data

Describing the histotypes and fates of cells is the most difficult aspect of the analysis of time lapse or transplantation, requiring care and time. For each animal produced, the cells and their histotypes must be recorded at the end of the experiment. This is usually the most time-consuming part of the experiment; a transplantation takes a minute or so but its analysis at 30h takes about 20–45 min, and sometimes much longer.

At each time point, we typically hand record the location and morphology of each labelled cell in the embryo, recording hand notes on a template to help interpret—and find—the images at later dates. This is shown in *Figure 9* (see also *Plate 11*). We also document each group of cells on video or computer files, recording data as separate images for white light and UV epilumination, usually at many different focal planes. Occasionally we will re-check complicated sections using a confocal microscope. Also, pay attention to dead and dying cells. In both lineage and transplantation experiments, dead cells can indicate overinjection or photodamage to the cells during the experiment. In transplantation experiments testing cell autonomy, dead cells may be an indication that the mutations are cell lethals.

One minimal strategy for data management is to score only neurons and muscle cells, which together comprise about 50% of the embryo. These data, expressed as a ratio, can reveal subtle changes in fate, especially when measured and averaged over many embryos. For example, in such an experiment, a given distance from the blastoderm margin should give a somewhat reproducible ratio between ectoderm and mesoderm; a change in this ratio, in a mutant or experimental embryo, may indicate changes in fate.

Hopefully, a figure can be prepared from the collected data. We have given an number of examples in our figures. Most biotin-labelled preparations can be simply photographed on to film or captured using a colour video camera. Images captured with black and white cameras are more of a challenge. Typically, a single image will be prepared from a DIC image, and 1–3 epifluorescence images. First, we process the individual images to match the black background to the same level, trying to avoid overediting. Then, to fuse these images into one red–green–blue (RGB) image, we place the rhodamine image into the red channel, the fluorescein image into the green channel and the DAPI image into the blue channel. (Unused channels are filled with a black background for most programs.) The resulting composite image with a black background is often the best image, as shown in *Figure 5e* (*Plate 7e*) or *Figure 7* (*Plate 9*). Alternatively, the epifluorescent image is fused or masked over the DIC image, such as in *Figures 3, 4e, 4f,* and *6* (*Plates 5, 6e, 6f,* and *8*).

Good luck.

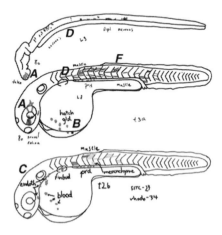

Figure 9 (see Plate 11) Examples of laboratory notes on a cutaway template. The top portion is a drawing of the medial regions of the 30-h fish, showing the central nervous system and the notochord; the bottom halves are drawings of the outer portions of the fish. The capital letters indicate the associated photograph from *Figure 6* (*Plate 8*). The cell count was done on the microscope, where more cells could be seen than were actually drawn. (Figure contributed by R. M. Warga and D.A.K.)

References

1. Roosen-Runge, E. (1939). *Biol. Bull.*, **77,** 79–91.
2. Kimmel, C. B., Warga, R. M.. and Schilling, T. F. (1990). *Development*, **108** (4), 581–94.
3. Stainier, D. Y., Lee, R. K., and Fishman, M. C. (1993) *Development*, **119** (1), 31–40.
4. Woo, K., Shih, J., and Fraser, S. E. (1995). *Curr. Opin. Genet. Dev.*, **5** (4), 439–43.
5. Woo, K. and Fraser, S. E. (1995). *Development*, **121** (8)**,** 2595–609.
6. Schilling, T. F., Walker, C., and Kimmel, C. B. (1996). *Development*, **122** (5), 1417–26.
7. Melby, A. E., Warga, R. M., and Kimmel, C. B. (1996). *Development*, **122** (7), 2225–37.
8. Felsenfeld, A. L., Walker, C., Westerfield, M., Kimmel, C., and Streisinger, G. (1990). *Development*, **108** (3)**,** 443–59.
9. Streisinger, G., Coale, F., Taggart, C., Walker, C., and Grunwald, D. (1989). *Dev. Biol.*, **131** (1), 60–9.
10. Strehlow, D., Heinrich, G., and Gilbert, W. (1994). *Development*, **120** (7), 1791–8.
11. Hammerschmidt, M. and Nüsslein-Volhard, C. (1993). *Development*, **119** (4), 1107–18.
12. Melby, A. E., Kimelman, D. and Kimmel, C. B. (1997). *Dev. Dyn.*, **209** (2), 156–65.
13. Girdham, C. H. and O'Farrell, P. H. (1994). *Methods Cell Biol.*, **44**, 533–43.
14. Cambridge, S. B., Davis, R. L., and Minden, J. S. (1997). *Science*, **277** (5327), 825–8.
15. Garcia-Martinez, V. and Schoenwolf, G. C. (1993). *Dev. Biol.*, **159** (2), 706–19.
16. Ho, R. K. and Kimmel, C. B. (1993). *Science*, **261,** 109–11.
17. Ho, R. K. and Kane, D. A. (1990). *Nature*, **348** (6303), 728–30.
18. Shih, J. and Fraser, S. E. (1996). *Development*, **122** (4)**,** 1313–22.
19. Oppenheimer, J. M. (1936). *J. Exp. Zool*, **72,** 247–69.
20. Sagerstrom, C. G., Grinbalt, Y., and Sive, H. (1996). *Development*, **122** (6)**,** 1873–83.

Chapter 5

Manipulating gene expression in the zebrafish

Darren T. Gilmour

Max-Planck-Institut für Entwicklungsbiologie, Abteilung Genetik, Spemannstr. 35, 72076 Tübingen, Germany

Jason R. Jessen

Lila Solnica-Krezel Lab, Vanderbilt University, Department of Biological Sciences, VU Station B 351634, Nashville, TN 37235–1634, USA

Shuo Lin

Department of Molecular, Cell and Developmental Biology, University of California, Los Angeles, 621 Charles E. Young Drive South, LS4325, PO Box 951606, Los Angeles, CA 90095–16061, USA

1 Introduction

The development of techniques to introduce and express recombinant DNA constructs is an important step in the establishment of any genetic model system. Such constructs can be used to express proteins transiently during development or to produce transgenic lines that express the construct in subsequent generations. Transformation allows one to assay the effects of expressing molecules in particular embryonic domains or at different time points during development. In addition, transgene constructs that express marker genes such as the green fluorescent protein (GFP) or lacZ under the control tissue of specific enhancers lead to the marking of particular cell types and greatly simplify the observation of these cells during morphogenesis (1). Such reporter constructs can also be used to identify cis-acting sequences and, in turn, the transacting factors that regulate tissue-specific transcription. Finally, given the great number of mutants that are available to the zebrafish community, one very important use for transformation will be in the rescue of embryonic phenotypes by the microinjection of candidate genomic regions or cDNAs. It is therefore likely that transformation techniques will play an increasingly important role in zebrafish research.

The aim of this chapter is to describe the methods most commonly used to transform zebrafish embryos, complete with experimental procedures and full descriptions of the equipment used. We will also cover, albeit briefly, the recently developed morpholino knockdown technique that has revolutionized zebrafish research. As well as providing protocols, we will explain why particular methods are used and what one can expect while following them.

2 Microinjection of zebrafish embryos

Although the use of other methods, such as retroviral infection (2), electroporation (3), and microprojectiles (4), has been reported, we will focus on microinjection as a technique for introducing nucleic acids into zebrafish embryos. Microinjection of newly fertilized eggs is relatively fast and, with a little practice, it is possible to inject hundreds of eggs in an hour. In addition, microinjection gives a high survival rate and reproducible results. As all of the methods covered by this chapter are dependent upon effective microinjection, this subject will be described in some detail. The techniques used for the microinjection of DNA constructs and mRNA are similar, the differences in the preparation of these samples will be discussed in Section 3.

2.1 Equipment required

The exact type of equipment used when microinjecting zebrafish eggs varies, but the basic requirements are the same. Appropriately pulled needles are required and some method of changing the pressure within the needle to deliver the injection solution is also needed. A manipulator is often used and the eggs are placed in a mould to hold them in position during injection.

2.1.1 Embryos

In addition to nucleic acid quality, the post-injection embryo survival rate is also influenced by the quality of zebrafish eggs obtained. To ensure that there is a sufficient number of healthy eggs for microinjection, dedicated stocks of healthy and highly productive breeding fish should be established. When young fish reach sexual maturity, a pre-selection can be made based on a pairwise mating test to determine which fish will provide the highest-quality eggs. The selected males and females are maintained separately and fed 3–4 smaller meals daily as opposed to two larger meals. We have observed that fish fed smaller meals more often tend to produce more eggs. The fish are fed a diet consisting of both brine shrimp and various flake foods. Tanks containing microinjection fish are labelled with the days of the week and the fish are used for mating only once a week. Fish often produce bad eggs after resting for several weeks without mating. Therefore, periodical mating of the microinjection fish is necessary to sustain the productivity of breeding stocks.

As it is desirable to inject the embryos at very early stages, these should become available when needed and therefore it is recommended that the matings should be staggered to produce embryos at a rate at which it is possible to inject. This is most simply done by separating single male and female fish the night before injection and placing a divider into the mating box, or putting them into separate mating boxes. On the morning of injection the fish are put together and usually they will mate within 10 minutes, often they lay three or more times over the course of an hour. The number of fish one puts together depends on the speed at which it is possible to inject. As it is difficult to identify unfertilized or low-quality clutches immediately after laying, mixing the clutches before injection will ensure that not too much time is wasted on unworthy embryos.

Some workers prefer to dechorionate embryos before injection as the chorion makes the eggs buoyant and more difficult to penetrate. However, it is not possible to disinfect eggs by bleaching once they have been dechorionated (see Chapter 1) and thus such injected eggs may pose a contamination risk. Although it is more laborious to inject eggs in their chorions, due to the increased difficulty of penetration, this also becomes simple with practice and therefore may be a choice for many labs.

2.1.2 Microinjection plates

As zebrafish eggs are fairly buoyant and move easily when touched, it is advised to use plates that hold them in position for injection. *Protocol 1* describes a simple method which uses a plastic mould to introduce a series of grooves in molten agarose. The grooves have a slope on one side to allow needle access and a perpendicular wall against which the embryos can be pressed. The exact dimensions of the plastic mould are shown in *Figure 1* (for further details contact D.T.G.). A simpler version of these plates can be generated by placing a glass slide into 1.5% molten agarose (prepared in double-distilled water (ddH$_2$O)) at a 10–20° angle.

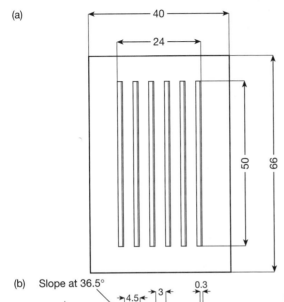

Figure 1 Diagram of injection plate mould. (a) Shows top view. (b) Shows side view. Dimensions are shown in millimetres.

Protocol 1

Preparation of injection plates

This protocol describes a quick method of preparing plates for microinjection. These plates are inexpensive, simple to prepare, and can be reused several times.

Equipment and reagents

- 94 mm Petri dish (Greiner, Cat. No. 633180)
- Agarose
- E2 fish buffer
- Injection slide mould (see *Figure 1*) or 5 × 7.5 cm glass slide

Method

(makes 2–3 plates)

1 Dissolve 1.5 g of agarose in 100 ml of E2 (see *Table 2* of Chapter 4) or distilled water (final concentration 1.5%) and bring to the boil in a microwave oven, swirl solution to remove bubbles.

2 Cover the base of a 94 mm Petri dish with agarose, 10–20 ml should suffice, and let it set. Then pour second layer of agarose and insert mould/glass slide.

3 When the agarose has hardened, remove mould/slide and cover with E2. When using a mould, take care not to trap bubbles between the grooves, by first inserting one edge and slowly lowering the other. These plates are then covered with E2 buffer or ddH$_2$O and stored at 4 °C until use. Plates can be reused several times if washed with 70% ethanol and distilled water (dH$_2$O) after use.

2.1.3 Microinjection needles

Microinjection needles should be thick enough in diameter to allow easy penetration of the chorion but should not be so thick as to damage the embryo. Microinjection capillaries (TW 120-F4, WPI; 1.2 mm) are pulled on a needle puller (DMZ Universal; Zeitz-Instruments) to give needles that are stubby and narrow quickly to a tip. Those that are pulled to produce very long, fine tips often bend when in contact with the chorion and are more prone to blockages. The end is then removed using fine forceps to give an opening of between 0.05 mm and 0.15 mm diameter. A needle that is broken too much will be too large to puncture the embryos effectively and is likely to rupture the embryos. Conversely, if a needle is broken too little, it will be more prone to clogging. It is not necessary to polish the needles before use. Once broken, the needle is easily filled using a pipette fitted with a tip designed for the loading of sequencing gels (Eppendorf Microloader Tips). The tip is placed down the centre of the capillary as close to the point as possible and the DNA solution is expelled, the use of capillaries with a filament ensures that the solution is pulled to the end of the needle. For injection of several hundred embryos 2 μl of injection solution are sufficient.

2.1.4 Microinjection apparatus

A typical microinjection set-up consists of a stereomicroscope, a micromanipulator, which is attached to a magnetic clamp stand, and a pneumatic microinjector, complete with air compressor (*Figure 2*). Stereomicroscopes that offer a long working distance, such as the Stemi range from Carl Zeiss, are recommended when using a micromanipulator as they allow more space for manoeuvre. The loaded needle is placed in a microelectrode holder (WPI, MPH415 1.2 mm) that is attached to the pneumatic microinjector with Teflon tubing (Narishige, CT-1). We tested several injection systems and found that the Pneumatic Picopump range from WPI

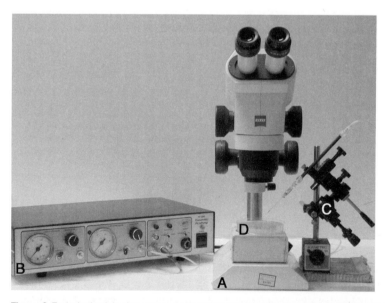

Figure 2 Typical microinjection set-up. A stereomicroscope that allows a large working distance is recommended (A, shown here, is a Stemi-2000 from Zeiss). Also shown are the pneumatic microinjector (B, PV820 from WPI) and micromanipulator (C, Narishige). D shows a microinjection plate (*Protocol 1*).

(PV820, PV830), although ugly, gives the most reliable injection volumes and is less prone to clogging than more aesthetically pleasing alternatives. We use the Narishige micromanipulators MN-151 (with joystick) and MN-152 (less expensive version without joystick).

Far less expensive systems can be used when finances are limited. Those with a steady hand may find that the micomanipulator is a mere luxury and can be omitted. In addition, there are cheaper alternatives to the microinjector and air compressor. The cheapest option is to 'mouth pipette': instead of attaching the microinjection needle and Teflon tubing to an expensive pneumatic microinjection apparatus, simply place a little plastic mouthpiece in the end and blow, with practice one can achieve almost the same level of control.

2.2 The microinjection procedure

2.2.1 How much to inject

DNA injection

Injection of DNA at a high concentration often leads to lethality or the generation of abnormal embryos, which often show a typical 'dorsalized' phenotype with a reduced trunk and tail. This is normal and does not depend on the sequence injected, nor does it always suggest bad microinjection technique. The secret is to find a balance between expression levels and survival rate, as abnormal embryos are often the ones that express highest levels. Therefore the aim seems to be to inject as much as DNA as possible. Concentrations ranging from 20 to 100 ng/μl will normally give good results, although it is possible to inject considerably higher concentrations. We have injected as much as 350 ng/μl of DNA and have achieved high survival rates, suggesting that the purity of the DNA, not the concentration, may be the limiting factor to embryo survival. The volume to inject depends on the concentration, we usually inject a bolus of approximately ⅓ of the total cell diameter (*Figure 3A*, see also *Plate 12a*). The addition of an injection tracer, such as Phenol Red (Sigma), helps visualize exactly the injection volume for each embryo. One can usually expect a lethality/abnormality rate of anything from 25 to 50%, and we find that it is often worthwhile to try a range of concentrations upon the first injection attempt.

mRNA injection

The total amount of RNA to inject depends, of course, on the potency of the molecules being assayed. For example, strongly ventralized embryos have been produced by injecting as little as 2 pg of BMP-2 mRNA (5), whereas other molecules were injected at higher concentrations before an effect was seen.

2.2.2 Where to inject

DNA injection

As discussed in Section 3.2, the microinjection of DNA constructs invariably leads to a expression pattern which can be described as highly mosaic when compared to the expression from mRNA. Although the exact cause of this mosaicism is unclear, it is probably at least in part due to the fact that injected DNA, unlike injected mRNA, must enter the cell nucleus and be transcribed efficiently. Indeed, it has been suggested that this mosaic pattern is due to the DNA being excluded from the nuclei of the non-expressing cells (6). However, when germ line transmission of the injected construct is the aim, it is important that the construct integrates into the host genome as quickly as possible. It is known from work in mice that it is important to inject the DNA solution directly into one of the two pronuclei to ensure a high

(a) (b) (c)

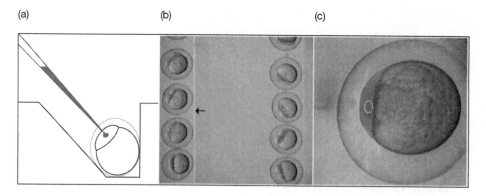

Figure 3 (see Plate 12) Microinjection of zebrafish embryos. (a) Side-view representation of a embryo sitting in an injection plate made using the mould described in *Figure 1*. The injection amount and position shown is approximately what we recommend.
(b) A series of two-cell stage embryos aligned within such an agarose microinjection plate. The walls of the injection plate trenches, against which the embryos are pressed, can be seen (arrow). (c) An embryo which has just been injected with a DNA solution. The injection tracer, which in this case is Phenol Red, has diffused but the original injected bolus size is indicated by a hatched circle. The injection needle can be seen at the left edge of the picture.

rate of transgenesis (11). Unfortunately it is not possible to visualize the zebrafish oocyte nucleus under a stereomicroscope. Nevertheless, injection into the forming cell is likely to increase the chances targetting the nucleus and, in our experience, this leads to a higher frequency of germ line transmission.

As the early cleavage divisions of zebrafish embryos are extremely rapid, it is advisable to inject as early as possible, which will reduce the mosaicism by allowing more time for integration of the construct into the host genome. Support for this idea comes from a recent paper by Higashijima and colleagues, which showed that injection of eggs only up until the first 20 min post-fertilization (pf) leads to a very high rate of transgenic founder fish, with some injections giving one transgenic for every four injected fish tested (7). Injection immediately after egg laying has the drawback that it is not possible to determine at which point on the egg's circumference the future cells will form, and it is difficult to know where to inject. In practice, we wait until the forming cell can be seen, as a crescent of clear ooplasm at one side of the opaque yolk, and then this is injected.

mRNA injection

Immediately after fertilization, zebrafish eggs undergo a process known as ooplasmic streaming, by which the egg cytoplasm segregates from the yolk and is transported to the newly forming cell. This process is capable of transporting large molecules such as high molecular weight dyes or mRNAs from the yolk to the blastoderm. Therefore it is possible to inject transcribed mRNAs even into quite vegetal positions within the yolk and these will be transported into the cytoplasm of the forming cells and efficiently translated. Indeed when mRNAs encoding lacZ or GFP are injected into the yolk cell, translated protein can be observed in many, if not all, embryonic cells (*Figure 4*, see also *Plate 13*, shows the result of injecting an embryo in the yolk cell with GFP mRNA). In addition, as the membrane which separates the yolk cell from the blastomeres does not begin to form until the 16-cell stage, it is possible to

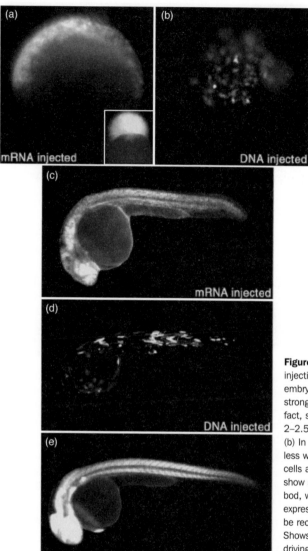

Figure 4 (see Plate 13) Comparing expression from injecting mRNA and DNA encoding GFP. (a) An embryo injected with 40 pg of GFP mRNA shows strong widespread expression by the shield stage. In fact, strong GFP fluorescence can be observed by 2–2.5 h (128–256-cell stage) after injection (inset). (b) In comparison, expression from DNA injection is less widespread and by shield stage only scattered cells are fluorescent. (c) By 24 h mRNA-injected larvae show strong GFP expression throughout the entire bod, whereas DNA injection gives a highly mosaic expression pattern and individual expressing cells can be recognized, in particular muscle cells (d). (e) Shows a transgenic embryo that carries a transgene driving GFP under the control of a ubiquitously expressed promoter (EF1-α)

inject into the yolk up until the 8-cell stage and still observe the mRNA in all blastomeres. Injection into the yolk cell is very simple and has become the standard method for mRNA injection. It has the further advantage that it can be done at a stage when unfertilized or low-quality embryos can be unequivocally identified, by their inability to cleave properly, and discarded. Please note that injection into the yolk cell is not recommended when attempting to generate stable transgenic lines as it reduces the likelihood of the injected DNA being integrated quickly into the host genome.

Protocol 2

Microinjection of zebrafish embryos

Equipment and reagents

- Male and female fish, up to 20 pairs
- Microinjection plates (*Protocol 1*)
- mRNA/DNA solution to be injected (*Protocols 3 and 4*, respectively)
- Phenol Red (0.5%, Sigma)
- 10 μl pipette and Microloader tips (both from Eppendorf)

- Microinjection needles (TW 120-F, WPI, see Section 2.1.3)
- E2 buffer supplemented with gentamycin (GibcoBRL) to 20 μg/ml
- Microinjection set-up: stereomicroscope (Stemi 2000, Zeiss), micromanipulator (MN-152, Narishige) and pneumatic microinjector (Picopump PV830, WPI)

Method

1. Male and female fish are separated the night before the injection experiment by placing them in single boxes. The next morning the fish are transferred to clean fish water, the divider is removed, and the fish are allowed to mate. Up to 10 pairs are placed together at any one time and the newly laid eggs are collected in E3 medium. Tranfer embryos to a microinjection plate which has been covered with E2 medium plus 20 μg/ml gentamycin and prewarmed by placing in an incubator at 28°C for 30 min before injection. Embryos are transferred with a clean 2.5 ml plastic pipette; however, a glass pipette should be used for transferring dechorionated embryos. The volume of buffer is reduced until it only just covers the eggs, thus preventing them from floating around.

2. Break the tip off a clean microinjection needle to give an opening of the appropriate diameter and load with 2–4 μl of injection solution using a pipette fitted with a gel-loader tip. Phenol Red acts as a non-toxic injection tracer when added at 0.05% (¹/₁₀ dilution final). Adjust the injection pressure and time to inject the right amount of solution into the rising cell, the first few embryos are usually sacrificed while finding the right settings.[a]

3. Once all embryos in the plate are injected, remove by flooding the plate with E2 buffer and dislodging the embryos by gentle swirling. Transfer to a sterile Petri dish filled with prewarmed E2 buffer plus gentamycin, and allow to recover at 28°C. Remove dead and unfertilized embryos 4–5 h later.

4. Remove dead and abnormal embryos the next day and, when injecting GFP constructs, embryos can be sorted on the basis of expression levels using a fluorescence stereomicroscope.[b]

[a] We find that injection of a bolus approximately ⅕ of the ooplasmic volume gives good results.

[b] Sorting on the basis of GFP expression should be done before disinfecting the embryos by bleaching, as this treatment makes the chorion autofluorescent (see Chapter 1 for a bleaching protocol).

3 Transient expression approaches

It is often desirable to express gene products transiently during embryogenesis without having to wait for germ line transmission and the propagation of transgenic lines. Transient expression from *in vitro* transcribed mRNA or DNA constructs can allow one to assay quickly the effects of misexpression of particular molecules and has been used to great effect by many zebrafish labs. Furthermore, in cases where ectopic expression has deleterious effects upon development, transient expression is often the only option. In addition, recent work has shown that transient expression from reporter gene constructs in injected zebrafish embryos can allow rapid dissection of *cis*-acting regulatory elements (8, 9, discussed in Section 3.4).

3.1 DNA constructs and mRNA transcripts give different expression profiles upon injection

Both capped mRNA transcripts and DNA constructs can be used to express high levels of protein transiently in early zebrafish embryos, but the behaviour of the expression from each is very different. Understanding how the expression differs in spatial, temporal, and quantitative terms is important in order to decide which approach to take and how to interpret the results. *Figure 4* (see also *Plate 13*) compares the GFP expression obtained from mRNA injection and from injection of a DNA construct which can drive GFP expression in every cell type (pESG). Also shown is an embryo from a stable line transgenic for pESG, which confirms the ubiquitous expression seen observed when using this promoter.

3.2 RNA injection

3.2.1 Misexpression by mRNA injection

Injection of *in vitro* transcribed mRNA is the quickest way to gain insight into the *in vivo* activity of newly isolated cDNAs and has been the most commonly used method to misexpress proteins in zebrafish embryos. The rapid success of this approach is mainly due to the fact that mRNA injections had been previously perfected by researchers working on *Xenopus laevis* and a virtually identical approach can be used in fish. Expression of functional GFP protein can be seen in the mRNA-injected embryo as early as the 256-cell stage, a time point which is before the initiation of zygotic transcription at the mid-blastula transition, which is between 3 and 4h in zebrafish (10). This shows that expression from mRNAs is very rapid indeed, especially as GFP is considered to be a slowly folding protein. One advantage of mRNA injection therefore is that it can be used to address questions of even very early developmental events. Indeed, there are a number of reports of RNA injections leading to very early phenotypes or changes in gene expression (11, 12), results which would have been impossible to obtain with DNA constructs. However, as mRNA is always injected in limiting amounts, it soon becomes diluted or degraded, leading to a rapid reduction in expression at later time points. In addition, expression from injected RNA is very widespread and appears to be ubiquitous, showing that most embryonic cells have received and translated the transcript. This relatively uniform reporter expression and low degree of mosaicism means that mRNA injection is well suited to studying effects on a large number of cells.

However, the fact that the expression is difficult to control in temporal and spatial terms, the message being translated almost immediately by all cells that inherit it, is one of the main drawbacks of this technique. Often it is more informative to look at the effects of misexpression in distinct embryonic regions or clones of cells, and several mRNA injection strategies attempt to allow this. The injection of individual blastomeres at slightly later developmental stages can deliberately increase the mosaicism. For example, Fürthauer and colleagues have demonstrated that injection of individual blastomeres can lead to distinct stripes of expressing cells which have different dorsoventral positions (13). Taking this idea further, mRNA can be introduced into individual embryonic cell types at very late stages using a technique known as iontophoresis which, although technically demanding, shows great promise. Iontophoresis has recently been used to introduce specific mRNAs into individual premigratory neural crest cells to affect their differentiation (14) and could be applied to many problems at such late developmental stages.

3.2.2 Preparation of capped mRNA for injection

In vitro transcribed mRNAs should possess several features, such as a 5′-methyguanosyl cap and a poly(A) tail, to ensure efficient translation upon injection, and several vectors have been developed specifically for this aim, such as pSP64T (15); and although these vectors were designed primarily for *Xenopus* experiments, they also work well in zebrafish. These vectors usually have promoters for two phage-derived RNA polymerases, such as SP6, T7 or T3, separated by a multiple cloning site to allow insertion of cDNA sequences. The result is that the inserted cDNA has an RNA polymerase promoter at either end, allowing the transcription of both the sense and the antisense strand. They often also use 5′ and 3′ untranslated regions (UTRs) which are known to increase translation rates, such as the UTRs from the human β-globin gene. Before transcription, the plasmid should be linearized at a site downstream of the cDNA insert to ensure that the polymerase 'runs-off' and produces transcripts of a constant length. The linearized DNA is then phenol/chloroform extracted to remove all RNases, and 1 μg of DNA is then transcribed in the presence of cap analogue using a commerically available kit. We find that the mMessage mMachine In Vitro Transcription Kits from Ambion work well for *in vitro* transcription, and the instruction manual has several helpful tips in addition to the necessary protocols.

Protocol 3

In vitro transcription

This protocol is essentially as described in the instruction manual that comes with mMessage mMachine kits from Ambion.

Equipment and reagents

- 5 μg RNase-free column-purified plasmid sample
- 10 mg/ml proteinase K (Boehringer Mannheim), phenol, chloroform, and RNase-free ethanol
- mMessage mMachine kit from Ambion

Method

1 Linearize 5 μg of plasmid DNA by digestion with the appropriate restriction enzyme for 2 h.[a] The DNA is then prepared for transcription by treating with 200 μg/ml proteinase K in the presence of 0.5% sodium dodecyl sulphate (SDS) at 50 °C for 1 h. The sample is then phenol/chloroform extracted, ethanol precipitated and resuspended in DEPC-treated H_2O.

2 Add 1 μg of the linearized sample to a RNase-free Eppendorf which contains 10 μl of 2× ribonucleotide mix, 2 μl of 10× reaction buffer and 2 μl of 10× enzyme mix, made up to a final volume of 20 μl with DEPC-treated H_2O. Incubate the tube at 37 °C for 1–2 h to allow complete transcription.[b]

3 The DNA template is then removed by adding 1 μl of the RNase-free DNase I provided with the Ambion kit and allowing a further 15 min incubation at 37 °C.[c] Terminate the reaction and recover the mRNA by LiCl precipitation. Add 30 ml of nuclease-free dH_2O and 25 ml of LiCl precipitation solution (i.e. ½ vol of LiCl solution) to the tube and chill for at least 30 min at −20 °C. Pellet mRNA by spinning at 4 °C for at least 15 min at maximum speed. The pellet, which should be clearly visible, is washed with 70% ethanol/30% DEPC-treated H_2O before being allowed to air dry.

4 Resuspend the mRNA pellet in a small volume of nuclease-free water so that the final concentration is high enough to be useful for RNA injection experiments.[d] The mRNA yield can then be estimated by running a fraction on a denaturing gel (16). When injecting a specific mRNA for the first time, use a range of concentrations to determine the optimum injection amount. The total amount of mRNA per embryo can be estimated by injecting the mRNA/Phenol Red mix into a drop of oil sitting on a calibration scale. Measuring the diameter of the fluid sphere allows the injected volume to be calculated using the equation $V = \frac{4}{3}\pi R^3$ (V = volume, R = diameter) and hence the total amount of mRNA injected.

[a] The plasmid should be linearized somewhere in the 5′ end to ensure that the antisense transcripts are of a defined size. Restriction sites in the polylinker 5′ of the insert are ideal.

[b] An hour is usually sufficient for the complete transcription of inserts of 0.3–5 kb. However, an additional hour will maximize yield and, in particular, SP6 reactions, which are known to be a little slower, will benefit from an additional hour incubation.

[c] This step is optional and is often not necessary as the transcribed mRNA is usually in vast excess.

[d] After resuspension of the pellet, the final mRNA concentration should not be so low as to require reprecipitation before use. 10 μl is usually fine.

Microinjection of mRNA is essentially the same as described in Section 2, with the exception that care has to be taken to prevent degradation of the sample by RNase. Therefore all solutions must be RNase free and the transcripts should be stored at −20 °C. Although it is common practice to treat microinjection needles to inactivate RNases before use, by baking at high temperature or treating with DEPC, in our experience this precaution is not necessary.

3.3 DNA microinjection and its applications

3.3.1 Transient expression from DNA constructs is highly mosaic

Transient expression from DNA constructs is very different from germ line expression. First, expression is not detectable until around the shield stage, a time point after the MBT (see *Figure 4*, and *Plate 13*). When the construct is finally expressed it is only seen in a limited number of cells of random distribution, and hence the expression from DNA constructs is highly mosaic (discussed below). More disturbingly, the number of cells that express the construct invariably falls dramatically over the first few days after injection, rendering DNA injection virtually useless for techniques such as lineage tracing. However, the expression from DNA constructs is sometimes very strong and can often be observed several weeks after injection, longer than is possible from injected RNA. Of course, one major difference between the two approaches is that DNA constructs can drive tissue-specific expression, even in transient assays. Nevertheless, mosaic expression is also observed when using very specific promoters.

Clean supercoiled plasmid samples can be prepared using any of several commercially available kits, and we find that Qiagen MaxiPrep Kits are well suited. Please note that it is not necessary to purify plasmids using CsCl gradients. Ideally the DNA for microinjection should be gel purified from vector sequences; however, this may not always be possible, and a number of reports have shown that linearized constructs containing plasmid backbone sequences drive efficient expression in transgenic zebrafish (7, 8). After digestion the fragments should be separated by running the reaction on an agarose gel and the appropriate band excised and gel purified using one of many commercially available kits. After purification, the DNA is diluted in an isotonic buffer and a tracer dye can be added to facilitate estimation of the amount injected. We have found that DNA dialysed on a floating dialysis membrane is very clean and produces a high survival rate.

Protocol 4

Preparation of DNA for microinjection

Equipment and reagents

- Qiagen brand or other plasmid DNA preparation kit
- GENECLEAN II kit (Cat. No. 1001–400, Bio101 Inc.)
- Floating dialysis membrane (Cat. No. VSWP 02500, Millipore)

- 0.5× TE: 5 mM Tris, 0.5 mM EDTA, pH 7.4
- KCl: 1 M in ddH$_2$O, autoclave and store as 1 ml aliquots at −20 °C

Method

1 Plasmid DNA is prepared, using a Qiagen kit or other plasmid DNA preparation procedure, as directed by the manufacturer.

2 DNA for microinjection is linearized with the appropriate restriction enzyme(s), but supercoiled DNA can be used for transient gene expression assays. If possible, insert DNA fragments should be purified free of vector sequences. Digested plasmid DNA is run on an agarose gel containing 0.5 µg/ml ethidium bromide and the desired DNA fragment is excised and purified using a GENECLEAN II kit.

3 DNA is dialysed against 2 litres of 0.5× TE overnight on a floating dialysis membrane.

4 The DNA solution is spun for 30 s using a microcentrifuge and ⅔ of the volume is transferred to a new 1.5 ml Eppendorf microcentrifuge tube. Care should be taken to avoid transferring any precipitate, which can later clog the microinjection needle during DNA loading. Check 2 µl of the DNA by electrophoresis and ethidium bromide staining to make sure that the DNA is pure and structurally intact.

5 For microinjection, the DNA should be diluted to between 50 and 300 ng/µl in 0.1 M KCl. A tetramethyl–rhodamine dye (0.125%) can be used as an internal control for monitoring the amount of injected DNA. Higher concentrations of DNA can increase gene expression levels but often result in lower post-injection embryo survival rates. The prepared DNA solution can be stored at 4 °C for up to 2 weeks or at −20 °C indefinitely.

3.4 Promoter analysis

By microinjecting DNA constructs containing a tissue-specific promoter ligated to the GFP reporter gene into single-cell zebrafish embryos, one can examine dynamic GFP expression patterns during early development (1, 7–9, 17). The generation of deletions and point mutations in promoter/enhancer regions can be used to identify *cis*-regulatory elements responsible for tissue- and developmental stage-specific expression patterns. Using the injection method described above, in a single day one person can generate more than 1000 microinjected embryos that will transiently express GFP. This makes it possible to analyse statistically data obtained from multiple reporter constructs.

3.4.1 Generation of plasmid-based GFP reporter constructs

The expression of a gene is controlled by transcription factors that recognize and bind to specific sequence motifs present in various locations, e.g. 5′ and 3′ flanking sequences and intervening sequences of a gene. To analyse a DNA sequence's regulatory activity, one can isolate a long stretch of sequence immediately upstream of the translation initiation codon and ligate it to a reporter gene. This ensures the preservation of basic promoter elements and potential regulatory elements in the 5′ untranslated region and also eliminates the laborious work of identifying the transcription start site for a particular gene.

To isolate promoter/enhancer sequences of a gene, a DNA fragment that contains the 5′ part of this gene is used as a probe to screen a genomic library. This probe can be obtained by digesting a cDNA clone with proper restriction enzymes or by PCR using primers specific to the 5′ region of the gene. Although genomic libraries constructed using lambda phage vectors are often used for this purpose, bacterial artificial chromosome (BAC) (18) or P1-derived artificial chromosome (PAC) (19) libraries are preferred. The identified positive clones are mapped by restriction analysis and a large fragment containing 5′ sequences is subcloned into a plasmid vector.

The 5′ flanking region is then amplified by PCR using a specific primer complementary to the cDNA sequence just 5′ of the translation initiation codon and a primer complementary to vector sequences. A *Bam*HI site is usually introduced into the specific primer to facilitate subsequent cloning. The PCR conditions depend on the length of the 5′ flanking region, melting temperatures of the primers, and other factors. If the promoter is over 6 kb long, the Expand™ Long Template PCR System (Cat. No. 1 681 842, Boehringer Mannheim) should be used. After digestion with *Bam*HI and another enzyme suitable for cloning, the amplified fragment is gel-purified and ligated to a GFP reporter gene.

3.4.2 Generation of deletion and point mutation constructs

To identify specific *cis*-acting DNA elements that are responsible for regulating tissue-specific gene expression patterns, constructs containing deletions in the 5′ flanking region can be generated. Naturally occurring restriction sites may be used to create a series of gross deletions. Alternatively, if the transcription start site is known, specific mutations can be made based on consensus transcription factor binding sites identified by computer analysis. Each construct generated should be individually microinjected into single-cell-stage embryos and examined. PCR technology can be used to create rapidly a deletion series within a promoter/enhancer region, and can confine a potential regulatory sequence motif to 20–30 base pairs. To determine the core sequence necessary for the activity of a tissue-specific *cis*-acting element, site-directed base mutations can also be generated by PCR. Primers containing 2–3 missense mutations are used in conjunction with a vector primer for PCR amplification of the target sequence using a promoter/enhancer–GFP construct as the template. The PCR products are gel purified and used for microinjection without further subcloning.

3.5 Co-injection

Several studies have shown that transgene constructs form large concatamers when injected into fish embryos (20). A co-injection technique has been developed by Müller and colleagues to exploit this feature and allow very rapid identification of tissue-specific enhancers without the need for labour-intensive cloning strategies (21). Genomic fragments that contain possible enhancer regions are gel purified and simply mixed with a reporter gene, in this case *lacZ*, fused to a minimal promoter. When the mix is injected into zebrafish embryos, concatamerization of DNA molecules ensures that the enhancers are in sufficient proximity to the promoter to drive tissue-specific expression of the marker. Often the co-injected mixes give stronger expression than constructs that have the enhancers fused to the reporter by ligation and, amazingly, this technique has been shown to work with even heterologous promoters. Co-injection therefore represents a powerful method for scanning large genomic clones for enhancer elements and is likely to be widely used in the future.

3.6 Rescue of mutants using transgenesis

The transfer of nucleic acids into zebrafish also provides a means of testing candidate genes for their abilities to rescue mutant phenotypes. This can be accomplished using either RNA transcribed from cDNA/plasmid templates or genomic DNA microinjection. When using RNA, protein products are ubiquitously expressed and the spatiotemporal expression pattern of the endogenous gene is not mimicked. Therefore, negative results can be misleading and should be interpreted with caution. In contrast, the expression of a gene injected as part of a genomic clone is controlled by its own promoter/regulatory elements and should accurately recapitulate the expression pattern of the endogenous gene. One caveat to this approach is that the chosen genomic fragment must include all of the regulatory elements required to direct gene expression in a tissue- and developmental stage-specific manner. For some genes this may necessitate the use of large genomic clones, such as artificial chromosomes.

It is also possible to monitor the expression of a candidate gene by creating fusion constructs in which a cDNA sequence is linked in-frame with the GFP reporter gene. Expression of such a construct can be controlled by either a gene's endogenous promoter or a heterologous one that will direct transcription in a certain tissue or cell type. This approach was used recently by Wang *et al.* (22) to link the uroporphyrinogen decarboxylase gene to a zebrafish mutant with a porphyria syndrome. In this work, the zebrafish *GATA-1* promoter was used to direct expression of a urod-GFP fusion protein specifically within circulating blood cells.

3.7 Artificial chromosome transgenesis in zebrafish

We have already discussed how genomic sequences can be subcloned and analysed for their ability to direct GFP reporter gene expression transiently and in germ-line transgenic zebrafish. It is often found that a few kilobases of 5' flanking region are sufficient to drive reporter gene expression in a pattern mimicking the endogenous gene. However, it is evident that the expression of many genes is controlled by *cis*-regulatory elements located beyond the limits that can be cloned into conventional plasmid-based vectors.

Since the early 1990s, researchers have circumvented this problem by generating transgenic mice using yeast artificial chromosome (YAC) transgenes (23). While effective in the mouse, application of the YAC system to zebrafish is not possible because YAC DNA cannot be purified at a concentration adequate for zebrafish transgenics. BACs and PACs are also high-capacity vectors capable of maintaining inserts of up to 300 kb. Since BACs and PACs are stably maintained in bacteria and are amenable to the same purification procedures as conventional plasmid DNA, they offer a significant advantage over working with YACs. Although standard cloning procedures have limited potential when working with such large inserts, a number of techniques have been developed that use homologous recombination in bacteria to introduce highly specific changes into BAC and PAC clones (24, 25). Here we describe one method, adapted from Jessen *et al.* (20) and illustrated in *Figure 5*, that can be used to insert a reporter gene into a BAC or PAC clone, and the procedures for purifying and microinjecting these large constructs into zebrafish embryos. Large artifical chromosome-based reporter constructs can be used to identify and analyse those regulatory elements located many kilobases from the transcription start site. In addition, by including all regulatory elements that drive gene expression, artificial chromosome transgenes may be useful for phenotype rescue experiments where accurate tissue- and developmental stage-specific gene expression is absolutely required.

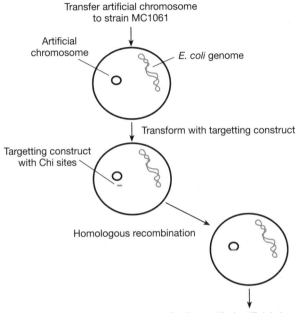

Figure 5 Modification of artificial chromosomes using Chi-stimulated homologous recombination in *Escherichia coli*. The artificial chromosome is transferred to a recombination proficient (RecBCD+) bacterial strain such as MC1061. Bacteria containing the artificial chromosome are transformed with a Chi-containing linearized homologous recombination targetting construct. Homologous recombination results in insertion of the GFP reporter gene and an antibiotic-selectable marker into the targetted genomic locus. Modified artificial chromosomes are directly purified from strain MC1061 for microinjection into zebrafish embryos.

Protocol 5

Modification of artificial chromosomes using Chi-stimulated homologous recombination

Equipment and reagents

- Targetting construct with Chi sites flanking the 5′ and 3′ regions of homology
- BAC or PAC genomic clone containing gene to be modified
- Electroporation-competent MC1061 *Escherichia coli* bacteria (ATCC)
- 0.2 cm electrode gap electroporation cuvettes (Cat. No. 1652086, Bio-Rad)
- Wizard plus minipreps DNA purification system (Cat. No. A7510, Promega) kb-100 (Cat.

No. kb-100, Genome Systems, Inc.) or other large-scale plasmid purification kit
- SOB medium: 2% (w/v) bacto tryptone, 0.5% (w/v) yeast extract, 10 mM NaCl, 2.5 mM KCl, 10 mM MgCl$_2$, 10 mM MgSO$_4$, pH7
- SOC medium: SOB medium plus 20 mM glucose
- Luria-Bertani (LB) or SOC medium and agar plates
- Ampicillin, choramphenicol, and kanamycin

Protocol 5 continued

Method

1   A typical genomic locus to be modified and the corresponding targeting construct containing 5′ and 3′ regions of homology flanking the GFP reporter gene and a selectable marker (kanamycin when modifying a BAC; chloramphenicol when modifying a PAC) are shown in *Figure 6a* and *Figure 6b*, respectively. Homologous recombinaton between the targetting construct and the BAC or PAC should result in replacement of the first coding exon with the GFP and selectable marker genes.

2   Prior to homologous recombination, the artificial chromosome is transferred into a recombination-proficient bacterial strain that contains an intact RecBCD enzyme pathway, such as MC1061. This is accomplished using electroporation in a 0.2 cm electrode gap cuvette at the following settings: 2.5 kV, 25 μF, and 400 Ω.

3   Transformants are grown on plates containing the proper selective medium (LB for bacteria containing BAC DNA and SOC for bacteria containing PAC DNA) and analysed for intact artificial chromosomes. Several colonies are each grown in 10 ml liquid cultures, and plasmid DNA is prepared using a Wizard plus minipreps kit following the manufacturer's protocol for *endA*-positive bacterial strains. Plasmid DNA is cut with a restriction enzyme and compared with the original plasmid DNA prior to electroporation using agarose gel electrophoresis. Only MC1061 bacteria containing intact artificial chromosome DNA (identical to original DNA) should be used in subsequent steps.

4   MC1061 bacteria containing the artificial chromosome DNA are made competent for transformation with the targetting construct using the $CaCl_2$-based method of Hanahan *et al.* (26). This method generates competencies of $>10^8$ transformants/μg which is absolutely critical for subsequent PAC modification. Lower competencies have worked for BAC modifications.

5   Prior to transformation, the targetting construct is linearized by removal of the origin of replication because the RecBCD enzyme requires free double-stranded DNA ends to function and, by removing the origin, the targetting construct will not replicate independently.

6   The $CaCl_2$-competent MC1061 bacteria containing the artificial chromosome are transformed with anywhere from 400 ng to 4 μg (may require optimization) of linearized targetting construct, following standard protocols (e.g. 50 min on ice, 1.5 min at 42 °C, 2 min on ice, 1 h at 37 °C in medium, and streaking on a selective plate). Transformants are grown on plates containing 12.5 μg/ml choramphenical and 25 μg/ml kanamycin to select for those bacteria having undergone recombination with the targetting construct.

7   Those colonies that are both choramphenicol and kanamycin resistant are subjected to ampicillin selection. Due to the design of the targetting construct, ampicillin resistance should be lost upon homologous recombination. Therefore, ampicillin-sensitive colonies most likely have undergone the correct recombinational event and should be analysed further using molecular methods.

8   To use molecular analysis to determine if a BAC or PAC has been modified, high-quality plasmid DNA is required. Plasmid DNA is prepared from choramphenicol- and kanamycin-resistant/ampicillin-sensitive colonies using a kb-100 or other large-scale plasmid purification kit. Each DNA prep and the original unmodified BAC or PAC clone are digested with a variety of restriction enzyme combinations. The enzymes are chosen based upon the sites present within the added DNA sequence (GFP and selectable marker genes) and known sites present outside of the boundaries of the targetting construct. By knowing the size of the restriction fragments that should be produced from a correctly modified BAC or PAC clone, they can be compared to those from an

Protocol 5 continued

unmodified clone using Southern hybridization. The correct probe to use is one that spans the region to be modified in the genomic locus (*Figure 6a*). PCR can also be used to detect homologous recombination using appropriately designed primers. A representative genomic locus after recombination is shown in *Figure 6c*.

9 Correctly modified BAC or PAC plasmid DNA is linearized for microinjection by digesting with *Not*I to release the genomic insert, followed directly by dialysis as described above. After digestion, the BAC or PAC DNA should be handled carefully and stored at 4°C to prevent shearing. This DNA should be very clean and can be injected at concentrations as high as 350 ng/μl. The procedures for injecting artificial chromosomes and for identifying germ line transgenic zebrafish are the same as for smaller constructs.

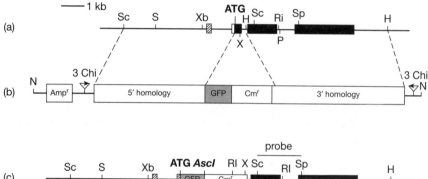

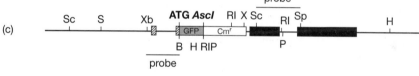

Figure 6 Schematic representation of the zebrafish *rag1* genomic locus before and after homologous recombination. (a) The region of the wild-type zebrafish *rag1* locus modified with the GFP reporter gene. Hatched boxes represent 5′ untranslated *rag1* exons, and black boxes, coding exons. (b) A targetting construct, containing properly oriented Chi sites (solid arrows), was designed to modify the *rag1* locus. Regions of 5′ and 3′ *rag1* homology, and their relationship with the *rag1* locus, are indicated by dashed lines. The targetting construct was digested with *Not*I (N, *Not*I) to linearize and simultaneously release the origin of replication prior to homologous recombination. (c) The zebrafish *rag1* locus after recombination with the targetting construct. Portions of the first coding exon and the following intron were replaced with the GFP gene and a chloramphenicol-resistance cassette (Cm^r). The probes used during Southern hybridization are designated. The added *Asc*I site 3′ of the GFP gene (in bold) facilitated the isolation of additional reporter constructs from the modified *rag1* artificial chromosome. Hatched boxes represent 5′ untranslated *rag1* exons, and black boxes, coding exons. Other restriction sites are designated as follows: B, *Bam*HI; E, *Eco*RI; H, *Hind*III; P, *Pst*I; Sc, *Sac*I; S, *Sal*I; Sp, *Spe*I; Xb, *Xba*I; X, *Xho*I.

4 Generation of stable germ line transgenic lines

Fertilized eggs can be injected using the above protocols and raised to adulthood to test for germ line transmission of the transgene to their progeny. Transgenic founders generated in this way may be as rare as 1–5% of all injected fish tested (although the rate of transgenic founders can be as high as 25%). The level of transgene expression can vary between lines and therefore it is worthwhile assessing the progeny from several independent lines before proceeding. This variability in observed expression levels is probably a feature of the transgene integration site or copy number. To ensure that sufficient numbers of transgenics are generated, we recommend that *at least* 100 fish are raised to adulthood for each injected construct. Furthermore, when screening potential founder fish please note that injected fish are often highly mosaic, with the transgene being present in only a fraction of the cells that give rise to the germ line. This germ line mosaicism results in founders giving rise to transgenic offspring at a lower frequency than the expected Mendelian frequency (50%) and, in some cases, transgenics can account for less than 1% of embryos. Therefore it may be worthwhile mating these fish more than once before concluding that they are not transgenic. However, as egg injection is much less labour-intensive and time-consuming than mating potential founders and screening of clutches, it is more effective to inject and raise a larger number of fish and set these fish up fewer times.

4.1 Identifying transgenic founder fish by PCR

Although it is possible to screen germ line transgenic zebrafish by examining expression of a reporter gene such as GFP or *lacZ*, PCR still represents the most commonly used method to detect germ line transmission of transgenes. Since the frequency of positive transgenic embryos in a clutch from a given founder fish can be very low (<0.5%), PCR provides a sensitive assay for the detection of transgenes. To reduce the labour involved in this process, eggs from multiple clutches can be pooled for the initial screening process.

Protocol 6

Identification of germ line transgenic zebrafish by PCR

Reagents

- DNA extraction buffer: 10 mM Tris pH 8.2, 10 mM EDTA, 200 mM NaCl, 0.5% SDS, and 200 µg/ml proteinase K
- Zebrafish ef1α and GFP primers
- ef1α (sense primer) 5′ TACGCCTGGGTGTTGGACAAA 3′
- ef1α (antisense primer) 5′ TCTTCTTGATGTATCCGCTGAC 3′
- GFP (sense primer) 5′ AATGTATCAATCATGGCAGAC 3′
- GFP (antisense primer) 5′ TGTATAGTTCATCCATGCCATGTG 3′

Method

1 Potential transgenic founder fish (2–4 months old) are mated to each other or to a nontransgenic wild-type fish. Since transmission of a transgene to the F_1 generation is mosaic (ranging from 0.5% to >50%), it is necessary to collect approximately 100–200 eggs from each founder fish. F_1 embryos from founder fish are grown for 24 hours at 28–30 °C and treated with pronase to remove their chorions, as described above.

2 Approximately 100 embryos are transferred to an Eppendorf tube containing 1 ml of DNA extraction buffer, vortexed briefly, and incubated with rotation at 55 °C for 2 hours. The tubes are vortexed briefly once during every hour of this incubation.

3 The samples are then centrifuged at 14 000 rpm for 10 min at room temperature, using a microcentrifuge. Ten samples are pooled by combining 20 µl of extraction solution from each sample in a new Eppendorf tube. The remaining DNA extraction solution from each sample should be saved at −20 °C until further use. The combined DNA mixtures are precipitated by adding 300 µl of 100% ethanol, washed with 70% ethanol, vacuum dried, and dissolved in 100 µl of 1 × TE (pH 8.0).

4 Using the isolated genomic DNA, standard PCR can be performed to identify pools of DNA samples that contain a transgene. The primer set corresponding to the GFP reporter gene (see Reagents) produces a 267 bp PCR product (1). A number of controls should be included in the PCR reactions. First, an internal control that detects an endogenous zebrafish gene should be included in each reaction to assess whether the PCR reaction worked at all. The primer set corresponding to the zebrafish *ef1α* gene (see Reagents) produces a 470 bp PCR product. Secondly, a positive control that contains one transgenic embryo per 500 embryos should be included to assure that the PCR conditions used are sensitive enough to detect such a low transgene concentration. This can be achieved by mixing a previously identified transgenic embryo with non-transgenic embryos, or mixing transgene DNA with DNA isolated from non-transgenic fish at appropriate ratios. Finally, a negative control should be included to address possible contamination. A typical PCR result is shown in *Figure 7*.

5 Once a positive pool is identified, PCR can be performed using the saved DNA extraction solutions from each individual sample to identify the transgenic founder fish. After a founder fish is identified by PCR, other methods, such as Southern blotting or visual examination of transgene expression, should be performed on the progeny to confirm the identity of founder fish.

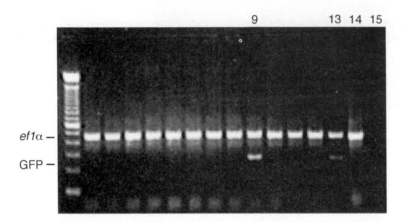

Figure 7 PCR identification of germ line transgenic zebrafish. Sexually mature zebrafish that had been microinjected with GFP reporter constructs were mated with wild-type fish. Genomic DNA was extracted from the progeny and analysed for the presence of the GFP gene using PCR. Lane 9 depicts the PCR results from a transgenic zebrafish and shows both *ef1α* and GFP bands. Lane 13 is a positive control PCR reaction performed using genomic DNA from a previously identified transgenic zebrafish. Lane 14 represents a negative control PCR reaction performed using wild-type genomic DNA. The PCR in lane 15 was performed without template DNA to assess for reagent contamination.

4.2 Identification of transgenic founder fish by reporter gene expression

The founder transgenic fish can also be identified based on the expression of a specific reporter gene such as GFP or *lacZ* in the founder's progeny (*Figure 8*, see also *Plate 14*). When GFP is used as the reporter gene, embryos derived from potential transgenic founder fish are screened for GFP expression at appropriate developmental stages using fluorescence microscopy. Stereomicroscopes with fluorescence capability, such as the Leica MZFLIII series,

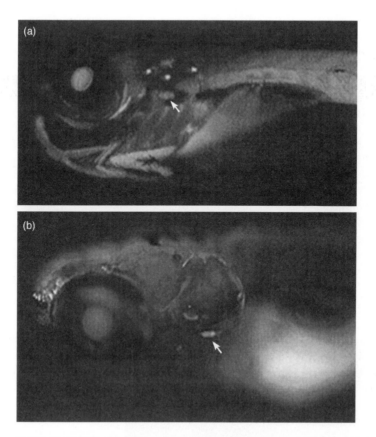

Figure 8 (see Plate 14) GFP expression in germ line transgenic zebrafish. (a, b) Lateral views with anterior to the left and dorsal to the top. (a) GFP expression in lymphoid (arrow indicates thymus) and several non-lymphoid tissues of a 7-day embryo containing a small *rag1* promoter construct transgene. (b) In contrast, germ line transgenic embryos containing the GFP-modified *rag1* artificial chromosome express GFP in a pattern representing the endogenous gene.

allow one to scan whole clutches rapidly for the presence of fluorescent embryos, without removing them from the Petri dish. This greatly facilitates the screening procedure, and the acquisition of such a microscope is highly recommended for any laboratory interested in using GFP in zebrafish embryos. Direct screening for reporter gene expression has two clear advantages over PCR screening. First, it identifies those embryos that actually express the transgene and can allow comparison of different lines based on transgene expression. PCR screening, on the other hand, only demonstrates the presence of the transgene sequence. Secondly, as the transgenic embryos can be selected while they are still alive, they can be grown up to found the transgenic colony without the necessity of additional rounds of crossing. However, this screening method can only be used if the transgene does indeed contain a reporter such as GFP, whereas PCR screening can be used to identify fish with any kind of transgene integration.

4.3 Generation of homozygous transgenic fish

The transgenic founder (F_0) fish can be bred to generate homozygous transgenic populations that can be maintained for generations without the loss of integrated transgenes. F_0 fish are crossed with wild-type fish and their progeny (F_1) grown to sexual maturity. The presence of a transgene can be detected by analysing GFP expression in F_1 fish, as described above. Alternatively, live transgenic F_1 fish can be identified by performing PCR on genomic DNA isolated from caudal fin clips.

Protocol 7

Preparation of fin-clip DNA for PCR screening

Equipment and reagents

- DNA extraction buffer: 10 mM Tris pH 8.2, 10 mM EDTA, 200 mM NaCl, 0.5% SDS, and 200 µg/ml proteinase K
- 0.16% Tricaine® in H_2O stock solution (Tricaine®; 3-aminobenzoic acid ethyl ester, Cat. No. A 5040, Sigma).

- Microcentrifuge
- Clean razor blades/scalpels

Method

1 Anaesthetize F_1 fish by placing in 0.16% Tricaine® H_2O stock solution diluted appropriately in fish water. The amount of Tricaine® to use varies with water volume and number of fish, but should be used sparingly so as not to kill the fish.

2 Cut off ½ to ⅔ of the caudal fin using a sterile razor blade/scalpel and place each in a separate Eppendorf tube containing 100 µl of DNA extraction buffer. Place fish in separate mouse cages numbered to correspond to the Eppendorf tube containing the fin clip.

3 Vortex tubes thoroughly and incubate at 55 °C for 2–3 hours. Vortex tubes periodically to ensure thorough mixing. After incubation, centrifuge at 14000 rpm for 20 min at room temperature to spin down debris. Dilute 100 nl–1 µl for PCR screening (see *Protocol 6*). If desired, multiple DNA samples can be pooled to minimize the number of PCR reactions.

Positively identified F_1 siblings are mated. An F_1 pair should normally give among their progeny (F_2) 25% transgenic homozygote positives, 50% heterozygotes and 25% wild-type homozygotes. The F_2 embryos are grown to sexual maturity and individual F_2 fish are mated to wild-type fish. The F_3 progeny are analysed for the transgene, either by assaying reporter gene expression or by analysing individual F_3 eggs by PCR, as described above. A homozygous F_2 fish should pass the transgene to 100% of its offspring when mated with a wild-type fish. Identified homozygous F_2 fish can be mated to each other to produce a large homozygous population.

5 Morpholino 'knockdown' of gene activity

Large-scale 'forward' genetic screens begin with the isolation of interesting mutant phenotypes and end with the molecular analysis of the genes underlying these phenotypes. Conversely, 'reverse' genetic approaches begin with interesting genes and end with the analysis of the phenotypes that result from their mutation. Such reverse genetic approaches are likely to be of increasing importance due to the vast number of recently isolated genes of unknown function resulting from genome sequencing projects. In terms of reverse genetics, the mouse is the vertebrate system *par excellence* as the use of homologous recombination in embryonic stem cells allows the researcher to introduce specific sequence changes into any genomic position and study the effects in living mice. Although attempts have been made to isolate such early pluripotent cell lines from fish embryos, it will be some time before there are reports of these being used to generate so-called 'knockout' fish (27).

This absence of reliable techniques for the inactivation of previously cloned genes was, until recently, a weak point in the zebrafish's armour. Although there were reports that RNAi and antisense RNA injection could be used to inactivate gene activity, both of these techniques proved too unreliable to be of general use (28). The breakthrough in zebrafish reverse genetics came with the recent discovery that a type of synthetic antisense oligo, known as a morpholino, can be used to efficiently 'knockdown' gene activity in zebrafish embryos (29). Morpolinos are chemically modified oligonucleotides that are able to bind to mRNAs with high affinity and efficiently prevent their translation. They consist of subunits that contain either of the four bases (A, C, G, or T) linked to a hexamer 'morpholine ring', hence the name, and these morpholine rings are connected via phosphorodiamidate intersubunit linkages. Unlike nucleic acids, they are uncharged and are apparently non-toxic, even when injected at high concentrations, and the ability to inject such large amounts may be the secret of their success.

Convincing proof of their specificy and efficacy comes from the fact that morpholino injection results in phenotypes that are identical to those observed from null mutations. July 2001 saw the publication of a special edition of the journal *Genesis* dedicated to papers describing morpholino-based studies. In this journal alone, over 20 previously identified zebrafish mutants were reported to be phenocopied by morpholino injection (30). Morpholinos can be used to eliminate the function of genes acting early in development, including maternally localized messages, and the effect can be completely penetrant through the first 2 days of development. Furthermore, weaker phenotypes could be observed by injecting lower amounts of morpholino, suggesting that the technique can be used to generate the equivalent of an allelic series. The morpholino reverse genetic approach not only complements forward genetic approaches, but it can help overcome some of the shortcomings of genetic screens, such as redundancy, by injecting morpholinos targetted to related genes into mutant embryos. Finally, it is likely that morpholino injection will be used increasingly to confirm or rule-out candidate genes thought to underlie mutant phenotypes.

Some batches of morpholinos demonstrate increased toxicity due to the presence of residual acid from the synthesis procedure, and injection of these results in a high lethality rate. However, such toxic batches can be easily be identified as the injection tracer we recommend, Phenol Red, is also a pH indicator and will turn yellow if the morpholino solution is acidic.

Protocol 8

Injection of morpholinos into zebrafish embryos

Equipment and reagents

- 1 × Danieau injection buffer (0.4 mM MgSO$_4$; 0.6 mM CaCl$_2$; 0.7 mM KCl; 58 mM NaCl; 5 mM Hepes, pH 7.6)
- Injection set-up (Section 2.1.4)
- Microinjection plates (*Protocol 1*)
- Phenol Red (0.5%, Sigma)

- 10 μl pipette and Microloader tips (both from Eppendorf)
- Microinjection needles (TW120-F, WPI, see Section 2.1.3)
- E2 buffer supplemented with gentamycin (GibcoBRL) to 20 μg/ml

Method

1 Dissolve lyophilized morpholino powder in injection buffer to give a final concentration of 10 mM. Normal injection concentrations range from 10 μM to 5 mM and it is recommended to try such a wide concentration range when injecting a morpholino for the first time.

Protocol 8 continued

2 Dilute with injection buffer that has 0.05% Phenol Red added (i.e. ¹⁄₁₀ dilution of Phenol Red solution from Sigma).

3 Inject as described in *Protocol 2* (and shown *Figure 3*). Unpublished results suggest that morpholinos injected into the yolk, like mRNA, are efficiently transported to the early embryonic cells.

References

1. Long, Q., Meng, A., Wang, H., Jessen, J. R., Farrell, M. J., and Lin, S. (1997). *Development*, **124**, 4105.
2. Lin, S., Gaiano, N., Culp, P., Burns, J. C., Friedmann, T., Yee, J. K., and Hopkins, N. (1994). *Science*, **265**, 666.
3. Muller, F., Lele, Z., Varadi, L., Menczel, L., and Orban, L. (1993). *FEBS Lett.*, **324**, 27.
4. Zelenin, A. V., Alimov, A. A., Barmintzev, V. A., Beniumov, A. O., Zelenina, I. A., Krasnov, A. M., and Kolesnikov, V. A. (1991). *FEBS Lett.*, **287**, 118.
5. Kishimoto, Y., Lee, K.-H., Zon, L., Hammerschmidt, M., and Schulte-Merker, S. (1997). *Development*, **124**, 4457–66.
6. Hogan, B. L .M., Beddington R. S .P., Constantini, F., and Lacey, E. (1994). *Manipulating the mouse embryo* (2nd Edition). CSH Press, New York.
7. Higashijima, S., Okamoto, H., Ueno, N., Hotta, Y., and Eguchi, G. (1997). *Dev. Biol.*, **192**, 289.
8. Meng, A., Tang, H., Ong, B. A., Farrell, M. J. and Lin, S. (1997). *Proc. Natl Acad. Sci. USA*, **94**, 6267.
9. Meng, A., Tang, H., Yuan, B., Ong, B. A., Long, Q., and Lin, S. (1999). *Blood*, **93**, 500.
10. Kane, D. A., and Kimmel, C. B. (1993). *Development.* **119**, 447–56.
11. Gritsman, K., Zhang, J., Cheng, S., Heckscher, E., Talbot, W. S., and Schier, A. F. (1999). *Cell*, **97**, 121–32.
12. Erter, C. E., Solnica-Krezel, L., and Wright, C. V. (1998). *Dev. Biol.*, **204**, 361–72.
13. Furthauer, M., Thisse, C., and Thisse, B. (1997). *Development*, **124**, 4253–64.
14. Dorsky, R. I., Moon, R. T., and Raible, D. W. (1998). *Nature*, **396**, 370–3.
15. Krieg, P. A. and Melton, D. A. (1984). *Nucleic Acids Res.*, **12**, 7057–70.
16. Sambrook, J., Fristsch, E. F., and Maniatis, T. (1989). *Molecular cloning, a laboratory manual*. Cold Spring Harbor Press, New York.
17. Jessen, J. R., Meng, A., McFarlane, R. J., Paw, B., Zon, L. I., Smith, G. R., and Lin, S. (1998). *Proc. Natl Acad. Sci. USA*, **95**, 5121.
18. Shizuya, H., Birren, B., Kim, U. J., Mancino, V., Slepak, T., Tachiiri, Y., and Simon, M. (1992) *Proc. Natl Acad. Sci. USA*, **89**, 8794.
19. Ioannou, P. A., Amemiya, C. T., Garnes, J., Kroisel, P. M., Shizuya, H., Chen, C., Batzer, M. A., and de Jong, P. J. (1994). *Nat. Genet.*, **6**, 84.
20. Stuart, G. W., McMurray, J. V., and Westerfield, M. (1988). *Development*, **103**, 403–12.
21. Müller, F., Chang, B.-E., Albert, S., Fischer, N., Tora, L., and Strähle, U. (1999). *Development*, **126**, 2103–16.
22. Wang, H., Long, Q., Marty, S. D., Sassa, S., and Lin, S. (1998). *Nat. Genet.*, **20**, 239.
23. Peterson, K. R., Clegg, C. H., Li, Q., and Stamatoyannopoulos, G. (1997). *Trends Genet.*, **13**, 61.
24. Zhang, Y., Buchholz, F., Muyrers, J. P., and Stewart, A. F. (1998). *Nat. Genet.*, **20**, 123–8.
25. Yang, X. W., Model, P. and Heintz, N. (1997). *Nat. Biotech.*, **15**, 859–65.
26. Hanahan, D., Jessee, J., and Bloom, F. R. (1991). *Meth. Enzymol.*, **204**, 63.
27. Ma, C., Fan, L., Ganassin, R., Bols, N., and Collodi, P. (2001). *Proc. Natl Acad. Sci. USA*, **98**, 2461–6.
28. Oates, A. C., Bruce, A. E., and Ho, R. K. (2000). *Dev. Biol.*, **224**, 20–8.
29. Nasevicius, A. and Ekker, S. C. (2000). *Nat. Genet.*, **26**, 216–20.
30. *Genesis* **3**, July 2001.

Chapter 6
Mutagenesis

Francisco Pelegri

Laboratory of Genetics, University of Wisconsin – Madison, 445 Henry Mall, Madison, WI 53706–1574, USA

1 Introduction

The ability to carry out genetic screens in the zebrafish is one of its most attractive features. The development of the genetic trickery of ploidy manipulation almost 20 years ago by Streisinger and co-workers underscored the genetic potential of the zebrafish and led to the identification of a number of zygotic patterning and behavioural mutations (1). More recently, the optimization of inbreeding strategies has allowed carrying out large-scale screens for zygotic mutations affecting development (2, 3). In this chapter we present the most practical methodologies used to perform mutagenesis and genetic screens in the zebrafish.

2 Mutagenic agents

The type of mutagen chosen for a particular screen will determine the efficiency of mutation induction as well as the type of mutations induced. A general comparison of the mutagenic treatments commonly used in the zebrafish is presented in *Table 1*.

2.1 Chemical mutagens: *N*-ethyl-*N*-nitrosourea

The monofunctional alkylating agent *N*-ethyl-*N*-nitrosourea (ENU) has been shown to be a powerful mutagen in a number of organisms, including the zebrafish (4–6). Mutations induced in germ cells at either postmeiotic or premeiotic stages can be obtained by raising offspring from males during or after, respectively, the first 2 weeks after the mutagenic treatment (5, 6). Mutagenesis of mature sperm cells has also been carried out by direct treatment with ENU in solution (4).

ENU results in modifications of a single DNA strand. Therefore, if induced in premeiotic sperm cells, ENU-induced mutations become fixed in both strands through DNA replication, and result in non-mosaic progeny. On the other hand, ENU-induced changes in postmeiotic germ cells do not become fixed until after fertilization and result in mosaic progeny (see Figure 1 in ref. 5 or, for a more general review, ref. 7). Of the more than 20 ENU-induced changes molecularly characterized so far, all those induced in premeiotic germ cells have been shown to be point mutations. In at least one reported instance, ENU mutagenesis of postmeiotic germ cells has also resulted in point mutations (8). However, a systematic analysis of five mutations induced by postmeiotic treatment with ENU indicate that it may induce primarily chromosomal rearrangements and deletions (9), in this study involving regions of the genome spanning less than 3 cM to greater than 20 cM.

ENU is toxic to adult fish. More than 50% lethality is induced by 1-hour treatments above 3 mM or 2-hour treatments at lower concentrations, leaving a 1-hour exposure to 3 mM as the

Table 1 Commonly used mutagenic treatments

Mutagen/target	Method of exposure	Specific locus rate ($\times\ 10^{-3}$)	Mutations/haploid-induced genome	Type of mutations	References
ENU/premeiotic male germ cells	Adult immersion, 3 mM, 3×1 h	0.6–4.8	1.2–1.7	Point mutations	5, 6 , 41
ENU/postmeiotic male germ cells	Adult immersion, 0.8 mM, 1 h	21 (mosaics)	9 (mosaic germ line)	Chromosomal rearrangements and deletions, point mutations	8, 9, 23
Gamma-ray/postmeiotic male germ cells	Irradiation of sperm in solution, 250 rads	5–10	1.0[a]	From few base pair changes to large chromosomal rearrangements and deletions	11, 12
Pseudotyped retroviral vectors	Injection of blastulae, 1×10^9 cfu/ml	n.d.[b]	0.1	Insertions	15, 16, 17

[a] Determined by irradiation of fertilized eggs.
[b] Presumably 0.03–0.5, but likely more variable due to insertional preferences.

maximal single exposure that permits adequate survival. Repeated exposures to ENU result in higher mutagenesis rates while allowing survival. Therefore, the standard protocol for ENU mutagenesis consists of 3–6 1-hour exposures to 3 mM ENU carried out at weekly intervals. At this ENU concentration, survival of the treated fish is variable. Low density of fish during treatment appears to be an important factor contributing to fish survival, especially at higher ENU concentrations. Changes in temperature or pH may affect the activity or stability of ENU, so that mutagenesis is carried out at a relatively constant temperature of 21–22.5 °C and in a buffered solution at pH 6.6. In addition, during the ENU treatment fish become very nervous, and a reduction in outside stimuli (noise, light) results in improved survival rates.

Protocol 1

ENU mutagenesis of premeiotic adult male germ cells

This protocol results in 10 males with mutations induced at premeiotic stages using 3 mM ENU. These males will allow the production of 5000 F_1 progeny carrying mutations in a non-mosaic germ line.

Equipment and reagents

- A chemical hood at a temperature of 21–22.5 °C. Line inside surfaces with disposable bench paper. In order to avoid exposing fish to light and noise, cover the hood door with cardboard and keep the room quiet during and for several hours after treatment

- Male fish: Due to ENU-induced lethality, treat 50 (10 groups of 5) males to obtain 10 mutagenized males. Preselect males for fertility through pair matings during the weeks immediately prior to treatment

Protocol 1 continued

- ENU stock solution: handle ENU in a chemical hood with gloves and appropriate protective clothing. Dissolve 1 g ENU powder in an isopack bottle (Sigma, Cat. No. N 3385) by injecting 100 ml of 10 mM acetic acid through the unopened rubber cap of the bottle. Using a 25 ml syringe with an 18G needle, inject first 10 ml of the acetic acid solution and gently mix the bottle contents so that the ENU powder becomes wet and less likely to disperse through air. Then insert through the bottle rubber cap a second, similar, syringe but with its stopper removed to release the pressure caused by the injected fluid. (To prevent ENU powder or solution from being dispelled into the air, fill this second syringe with gauze and do not allow its tip to contact the ENU solution.) Use the first syringe to inject an additional 90 ml of 10 mM acetic acid solution. Remove both needles, wrap the rubber cap with Parafilm and shake the bottle in a shaking platform until the ENU dissolves (0.5–2 h). The resulting ENU solution will have a concentration near 85.5 mM ENU. Because the ENU concentrations used are near the lethal dose, determine spectrophotometrically the precise final ENU concentration. Dilute 5 μl of ENU stock into 5 ml of 10 mM sodium phosphate buffer at pH 6.0. Using a standard quartz cuvette specifically assigned to measure ENU solutions, determine the absorbance (A) at lambda = 238 nm and pH 6.0.[a] Unless used immediately, store ENU in aliquots in Falcon tubes at −20 °C and thaw several hours before treatment. Confirm the concentration after prolonged storage through a new absorbance reading

- For each of the 10 groups of males, three 1-litre round plastic containers (e.g. Semadeni, Cat. No. 1749), one inner cylinder which fits within the former and with a fine mesh covering its bottom opening (Schwarz, 100 mm diameter circular box with fine grate insert), and a 94 mm Petri plate lid that covers the top opening of the inner cylinder

- 6 litres 10 mM sodium phosphate buffer, pH 6.6. Make a stock solution of 0.5 M phosphate buffer at pH 6.0, and dilute 1:50 with fish water to produce a working concentration of 10 mM at pH 6.6

- 3 litres fish water

- ENU inactivating solution: 8 litre 1 M NaOH. Place open NaOH solutions under the hood and handle with gloves and protective clothing. On the day of the ENU treatment, set up 1 litre 1 M NaOH in a 5-litre beaker and immerse all materials that were in contact with ENU during its preparation. On the same day, and after the fish have been removed from the containers in the hood, treat the solutions in each container by adding 100 ml 1 M NaOH to each of them (leave the inner cylinders in the containers so that they too get treated). Let all contaminated items and solutions rest overnight and then discard solutions and disposable items. Completely decontaminate containers and inner cylinders by immersing them, with the aid of long forceps, for several seconds into 4 litres 1 M NaOH in a 5-litre container. Containers and inner cylinders can be reused after rinsing them thoroughly with water and washing

Method

1 Arrange the plastic containers within the hood in three rows. Fill the round containers in the first two rows with 300 ml of 10 mM sodium phosphate buffer, pH 6.6, and those in the third row with 300 ml of fish water.

2 Place the inner cylinders inside the third-row containers with fish water. Transfer males in groups of five into these inner cylinders. Cover the inner cylinders with the plastic lids.

3 Add 10.9 ml of 85.5 mM ENU stock[b] to each first-row container, to make a final solution of 3 mM ENU in sodium phosphate buffer.

4 Begin the ENU treatment by gently transferring the covered inner cylinders with the males into the first-row containers with ENU solution. Leave undisturbed for 1 hour.

Protocol 1 continued

5 Gently transfer the covered inner cylinders to the second-row containers with phosphate buffer. Leave undisturbed for 1–2 hours.

6 Gently transfer the covered inner cylinders to the third-row containers with fish water. Leave undisturbed for 1–2 hours.

7 Transfer each group of fish to a tank with 3–4 litres fish water by removing the lid cover and 'pouring' the fish through the opening of the inner cylinder.

8 On the next day, connect fish to the water system and allow them to rest for a week with regular feedings.

9 Repeat steps 1–8 weekly two more times. If the survival of the males is good, the treatment can be extended up to a total of six 1-hour exposures.

10 One week after the last mutagenic treatment, set up the surviving males in pair matings.[c]

11 One week later, mate the males again.[d]

12 One week later, or 3 weeks after the last mutagenesis, give each mutagenized male an identifying label (e.g. a letter or combination of letters) and keep them separately. Verify the induction of new mutations by crossing mutagenized males to females homozygous for a recessive pigment marker such as *golden*, *albino*, or *sparse* in a locus-specific test. Test at least 3000 embryos.[e]

13 After the locus-specific tests have been performed, mate males to females of the desired genetic background and grow the resulting F_1 fish to adulthood.

14 After mutagenized males have produced the desired number of F_1 progeny, freeze them so that they can be used as a source of parental DNA for future cloning purposes. Anaesthetize males in MESAB (ethyl-*m*-aminobenzoate methanesulphonate) (as in *Protocol 6*), briefly rinse them in fish water, briefly blot dry them on a paper towel, insert them into a properly labelled 1.8 ml cryotube, quick-freeze them in liquid nitrogen and store them at $-70\,°C$.

[a] Under these conditions, the extinction coefficient (E) of ENU is 5830 m^{-1} cm^{-1}, and its molarity = $A/(E \times d)$, where d is the length of the field in the cuvette, usually 1 cm.

[b] Adjust this volume accordingly if the concentration of the ENU stock is not 85.5 mM.

[c] The resulting clutches should exhibit a reduced fertilization rate (30–40% lower than those of untreated clutches) and a drastically reduced fraction of normal embryos at 24 h (about 1% compared to untreated clutches). If the females are homozygous for a pigment mutation (e.g. *golden* or *albino*) a fraction of the progeny will be mosaic for the mutation (about 6–10 $\times 10^{-3}$).

[d] Fertility rates and the fraction of normal embryos at 24 h should now begin to approach normality, and the fraction of mosaics should decrease.

[e] Progeny embryos exhibiting the recessive pigment phenotype are expected at frequencies of 1–3 $\times 10^{-3}$. Regard only morphologically normal, non-mosaic embryos as potential carriers of new pigmentation mutations. If desired, allelism can be fully confirmed by growing the abnormally pigmented embryos to adulthood, crossing them to females homozygous for the same mutation, and showing the absence of segregating wild-type pigmented embryos in the resulting progeny.

Protocol 2

ENU mutagenesis of postmeiotic adult male germ cells

This protocol results in 10 males with mutations induced at postmeiotic stages using 0.8 mM ENU. These males will allow the production of 4000–8000 F_1 progeny carrying mutations in a mosaic germ line.

Equipment and reagents

- A chemical hood at a temperature of 21–22.5 °C as in *Protocol 1*
- Male fish: 14 (2 groups of 7) males. Preselect males for fertility through pair matings during the weeks immediately prior to treatment
- ENU stock solution as in *Protocol 1*
- For each of the two groups of males, three plastic containers, one inner cylinder and lid, as in *Protocol 1*

- 1.2 litres 10 mM sodium phosphate buffer, pH 6.6 as in *Protocol 1*
- 0.6 litre fish water
- ENU inactivating solution: 6 litres 1 M NaOH. Handle and treat items and solutions in contact with ENU as in *Protocol 1*

Method

1. Array the plastic containers within the hood in three rows. Fill the round containers in the first two rows with 300 ml of 10 mM sodium phosphate buffer, pH 6.6, and those in the third row with 300 ml of fish water.

2. Place the inner cylinders inside the third-row containers with fish water. Transfer males in groups of seven into these inner cylinders. Cover the inner cylinders with the plastic lids.

3. Add 2.8 ml of 85.5 mM ENU stock[a] to each first-row container to make a final solution of 0.8 mM ENU in sodium phosphate buffer.

4. Perform the mutagenesis treatment as in *Protocol 1*, steps 4–7.

5. On the evening of the same day, set up the mutagenized males against females of the desired genetic background and grow the resulting F_1 progeny. Repeat every other evening for a maximum of 2 weeks until a total of four outcrosses are obtained. Verify the induction of new mutations by crossing in a similar manner three of the mutagenized males to females homozygous for a pigmentation marker (e.g. *golden* or *albino*).[b]

6. After mutagenized males produce the desired number of F_1 progeny, freeze them so that they can be used as a source of parental DNA for future cloning purposes, as in *Protocol 1*, step 14.

[a] Adjust this volume accordingly if the concentration of the ENU stock is not 85.5 mM.

[b] The resulting progeny should exhibit somatic mosaicism (e.g. in the retina) in 4–8% of the morphologically normal embryos.

2.2 Radiation sources

Ionizing radiation induces double-strand DNA breaks and allows the production of chromosomal rearrangements. These, particularly deletions and translocations, facilitate genetic analysis and positional cloning of genes. PCR analysis of multiple gene markers after gamma-irradiation of mature sperm showed that the majority (13/15) of the induced mutations were complex rearrangements involving regions of the genome spanning the range of 7 to 160 cM, while the rest of the analysed changes (2/15) were deficiencies involving smaller regions of <10 cM (10). The induction of complex rearrangements by exposure to X-rays is also suggested by the low recovery of lethal mutations in a family inbreeding scheme, presumably by their inability to segregate through the germ line, in spite of a high mutation rate in a specific locus assay (6).

Exposure to gamma-radiation is most efficient when performed on mature sperm (11) (*Protocol 3*). Mutagenesis of mature sperm cells with this agent results in both non-mosaic and mosaic progeny (11, 12). Other radiation treatments, such as irradiation of blastula embryos or irradiation of adult males to target premeiotic germ cells, are described in (11).

Protocol 3

Gamma-irradiation of mature sperm

This protocol results in the mutagenesis of sufficient mature sperm to generate about 2000–4000 adult F_1 fish carrying gamma-ray-induced mutations.

Equipment and reagents

- A caesium-137 gamma-ray source (e.g. Mark I Irradiator, JL Shepherd and Associates)[a]
- 1 ml of sperm solution (*Protocols 6 or 7*) in a 2 ml glass vial
- One 250 ml glass beaker packed with ice

- Females to produce 40 freshly stripped egg clutches (*Protocol 8*)
- E3 embryonic medium (*Protocol 10*)

Method

1 Place the sperm solution in a beaker with ice and expose to a total dose of 200–300 rad.[b]

2 Use 25 µl sperm solution to fertilize each freshly stripped egg clutch, as in *Protocol 10*, and grow the resulting F_1 progeny to adulthood. 25–50% of the embryos will survive to adulthood.

[a] If necessary, use an ionization dosimeter (e.g. Victoreen) placed at the location where the sample will be located to calibrate the radiation source.

[b] Control the total exposure by adjusting the time of exposure. Rates of mutagenesis of zebrafish sperm are not significantly different in the dose range of 50–1000 rad/min (12).

Protocols for mutagenesis of mature zebrafish sperm using irradiation with ultraviolet light in the presence of psoralen have been reported recently (13, 14) and shown to induce rearrangements and deletions similar to those produced by gamma-irradiation (14).

2.3 Insertional mutagenesis

An inserted sequence acts as a tag that allows the rapid cloning of mutated genes. Pseudo-typed retroviral vectors are being used for mutagenesis in large-scale screens (15–17 and A. Amsterdam and N. Hopkins, personal communication). Injecting the pseudotyped virus into blastula embryos mutagenizes germ-cell precursors, resulting in mosaic founder fish with up to 25–30 insertions in their germ line. Intercrossing such founder fish results in F_1 non-mosaic fish, of which about 20% carry five or more insertions per fish. Founders carrying multiple insertions can be identified by a combination of quantitative PCR and Southern analysis. F_1 carriers, preferable covering non-overlapping sets of insertions, are intercrossed to generate F_2 families, which in turn are inbred and screened as in a standard family inbreeding scheme. Retroviral insertions produce about one embryonic lethal mutation per 85 insertions, so that the multi-insertional method is estimated to yield five- to tenfold fewer mutations per family than ENU mutagenesis. However, in about ⅔ of insertional mutants, the mutated genes can be cloned at a rate of 1–3 mutant genes per week by a single individual.

Protocol 4

Production of fish carrying multiple retroviral insertions

This protocol allows for the production of about 200 F_1 adults, each carrying five or more insertions which can then be utilized to produce 50 or more F_2 families to be tested as in *Protocol 5*. Methods for the production of pseudotyped VSV-G virions and the cloning of tagged genes are beyond the scope of this chapter and the reader is directed to the original sources (15–18). Handling and microinjection of the virus needs to be performed under a hood in an appropriate biohazard facility.

Equipment and reagents

- 20 µl VSV-G pseudotyped virion stock (1×10^9 cfu/ml) containing 8 µg/ml polybrene
- 1000 embryos fertilized within 20 min
- Injection needles and equipment for microinjection (Chapter 5)

- Materials for 1000 fin clips, genomic DNA isolation and Southern blotting analysis (Chapter 5)
- Materials for PCR analysis of single embryos and internal viral vector primers (optional)

Method

1 Align embryos in the microinjection plate while they age.

2 When the embryos reach the 512–2000-cell stage, inject a total of 10–20 nl of viral stock into 4–5 different locations among the cells of each embryo (see Chapter 5 for the general injection procedure). Grow the surviving embryos to adulthood (about 15–50% of injected embryos).

3 When founders reach 4 months of age, mate them against each other to produce 70 or more clutches of F_1 fish. Grow 30 embryos per cross. Reset those founders that do not mate once more after a 2-week rest period to produce more F_1 clutches.

4 When F_1 fish reach 8–12 weeks of age, perform fin clips as in Chapter 5 and keep the corresponding fish separated in single boxes. Isolate genomic DNA from each fin clip (Chapter 5). Use a small amount of these genomic DNA samples and quantitative PCR using internal proviral primers to choose the eight fish of each family with the largest amount of proviral DNA (see ref. 17 for details). Use the rest of the genomic DNA from these eight fish for Southern blot analysis using the viral vector as a probe.

5 Select F_1 fish with more than five unique insertions.[a] Discard the rest of the F_1 fish.

6 Mate multi-insertion F_1 founders, preferably carrying non-overlapping sets of insertions, to each other to produce F_2 families. Grow 65 embryos per family. Discard F_1 fish that do not mate after two attempts.

7 Perform test crosses between sibling F_2 fish as in *Protocol 5*, steps 3–6.[b] If a phenotype is observed in the F_3 generation, perform pair matings between sibling fish to identify pairs of fish that show the same phenotype in their offspring. Perform Southern analysis on genomic DNA extracted from all F_2 fish that yielded F_3 crosses. Two observations indicate a potential linkage between the observed phenotype and a given insertion. First, all fish in pairs that, when mated, produce mutant embryos should contain the insertion. Secondly, in pairs that, when mated, produce wild-type progeny, that same insertion should not be present in both members of the pair.[c] Additional linkage data can be obtained by Southern blot analysis of DNA, using proviral DNA as a probe, from individual mutant and wild-type embryos from a given cross. However, high-resolution linkage data is best obtained after the cloning of genomic fragments adjacent to the insertion of interest, since they can be used to create co-dominant markers that allow both detecting a single insertion and distinguishing tagged and non-tagged alleles.[c] Propagate insertions linked to mutant phenotypes as in Chapter 1.

Protocol 4 continued

^a About three F_1 fish per family. For high-throughput screening, it is also possible to omit the Southern analysis and choose the three F_1 fish per founder that have the highest numbers of insertions, as determined by quantitative PCR, as long as the F_1 fish carry more than five insertions. Typically, this streamlined procedure results in the selection of 2.5–3 F_1 fish per founder, which carry an average of 7–8 inserts per fish (A. Amsterdam and N. Hopkins, personal communication).

^b In this case, however, if six test crosses are not obtained during the first trial for a given family, the family should be set up once more after a 2-week rest period.

^c Co-dominant markers linked to the insertion of interest can be detected through Southern analysis using the genomic fragment as a probe, or by multiplex PCR with one primer on the virus and primers in the genomic sequence on both sides of the insertion site (see ref. 17 for details).

Transposable elements, specifically those of the Tc1/mariner family, are also currently being tested for insertional mutagenesis (19, 20).

3 Selection of background genetic lines

3.1 General criteria for selecting genetic backgrounds

The maintenance of genetic variation in a line should always be promoted to preserve the robustness of the line (see Chapter 1). In the case in which inbred lines are desired (Section 3.2), several lines with the desired characteristics can be combined into a more robust hybrid background.

The genetic background should not interfere with the screening of desired phenotypes. When possible, visible markers should be used to facilitate the logistics of the genetic scheme, such as to detect incomplete paternal DNA inactivation (see *Protocols 11* and *12*) or to safely distinguish newly induced mutations in an allele screen (*Protocols 14* and *15*). Finally, it is important that mutations be induced in a background that is polymorphic with respect to at least one other zebrafish strain so that this variation can be subsequently used for genetic mapping.

3.2 Selection of lines amenable for *in vitro* fertilization and parthenogenesis

In vitro fertilization and parthenogenesis (see Section 4.2.1) require foremost that eggs be manually stripped from a female, and only a minority of genetic strains, such as the AB or AB* lines (21) or the golden-derived Tübingen lines (22), can yield eggs even in the absence of males. In gynogenesis-based methods, the additional hurdle of producing viable diploids is added to the procedure. In addition, most gynogenetically derived adults show a strong tendency towards maleness.

It is possible, however, to select for genetic backgrounds amenable to these procedures (see for example, ref. 22). A simple recipe to select for such special lines, is to begin with a large population (e.g. 50–100 females) with as much genetic heterogeneity as possible and attempt to propagate the line through the procedure that wishes to be facilitated. For example, lines that readily yield eggs in the absence of males can be selected by propagation through *in vitro* fertilization (*Protocol 10*), and lines that produce viable gynogenetically-derived adults with a greater proportion of females can be selected by propagating the fish through early pressure or heat shock (*Protocols 11* and *12*, respectively). As a general rule, in the initial round of selection the appropriate genetic background is found in a low fraction (as low as 5% or less) of the population. Once this first selection step is carried out, propagation through subsequent generations is more efficient.

3.3 Selection of lines free of lethal or sterile mutations

The standard selection against embryonic lethals during normal stock maintenance (Chapter 1) does not select against mutations affecting late viability or fertility. A rapid method to produce lines lacking lethal mutations at any developmental stage, as well as sterile mutations, is to produce a large number (at least 50) of heat-shock-derived clutches (*Protocol 12*). Because heat-shock-derived progeny are homozygous at every locus (Section 4.2), those which become viable and fertile adults (about 12% of viable day 5 fish, or 1% of fertilized eggs) (22) can be used to initiate lethal/sterile-free lines.

4 Genetic screening strategies

The combination of a variety of mutagenic methods with different breeding techniques allows for a large number of possible mutagenesis schemes. Here we present those genetic schemes that we consider most effective for a given purpose (summarized in *Table 2*).

4.1 Inbreeding of families carrying non-mosaic germ lines

In this scheme, non-mosaic F_1 individuals that carry induced mutations are intercrossed to form F_2 families, which in the following generation are tested through random sibling F_2 crosses (*Figure 1*). If one of the original F_1 fish carried a mutation, 25% of the crosses between F_2 siblings lead to phenotypes in 25% of the F_3 embryos.

Family inbreeding, although laborious, is technically simple. It allows testing traits at all stages of development with minimum background abnormalities. In addition, the expected fixed fraction of mutant progeny facilitates distinguishing defects in Mendelian traits from epigenetic abnormalities. For these reasons, family inbreeding has been chosen for the large-scale screens for zygotic mutations affecting embryonic development (2, 3).

Table 2 Screening strategies

Strategy	Mutagen	Breeding technique	Fraction mutants	Background defects	Stage restrictions	Space requirement	Other comments	Application
Family inbreeding	ENU (pre), retrovirus	Natural crosses	Fixed 0.25	Low	None	High		General screening with multiple tests
Non-mosaic haploid	ENU (pre)	Haploids	Fixed 0.5	High	n.a. >3 days pf	Low		Screening with few tests
Mosaic haploid	ENU (post), radiation[a]	Haploids	Variable 0-0.5	High	n.a. <24 h or >3 days pf	Low		Screening with 1–2 tests
EP	ENU (pre)	EP	Variable 0-0.5	High <24 h pf, moderate >24 h pf	n.a. <24 h pf	Low	Discrimination against distal genes	Screening with 1–2 tests
Allele screen (non-mosaic)	ENU (pre) radiation[a]	Natural crosses	Fixed 0.25	Low	None	Moderate	Suggested use of unselected tester fish	Identification of new alleles in multiple genes
Allele screen	ENU (post) radiation[a]	Natural crosses, haploids/EP	Variable 0-0.25	Screen: low; recovery: high/ moderate	None	Low	Suggested use of preselected fish	Identification of new alleles in single or multiple genes

[a] Radiation induces a significant fraction of germ line mosaicism.

Abbreviations: EP, early pressure; pre, mutagenesis at premeiotic stages; post, mutagenesis at postmeiotic stages; n.a., not applicable; pf, postfertilization.

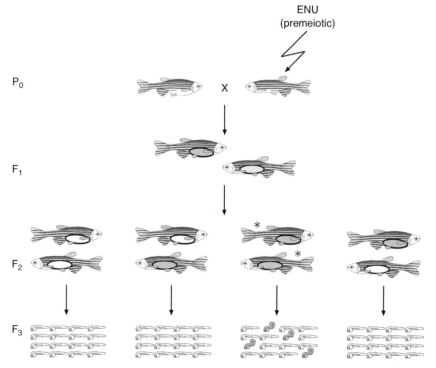

Figure 1 Family inbreeding screen. Males mutagenized in their premeiotic cells are used to produce non-mosaic F_1 carriers. These are intercrossed to produce F_2 families. Crosses between F_2 siblings produce F_3 clutches that can be screened for new phenotypes. Homozygosity for newly induced mutations occurs in 1/4 of the embryos in 1/4 of the F_3 crosses. For simplicity, only one of the two mutations segregating in the family is shown in the F_2 and F_3 generations. In *Figures 1–3*, ovals within fish represent their germ line, which are partially (mosaic germ lines) or completely (non-mosaic germ lines) filled with different patterns to represent heterozygosity for different segregating mutations. Embryos exhibiting the mutant phenotype are also drawn with the corresponding hatching pattern. Asterisks indicate fish that can be used to recover the mutation through natural crosses.

Family inbreeding requires large amounts of labour, space, and generation-time, and therefore it is most effectively used when combined with a high-efficiency mutagen such as ENU and multiple types of screen tests on the F_3 progeny. A modified inbreeding strategy using backcrosses of F_2 fish to F_1 fish carrying ENU-induced mutations in a mosaic germ line has also been reported (23). Family inbreeding is also used in combination with insertional mutagenesis (Section 2.3).

Protocol 5

Family inbreeding screen

This procedure will result in the screening of about 1500 F_2 inbred families, or the equivalent of 2600 haploid genomes.[a] Beginning at the production of the F_2 families, each week a suggested number of 50 F_2 clutches is produced. This cohort of 50 clutches is processed together in a treadmill-like fashion, all of them being transferred to tanks, tested and discarded at the same time. The suggested genetic scheme (excluding mutagenesis, retests, and recovery) requires about 66 weeks (30 weeks for the screening phase).[b]

Protocol 5 continued

Equipment and reagents

- 10 males mutagenized with ENU in their *premeiotic* germ cells (*Protocol 1*)
- Fish facility with minimum nursery and adult tank capacities of 500 and 450, respectively

- 800 pair mating set ups (Chapter 1)
- 400 1-litre single fish boxes with a lid (Chapter 1)

Method

1 Three weeks after ENU mutagenesis of premeiotic germ cells (*Protocol 1*), cross the mutagenized males to wild-type females to produce F_1 fish carriers of new mutations. Raise a maximum of 500 adult F_1 progeny for each mutagenized male.[c] This will produce a total of about 5000 F_1 fish. To help verify the independent induction of alleles, keep track of each set of F_1 and F_2 fish originating from the same mutagenized male by including the identifying label of each mutagenized male in the tags of all tanks with fish derived from it. After producing the desired number of F_1 progeny, freeze the mutagenized males (*Protocol 1*, step 14).

2 After the F_1 fish reach 4–5 months of age, set up each week 200 pair matings (where both F_1 parents are from the same mutagenized male, to simplify keeping track of potential clones of mutations) to grow 50 F_2 families/week, and process the weekly clutches together as a cohort. Each set of F_1 fish (originating from the same mutagenized male) should produce about 150 F_2 families (to yield a total of about 1500 F_2 families). Label each F_2 family with the identifying label of the mutagenized male, followed by a serial number, starting at 1, for each set of F_2 families originating from the same mutagenized male. Discard F_1 fish that produce progeny and pool those that do not mate without mixing sets of different origin. Repeat this step twice more and discard all F_1 fish.

3 When the F_2 fish reach 4 months of age, set up as many pair matings between F_2 siblings as possible for each family.[d]

4 At noon on the following 2 days, collect the laid clutches, labelling them and the fish pair with duplicate tags containing a serial test number and the F_2 family serial number (including the label identifying the mutagenized male). Transfer the fish pair and their tag to a fresh single box and cover it with a lid. These fish can be kept outside of the running water system and without feeding until the screen tests of their respective clutch are finished (up to 10 days). Sort fertilized eggs in separate groups corresponding to the different screen tests that will be applied. Incubate embryos at 27–29 °C.

5 Perform screen tests on clutches. Record the number of tests carried out for each family. If the family does not carry a mutant phenotype, discard all members of the family.[e]

6 If a mutation is present in the family, one-quarter of the embryos in one-quarter of the test crosses will exhibit the mutant phenotype. Keep the pair(s) that produced the mutant phenotype, separate male and female, and allow them to rest and feed for 2 weeks.

7 Cross both the male and female carriers of the mutation to wild-type fish to produce two outcrosses of 65 embryos each. The mutation can be re-identified in the next generation by random sibling crosses and maintained as in Chapter 1.

[a] The number of screened mutagenized genomes tested contributed by each F_2 family is $= 2(1 - (0.75)^n)$, where n is the number of successful F_2 test crosses in a given family.

[b] All calculations are based on the following estimates: average clutch size, 150 fertilized eggs; number of weeks in the growing facility, 10; number of weeks required to reach sexual maturity:

Protocol 5 continued

18; average sex ratio, 50%; average number of male/female pairs in a 50 fish tank, 15; laying rate, 50%. The number of progeny is given as number of adults. In our facility, about 50 adults are produced from a tank with an original number of 65 viable day 5 embryos.

c This reduces the possibility of transmitting clones of the same mutation, since the number of the stem germ cells in adult males is 500–1000 (6).

d This means setting up each week about 750 pair matings and collecting and sorting the resulting eggs, which requires two full-time workers.

e With respect to the number of screened genomes, it is more labour-efficient to discard this F_2 family and set up a different one than to attempt to obtain additional tests from the same family.

4.2 Screens involving ploidy manipulation

Ploidy manipulation allows bypassing one generation in a genetic scheme by directly uncovering recessive mutations through hemizygosis (haploid progeny) or homozygosis (parthenogenetic progeny), resulting in greatly reduced efforts in terms of generation time, space, and labour. Ploidy manipulation schemes are therefore most appropriately used when the size of the fish facilities is limiting and when a single or small number of screen tests are performed.

On the other hand, ploidy manipulation requires greater technical skill and especially selected lines, restricts the types of mutations that can be isolated, and produces a higher incidence of background abnormalities. Because of this latter aspect, it is essential to test the robustness of the trait under the ploidy manipulation condition used, by carrying out screen tests on a large number (50–100) of clutches from unmutagenized females. Ploidy manipulation schemes are often combined with screening with molecular markers, which generally allows a better visualization of specific wild-type patterns and a more clear distinction between new phenotypes and background abnormalities.

4.2.1 *In vitro* fertilization and commonly used ploidy manipulation techniques

Zebrafish oocytes, which are normally arrested at metaphase of the second meiotic division, can be extruded from anaesthetized females by applying gentle pressure upon their abdomens. Fertilization with sperm whose genetic material has been inactivated by irradiation with UV light allows haploid development for up to 4 days. Parthenogenetic haploids can be induced to double their genetic content through treatments that inhibit cytokinesis. There are two main methods to produce gynogenetic (maternally derived) offspring, early-pressure (EP) and heat-shock (HS) (21). EP consists in the application of hydrostatic pressure to inhibit completion of the second meiotic division, therefore producing diploid oocytes. HS consists in a heat pulse at a slightly later stage that inhibits cytokinesis of the first mitosis, thereby transforming haploid embryos into diploid ones. It is also possible to produce androgenetic (paternally derived) haploids by fertilizing irradiated oocytes with unirradiated sperm, and androgenetic diploids by treating such haploids with heat shock (24, 25). Gynogenesis is more widely used than androgenesis because it is more efficient, less cumbersome, and produces fewer background abnormalities. Here we provide the basic methods for *in vitro* fertilization and gynogenetic methods.

Protocol 6

Sperm collection using dissected testes

Sperm solutions can be made either by shredding dissected testes (this protocol), which we find to be more reliable and less laborious, or by collecting milt from anaesthetized males (*Protocol 7*). This protocol produces sperm solutions to fertilize about 40 egg clutches. Avoid contact of sperm with water to prevent its premature activation.

Equipment and reagents

- 100 ml of MESAB working solution in a 250 ml beaker: 7 ml stock solution per 100 ml fish water. (MESAB stock solution: 0.2% ethyl-*m*-aminobenzoate methanesulphonate. Adjust to pH 7.0 with 1 M Tris, pH 9.0. Keep at 4°C)
- 1 ml of Hank's solution in a 1.5 ml microfuge tube. Hank's stock solutions: Solution 1: 8.0 g NaCl, 0.4 g KCl in 100 ml double-distilled (dd) H_2O. Solution 2: 0.358 g anhydrous Na_2HPO_4, 0.60 g KH_2PO_4 in 100 ml ddH_2O. Solution 4: 0.72 g $CaCl_2$ in 50 ml ddH_2O. Solution 5: 1.23 g $MgSO_4.7H_2O$ in 50 ml ddH_2O. Hank's Premix: Combine the following in order: 10.0 ml Solution 1, 1.0 ml Solution 2, 1.0 ml Solution

4, 86.0 ml ddH_2O, and 1.0 ml Solution 5. Store stock solutions 1, 2, 4, and 5 and Premix at 4°C. Prepare fresh Solution 6: 0.35 g $NaHCO_3$ in 10 ml ddH_2O. Hank's (Final): combine 990 μl Hank's Premix and 10 μl Solution 6
- One tablespoon
- Ice bath
- Ten males, preferably 6–12 months old
- Paper towels
- One scalpel
- Two pairs of fine forceps
- Pellet pestle which fits snugly into a microfuge tube

Method

1 Anaesthetize the males in a 250 ml beaker containing 100 ml of MESAB working solution until they stop their gill movements (about 4 min).

2 With the aid of a spoon, rinse the males in fish water and place them on paper towels to remove excess moisture.

3 Using the scalpel, section the males, first transversely at a level immediately posterior to the pectoral fins, and subsequently longitudinally, along the ventral side from the plane of the first section to the anal region.

4 With the aid of forceps and under the dissecting microscope, remove the internal organs of the digestive track and the swim bladder. The testes will become visible as two long, whitish structures longitudinally attached to the body walls.

5 Transfer the testes to Hank's solution in a 1.5 ml microfuge tube on ice.

6 Release the sperm from the testes by gently grinding them using the pellet pestle. Alternatively, shear the testes by breaking them with a small spatula and gently pipetting the testes suspension several times using a 1000 μl Pipetteman. Allow debris to settle and transfer supernatant to a new tube[a].

a Sperm solution on ice is effective for about 2 h.

Protocol 7

Sperm collection using live anaesthetized males

The following protocol produces sperm solutions to fertilize about 40 egg clutches. Avoid contact of sperm with water to prevent its premature activation.

Equipment and reagents

- 100 ml of MESAB working solution in a 250 ml beaker (*Protocol 6*)
- 1 ml of Hank's solution in a 1.5 ml microfuge tube (*Protocol 6*)
- One tablespoon
- Ice bath
- 40–60 males, preferably 6–12 months old
- Paper towels and soft tissues
- Sponge with a slit to hold fish during sperm collection

- One pair of flat forceps
- 10 µl capillary tubes. Calibrate one capillary tube by making marks from the end corresponding to known volumes (e.g. 0.5, 1.0, 2.0, 4.0 µl) and keep as a measuring standard
- Mouth pipette apparatus with a capillary tube adapter

Method

1 Anaesthetize the males in a 250 ml beaker containing 100 ml of MESAB working solution until they reduce their gill movements (2–4 min).[a]

2 With the aid of a spoon, rinse the males in fish water and place them on paper towels to remove excess moisture.

3 Transfer the males to the sponge so that they are held, ventral side up, by the slit of the sponge. Under a dissecting microscope, gently push aside the anal fins to expose the cloaca. Gently dry this area with a soft tissue.

4 Using a pair of flat forceps, gently squeeze the sides of the fish just anterior to the base of the anal fins while applying gentle suction with the mouth pipette apparatus and collecting the sperm in the capillary tube. Sperm should have a milky, white appearance.

5 Using the calibrated capillary tube, estimate and keep count of the collected number of microlitres of sperm.

6 Gently expel the sperm into the Hank's solution on ice. Return the male to a tank with fish water. Repeat this procedure until a total of 20–25 µl sperm/ml of Hank's is collected.[b]

[a] Overexposure to MESAB will impede the recovery of the fish. MESAB working solution may have to be boosted through time with more stock solution in 0.5–1 ml increments.

[b] Sperm solution on ice is effective for about 2 h.

Protocol 8

Stripping females of eggs

For those lines which readily release eggs in the absence of males, females are separated overnight in clean tanks with running water (1–10 females per 2-litre tank) and stripped of eggs during the first 4 h after the start of the light cycle of the following day. About 50% of females from such lines will yield eggs in a given morning. For lines which no not readily yield eggs, females should be set up

Protocol 8 continued

together with males in pair matings overnight. Manual stripping may be attempted at the beginning of the following light cycle, with varying success. Waiting for the first eggs to be released during normal mating and immediately separating the female allows to manually release eggs from most females. This latter procedure, however, is too cumbersome to carry out on a large scale. Therefore, for large-scale approaches, such as genetic screens, females should be of a line that readily yields eggs (AB or golden-Tübingen).

Equipment and reagents

- Females separated into clean 2-litre tanks in the running water system or set up with males in pair matings. Set up twice as many females as clutches are desired
- 100 ml of MESAB working solution (*Protocol 6*)
- One tablespoon
- Paper towels and soft tissues
- One small spatula

Method

1 Anaesthetize females in MESAB working solution until they reduce their gill movements (2–4 min).[a]

2 With the aid of a spoon, rinse a female in fish water and place her on several paper towels to remove excess moisture.

3 Place the female on the bottom half of a Petri plate. With a soft tissue, gently dry further the anal fin area.[b]

4 Slightly moisten the index fingers of both hands.[c] With one finger support the back of the female, and with the other gently press her abdomen. Females that can be stripped of eggs will release them upon gentle pressure. Healthy eggs have a translucent, yellowish appearance.

5 Separate the eggs from the female with a dry, small spatula and place the female in an appropriate tank. Activation of egg clutches can be delayed for several minutes. In this case, cover the clutches with a Petri plate lid to reduce their drying.[d]

[a] Overexposure to MESAB will impede the recovery of the female, and fish should be placed in fresh water if they are not going to be used within 1 min after they reduce their gill movements. MESAB working solution may have to be boosted through time with more stock solution in 0.5–1 ml increments.

[b] Excess water may prematurely activate the eggs.

[c] Dry hands will stick to the skin of the fish.

[d] Longer delays before egg activation are possible using rainbow trout or coho salmon ovarian fluid (26) or Hank's saline supplemented with 0.5% BSA (27).

Protocol 9

UV inactivation of sperm

Equipment and reagents

- Sperm solution (*Protocols 6* or *7*)
- UV lamp: Sylvania 43 cm 15 W germicidal lamp
- One glass watchglass
- Ice bath

Protocol 9 continued

Method

1 Transfer 0.5–1.0 ml of sperm solution to a watchglass. Avoid pieces of debris, as they may shield sperm from the UV light.

2 Place the watchglass on ice at a distance of 30 cm directly under the UV lamp. Irradiate for 2.5 min (*Protocol 6*) or 2.0 min (*Protocol 7*) with gentle stirring every 30 s.

3 Using a clean pipette tip, transfer to a new Eppendorf tube and keep on ice.[a]

[a] UV-treated sperm solution on ice is effective for about 2 h.

Protocol 10

In vitro fertilization/haploid production

Equipment and reagents

- Sperm solution: untreated (*Protocols 6* or *7*) for *in vitro* fertilization, or UV irradiated (*Protocol 9*) for haploid production
- Freshly stripped eggs (*Protocol 8*)
- E3 medium: 5 mM NaCl, 0.17 mM KCl, 0.33 mM $CaCl_2$, 0.33 mM $MgSO_4$, 10^{-5}% Methylene Blue

Method

1 Add 25 µl of sperm solution to the egg clutch.

2 Mix the sperm and eggs by moving the pipette tip without lifting it from the Petri plate (to minimize damage to the eggs).

3 Add 1 ml of E3 medium to activate the eggs.

4 After 1 min, fill the Petri plate with E3 medium. Incubate at 27–29 °C.

5 After several hours, transfer fertilized eggs into new Petri dishes. For *in vitro* fertilization: proceed with fish-raising protocols (Chapter 1). For haploid production: because of the high level of embryonic lethality present in these clutches, incubate embryos at a low density of 80 embryos maximum per 94-mm plate and remove lysing embryos regularly (every 12 h for the first 36 h and every 24 h afterwards).[a]

[a] When producing haploids, the presence of viable survivors (e.g. 5-day-old embryos with an inflated swim bladder) indicates incomplete sperm inactivation.

Protocol 11

Gynogenesis through early pressure

Because a limited number of vials fit within a single pressure cell, in those cases where it is desired to maximize the number of clutches that can be produced, it is necessary to work on cycles in which up to four clutches in separate vials are inserted within the pressure cell. For this, we typically anaesthetize 6–12 females. Once four healthy-looking clutches are obtained, the females that have not yet been stripped of eggs are transferred to fresh fish water until they completely recuperate. It works well to begin to anaesthetize females for the next EP cycle at around minute four within a current cycle.

Protocol 11 continued

Equipment and reagents

- UV-irradiated sperm solution (*Protocol 9*)
- Freshly stripped eggs (*Protocol 8*)
- E3 saline (*Protocol 10*)
- EP vials: disposable glass scintillation vials, with plastic caps (3.2 cm height and 2.2 cm diameter, Wheaton) or similar vials. Only two vials of this type can fit at once in a pressure cell. In order to fit four vials in one cell, we have custom-built shorter plastic vials (1.8 cm height, including cap, 2.5 cm diameter, 0.3 mm wall thickness) which fit the plastic

caps from the scintillation vials. Perforate the plastic caps several times with a needle to allow better exposure to the hydrostatic pressure
- French Press cell, 40 ml (Thermo Spectronic)
- French Pressure Cell Press (Thermo Spectronic) or Hydraulic Laboratory Press (Fisher)

Method

1 Add 25 µl of UV-irradiated sperm to the egg clutch.

2 Mix the sperm and eggs by moving the pipette tip without lifting it from the Petri plate (to minimize damage to the eggs).

3 Activate up to four clutches simultaneously by adding 1 ml of E3 saline to each clutch and start the timer.[a]

4 After 12 s, add more E3 medium. A squirt to the side of the Petri plate will make the fertilized eggs collect in the middle of the plate.

5 With a plastic pipette, transfer the fertilized eggs to an EP vial. Fill the vial with E3 and cap it with the perforated plastic lid. Avoid large air bubbles. Place the vials inside the pressure cell, ensuring that no air remains trapped inside it. Record the relative position of the clutch within the pressure cell by placing tags labelling each clutch in the corresponding order on a dry surface. Fill the pressure cell with E3 and close it, allowing excess E3 to be released from the side valve. Close the side valve without overtightening. Insert the entire assembly on the French Press apparatus and apply pressure to 8000 lb/sq. in. by time 1 min 20 s after activation.[b]

6 At 6.0 min, release the pressure and remove the pressure cell from the French Press apparatus. Maintaining the relative order of the vials, remove the vials from the pressure cell, dry them with a towel and label them with their corresponding tags. Place the vial in a 27–29 °C incubator.

7 After all EP cycles have been completed, allow the embryos to rest in the vial for at least 45 min but not more than 4 h. Transfer embryos with their corresponding tags to Petri plates. Incubate at 27–29 °C. Because of the high level of embryonic lethality present in these clutches, incubate embryos at a low density of 80 embryos maximum per 94-mm plate and remove lysing embryos regularly (every 12 h for the first 36 h and daily afterwards).[c]

[a] At least two people are required to timely manipulate four clutches.

[b] For different strains and/or presses, different pressure values may be optimal (see, for example, ref. 28).

[c] The use of genetically marked lines can serve as a convenient safeguard against incompletely inactivated sperm. For example, if UV-irradiated sperm from wild-type males is used to fertilize eggs from females homozygous for a pigment mutation (e.g. *golden* or *albino*), embryos with a wild-type pigment pattern derive from incompletely inactivated sperm and can be discarded.

Protocol 12

Gynogenesis through late heat shock

Equipment and reagents

- UV-irradiated sperm solution (*Protocol 9*)
- Freshly stripped eggs (*Protocol 8*)
- E3 medium (*Protocol 10*)
- Heat-shock baskets: construct these by cutting off the bottom of ultracentrifuge tubes (Beckman Ultra-clear, 13 × 51 mm), and

- heat-sealing a fine wire mesh to the bottom edge of the tube
- Timer
- Stirring water baths of E3 saline at 28.5 °C
- Stirring water bath of E3 saline at 41.4 °C
- Paper towels

Method

1 Add 25 μl of UV-irradiated sperm to the egg clutch.

2 Mix the sperm and eggs by moving the pipette tip without lifting it from the Petri plate (to minimize damage to the eggs).

3 Add 1 ml of E3 saline to activate the eggs. In a premade chart, and using a timer, record for each clutch the time of activation (*t*) to the second and calculate and record the times to begin (*t* + 13 min) and end (*t* + 15 min) the heat-shock treatment.

4 Add more E3 after 30 s. Transfer the eggs to a heat-shock basket. Immerse the basket in a water bath with stirring and E3 saline at 28.5 °C.

5 At time = *t* + 13.0 min, blot the bottom of the basket briefly on to a stack of paper towels and transfer the basket to a water bath with stirring and E3 saline at 41.4 °C.

6 At time = *t* + 15.0 min, blot the bottom of the basket briefly and transfer the basket back to the 28.5 °C E3 bath.

7 Allow the embryos to rest for about 45 min and transfer to a Petri plate. Incubate at a low density (as in *Protocol 11*, step 7) at 27–29 °C.[a]

[a] The use of genetically marked lines can serve as a convenient safeguard against incompletely inactivated sperm (see *Protocol 11*, footnote c).

4.2.2 Haploid screens

Haploid screens from non-mosaic germ lines

The production of haploids from F_1 females heterozygous for newly induced mutations results in the uncovering of zygotic mutations. The haploid phenotype is usually similar to the homozygous diploid phenotype, although for some mutations the latter is weaker or even absent (11).

If the F_1 females are not mosaic, 50% of haploid embryos exhibit the mutant phenotype (*Figure 2a*; *Protocol 13*). Such an approach has been applied to identify mutations in early development (29), early motility (30, 31), and brain patterning (32). Haploid embryos exhibit a characteristic continuum of background abnormalities, and only a fraction of embryos (50–70% in selected lines) appear relatively normal after 24 h of development (11). These better-developed haploids nevertheless exhibit a reduction in the extension of the axis, and show a reduced robustness in developmental patterns, such as defects in some aspects of brain development and organogenesis (11, 33, 34). Because haploids do not survive past day 4 after fertilization, haploid screens are restricted to at most the first 3 days of development. Background abnormalities present in haploids may obscure some phenotypes, especially

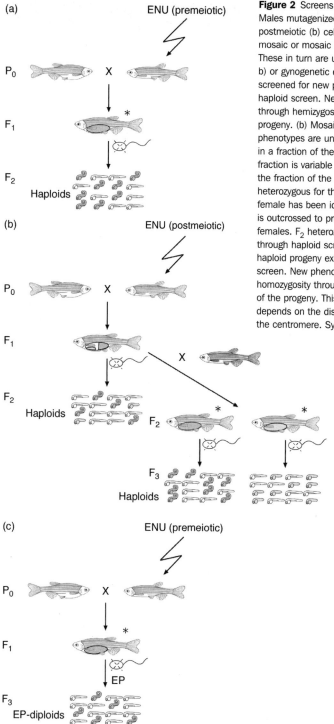

(a)

ENU (premeiotic)

P_0 X

F_1 *

F_2
Haploids

(b)

ENU (postmeiotic)

P_0 X

F_1

F_2
Haploids F_2 *   *

F_3
Haploids

(c)

ENU (premeiotic)

P_0 X

F_1 *

EP

F_3
EP-diploids

Figure 2 Screens based on ploidy manipulation. Males mutagenized in their premeiotic (a, c) or postmeiotic (b) cells are used to produce non-mosaic or mosaic F_1 females, respectively. These in turn are used to produce F_2 haploid (a, b) or gynogenetic diploid (c) clutches, which are screened for new phenotypes. (a) Non-mosaic haploid screen. New phenotypes are uncovered through hemizygosity in 50% of the F_2 haploid progeny. (b) Mosaic haploid screen. New phenotypes are uncovered through hemizygosity in a fraction of the F_2 haploid progeny. This fraction is variable and corresponds to one-half the fraction of the mosaic germ line heterozygous for the mutation. Once a mosaic F_1 female has been identified to carry a mutation, it is outcrossed to produce non-mosaic F_2 females. F_2 heterozygous carriers are identified through haploid screening, since 50% of their haploid progeny exhibit the phenotype. (c) EP screen. New phenotypes are uncovered through homozygosity through early pressure in a fraction of the progeny. This fraction is variable and depends on the distance of the mutated gene to the centromere. Symbols as in *Figure 1*.

during the first 24 h of development. However, the expected 50% mutant frequency is a useful aid in the recognition of new mutations.

Haploid screens from mosaic germ lines

Haploid screening has also been carried out in embryos from F_1 females with mosaic germ lines (derived from sperm mutagenized during postmeiotic stages) (35) (*Figure 2b*; *Protocol 13*). In this mosaic-haploid approach, mutant phenotypes are obtained in a variable fraction of the haploid F_2 embryos (corresponding to one half of the fraction of the germ line heterozygous for the induced mutation). Mosaic germ lines derived after postmeiotic treatment with ENU can carry a higher mutational load than non-mosaic germ lines (23), so that, in principle, the mosaic haploid approach allows isolating mutations more efficiently than a standard haploid approach. However, a majority of mutations induced by postmeiotic mutagenesis with ENU appears to be caused by large chromosomal rearrangements or deletions (9), so that the ENU/non-mosaic haploid approach is not appropriate to efficiently isolate single gene lesions. Another disadvantage of a non-mosaic haploid approach is that the variable, and in some cases, low fraction of embryos exhibiting new phenotypes increases the difficulty of distinguishing them from background abnormalities, and therefore only traits robustly expressed in haploids can be screened using this method.

4.2.3 Parthenogenesis-based methods

Parthenogenetic diploids can survive through embryogenesis and even reach to be fertile adults. Therefore, parthenogenesis, especially gynogenesis, extends the window of developmental stages past that which can be screened using haploids. In addition, gynogenetic diploids also show more reliable developmental patterns than haploids (11, 33, 34).

Screens based on early pressure-induced gynogenesis

F_1 females carrying newly induced mutations can be treated to produce partially homozygous embryos through EP (*Figure 2c*; *Protocol 13*) and the resulting diploids screened for lethal phenotypes caused by homozygosity for recessive mutations. EP is a relatively efficient method of diploidization (producing about 21% viable day 5 embryos) and has been used in general screens to identify mutations affecting patterning and morphogenesis (1, 36, 37), as well as specific screens to identify mutations affecting the neural crest (34) and maternally provided factors (22).

The degree of homozygosity in EP-derived progeny depends on the distance of a given locus to the centromere. For centromere-linked loci, the fraction of embryos homozygous for a given allele is 50%. As the centromere–locus distance increases, however, the fraction of homozygous progeny decreases and the fraction of heterozygous progeny increases, approaching values close to 0 and 100%, respectively. Therefore, EP has the disadvantage that it is biased against the identification of distal genes. Observed frequencies of homozygosity vary from 50 to 4%, with an average value of 23% (38, 39), suggesting that the majority of the genome is accessible to its identification through EP. Indeed, mutations resulting in mutant phenotypes represented by fractions as low as 5% have been identified using this method (34).

Another disadvantage of EP-based screens is that the expected fraction of embryos exhibiting new mutant phenotypes is not fixed. The lack of a fixed expected mutant fraction, combined with the relatively high incidence of syndromes induced by the gynogenetic treatments themselves (especially axial deficiencies) (40), increases the difficulty in distinguishing phenotypes with a genetic cause from epigenetic defects.

Protocol 13

Common protocol for gynogenesis (non-mosaic haploid, mosaic haploid, and EP) based screens

This protocol allows the screening of 1500 genomes (non-mosaic haploid, EP) or mutagenized mosaic germ lines (mosaic haploid).[a] The suggested genetic scheme (excluding mutagenesis, retests, and recovery) requires about 43 weeks (non-mosaic haploid) or 52 weeks (EP screens and mosaic haploid).[b] The screening phase requires about 25 weeks. For simplicity, this protocol is only based on ENU mutagenesis. However, it is also possible to use it in combination with other mutagenic treatments, e.g. gamma-irradiation of mature sperm (*Protocol 3*; *Table 2*). It is essential to use parental males and females from lines that readily yield eggs, such as AB or golden-derived Tübingen.

Equipment and reagents

- 6–10 males mutagenized in their premeiotic cells (non-mosaic haploid, EP) or 10 males mutagenized in their postmeiotic cells (mosaic haploids)
- Fish facility with minimum nursery and adult tank capacities of 120 each
- 120 1-litre plastic boxes with lids
- 500 males, preferably 6–12 months old, for sperm production (20 males per haploid/EP session, total of 25 sessions)
- Materials for sperm collection in order to produce 2 ml of sperm in Hank's solution

- using dissected testes from 20 males (preferably *Protocol 6*)
- Materials for UV irradiation of sperm (*Protocol 9*). Irradiate the sperm in aliquots of 1 ml immediately prior to use
- Two working stations for stripping females of eggs (*Protocol 8*) and haploid production (*Protocol 10*)
- 60 EP vials (EP screens only), French Press cell and French Pressure Cell Press (*Protocol 11*)

Method

1 Cross each mutagenized male to females of the same (egg-yielding) genetic line to generate about 6000 adult F_1 progeny (about 3000 females, half of which will release their eggs). In all cases, freeze the mutagenized males (*Protocol 1*, step 14) after the desired number of adult F_1 fish is produced. Non-mosaic haploids and EP: perform mutagenesis of *premeiotic* cells with 3 mM ENU (*Protocol 1*) and up to 2000 adults per mutagenized male,[c] and keep track of the parental origin of each F_1 tank. Mosaic haploids: perform mutagenesis of *postmeiotic* cells with 0.8 mM ENU (*Protocol 2*).

2 After reaching sexual differentiation (2-3 months), separate the F_1 fish into tanks containing 20–30 F_1 females and five sibling males for each tank. Discard the rest of the F_1 males.

3 Once F_1 females become 4 (non-mosaic haploid) or 6 (mosaic haploid and EP) months old,[d] begin to produce haploid or EP-derived clutches as in *Protocols 10* and *11*, respectively. In each session, attempt to strip 120 females to produce about 60 egg clutches.[e] Sort up to 1×80 fertilized eggs (non-mosaic haploid screen), 2×80 fertilized eggs (mosaic haploid) and all fertilized eggs in groups of 80 per plate (EP screens) and incubate at 27–29°C. Keep F_1 females which produced clutches separate in a single box outside of the running water system until the screen test is finished. Label both clutches and parental F_1 females with duplicate tags. Return females that do not release their eggs to tanks (in non-mosaic haploid and EP screens, record the paternal origin of the F_1 females and do not mix females from different paternal origins) and attempt to strip them of eggs once more after performing the procedure once on all F_1 females. Discard females that do not release eggs after the second attempt.

Protocol 13 continued

4   Perform screen tests. Consider as putative mutant phenotypes those occurring in 50% of the embryos (non-mosaic haploids) or in clusters in a variable fraction between 5 and 50% (mosaic haploids and EP).

5   If no mutation is found in a clutch, discard the corresponding F_1 female.

6   If a putative mutation is found in a clutch, return the female to the system and allow her to rest and feed for 2 weeks.

7   Propagate the mutation by crossing the F_1 female to a wild-type male to produce an F_2 generation.[f] Repeat this outcross on a weekly basis to produce two (non-mosaic haploids and EP) or four (mosaic haploids) F_2 crosses of 65 embryos each.

8   Three weeks after the last outcross, produce another haploid or EP clutch from the putative carrier F_1 female to verify the presence of the phenotype. If the female produces a clutch, and a cluster of embryos exhibiting the original phenotype is not observed, discard the F_1 female and corresponding growing F_2 clutches. If a cluster of embryos exhibiting the original phenotype is observed, or the female does not produce a clutch, continue growing the corresponding F_2 clutches.

9   Identify heterozygous carriers in the next generation. Non-mosaic haploids and EP: retest and recover the phenotype through random incrosses of the adult F_2 progeny. One-quarter of random incrosses should exhibit the mutant phenotype in one-quarter of the embryos. Mosaic haploids: once the F_2 generation has reached 4–5 months of age, produce and screen haploid clutches from as many F_2 females as possible or until 4–5 heterozygous carriers are identified.[g] Propagate the mutation as in Chapter 1.

[a]   The number of mutagenized genomes (non-mosaic haploid, EP) or mosaic germ lines (mosaic haploid) screened is equal to the number of (haploid or EP-derived) clutches tested. In EP-based screens, however, heterozygosity in distal regions results in a decrease in the actual number of screened genomes per clutch tested (to a value equal to $(1 - (1 - F_m)^n)$, where F_m is the frequency of homozygosity of loci in those regions after EP (expressed as a fraction, ranging between 0.5 and 0) and n is the number of diploid embryos screened in the clutch).

[b]   Different time requirements depend on the age of the stripped F_1 females (step 3). All other calculations are estimated as in *Protocol 5*, footnote b.

[c]   As in *Protocol 5*, footnote c, and considering that only 25% of F_1 fish will be females that produce egg clutches.

[d]   In order to produce larger clutches, 6-month-old (or older) females should be used in mosaic haploid and EP screening.

[e]   During a session, at least two people are required to manipulate this number of females and haploid or EP clutches (preferably three in the case of EP).

[f]   Females sometimes do not mate after stripping procedures. For this reason, and although it is more laborious, some workers prefer to set up outcrosses 10 days before the haploid or EP clutch production, i.e. prior to step 3. This also serves to preselect females that are fertile. In this case, the screening procedure should be finished within 10 days after clutch production, so that the growing babies do not need to be connected to the running water system. In mosaic haploid screens, the male used for this outcross should be from a egg-yielding genetic line.

[g]   Due to the mosaicism of the F_1 germ line, only a fraction of the F_2 females will be heterozygous for the mutation (this fraction should be similar to the fraction of mutant embryos in the original screen tests). Heterozygous carrier F_2 females will contain a non-mosaic germ line, so that the phenotype will be exhibited by 50% of the haploid progeny.

Heat shock-induced gynogenesis

HS-derived progeny are homozygous at every single locus, and therefore 50% of HS-derived progeny will be homozygous for any mutation present in a heterozygous mother, regardless of its chromosomal location. Screening of HS-derived progeny from F_1 females carrying newly induced mutations has been used to estimate mutagenesis rates and isolate a number of mutations (4). However, HS induces diploidization at lower frequency than EP (about 9% viable day 5 embryos) and HS-derived embryos are generally less robust than EP-derived ones. Therefore HS is not commonly used in genetic screens, although it is useful for the rapid production of lethal-free lines (Section 3.3 and *Protocol 12*).

Androgenesis-based methods

Irradiation of eggs required for androgenesis often results in cellular damage that precludes screening of embryos at the phenotypic level. However, haploid androgenesis has been applied to PCR-based screens of genomic DNA to detect mutations in particular cloned genes (see ref. 25).

4.3 Screens for recessive mutations affecting adult traits and maternal-effect mutations

Screening for mutations in adult traits and fertility requires growing homozygous fish for weeks or months until the appearance of the trait. If carried out solely using natural crosses, the generation of adult homozygous fish (i.e. growing F_3 crosses to adulthood in *Figure 1*) requires large amounts of labour, generation time, and space. For this reason, the application of EP-induced gynogenesis (i.e. growing F_2 gynogenotes in *Figure 2c*), which allows bypassing one generation to produce homozygotes for newly induced mutations, is an attractive alternative to carrying out adult screens (see ref. 22).

However, the number of viable offspring produced by EP is much lower than that produced by natural crosses (on average, about 30 and 150, respectively, at day 5 of development) and EP results in variable fractions of phenotypes. Therefore, the choice between a family inbreeding and an EP-based approach depends largely on the robustness of the trait at the individual level. Certain traits with large individual variations, such as behaviour, are more appropriately tested using a scheme based solely on natural crosses. On the other hand, an EP-based approach is particularly well-suited for a screen for maternal-effect mutations, since a single EP-derived female can produce several hundred eggs, most of which develop normally (22).

4.4 Screens for dominant mutations

Viable or subviable dominant zygotic mutations can be screened for in their own right or in conjunction with a genetic screen for recessive mutations. However, since each generation in a genetic screen might select against a subviable dominant mutation, it is most efficient to screen for dominant mutations in the first generation comprising of non-mosaic individuals. In our laboratory we have identified and recovered viable dominant zygotic mutations as early as in the direct progeny derived from germ cells mutagenized during premeiotic stages.

4.5 Allele screens

The identification of new alleles of particular mutations can be of valuable help for further functional genetic analysis and as an aid to the molecular cloning of the affected gene. Fish carriers of newly induced mutations are crossed to tester fish carrying the original mutation, and the appropriate phenotype is screened for in their progeny.

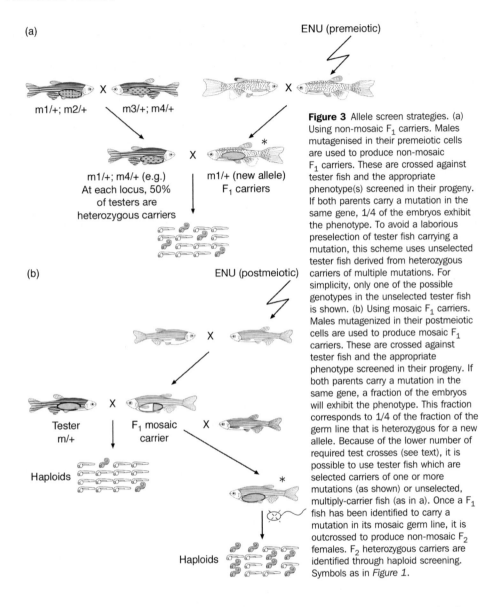

Figure 3 Allele screen strategies. (a) Using non-mosaic F_1 carriers. Males mutagenised in their premeiotic cells are used to produce non-mosaic F_1 carriers. These are crossed against tester fish and the appropriate phenotype(s) screened in their progeny. If both parents carry a mutation in the same gene, 1/4 of the embryos exhibit the phenotype. To avoid a laborious preselection of tester fish carrying a mutation, this scheme uses unselected tester fish derived from heterozygous carriers of multiple mutations. For simplicity, only one of the possible genotypes in the unselected tester fish is shown. (b) Using mosaic F_1 carriers. Males mutagenized in their postmeiotic cells are used to produce mosaic F_1 carriers. These are crossed against tester fish and the appropriate phenotype screened in their progeny. If both parents carry a mutation in the same gene, a fraction of the embryos will exhibit the phenotype. This fraction corresponds to 1/4 of the fraction of the germ line that is heterozygous for a new allele. Because of the lower number of required test crosses (see text), it is possible to use tester fish which are selected carriers of one or more mutations (as shown) or unselected, multiply-carrier fish (as in a). Once a F_1 fish has been identified to carry a mutation in its mosaic germ line, it is outcrossed to produce non-mosaic F_2 females. F_2 heterozygous carriers are identified through haploid screening. Symbols as in *Figure 1*.

A large-scale allele screen has been performed in combination with mutagenesis of premeiotic cells using ENU (*Figure 3a*) (41). This treatment induces mutations at frequencies of about 0.5–1 $\times$ 10^{-3} tested genomes. Therefore, to identify two new alleles, for example, requires about 2000–4000 successful test crosses (or about 4000–8000 pair matings). These, in turn, would require a large number of tester fish carrying the original mutations (in this same example, 200–400 tester fish repeatedly set up over 20 weeks). For most lethal mutations the identification of such a large number of heterozygous carriers through random crosses constitutes in itself a large effort. van Eeden *et al.* (41) have circumvented a preselection for tester fish altogether by increasing the number of mutations segregating amongst the tester fish (to four mutations in their screen), and therefore increasing the chances that a given unselected tester fish carries a mutation.

Protocol 14

Allele screens using non-mosaic mutagenized germ lines and unselected tester fish

This protocol allows isolating about two new alleles at an average locus.[a] The suggested genetic scheme (excluding mutagenesis, retests, and recovery) requires about 34 weeks (18 weeks for the screening phase).[b] Germ lines mutagenized using other mutagens (e.g. gamma-rays) may also be screened using this procedure.

Equipment and reagents

- About 25 mutagenized males carrying mutations in their premeiotic germ cells (*Protocol 1*, adjusting for the amount of starting males and materials). These should be marked with appropriate visible markers (e.g. *leopard-Long Fin*) to distinguish them from fish carrying the original mutations
- 3–4 pairs of fish known to be doubly heterozygous for each of two sets of different mutations. Produce these by crossing known

single heterozygous carriers, and selecting double heterozygotes through random sibling crosses in the next generation
- Fish facility with minimum nursery and adult tank capacities of 200 tanks each
- 1100 pair mating set-ups (Chapter 1)
- 600 single fish boxes with lids (Chapter 1)

Method

1 Cross heterozygous carriers for multiple mutations (e.g. heterozygous carriers for two mutations against heterozygous carriers for two other mutations) to produce about 1800 progeny tester adults. Dominant effects on viability caused by the mutation can significantly decrease the expected allele frequency of 0.5. Therefore, test the actual allele frequencies by carrying out 100–200 random crosses between tester fish.[c]

2 On the week immediately after these crosses have been performed, begin crossing mutagenized males with mutations induced premeiotically to females of the same (marked) genetic background, to produce a total of about 12 000 fish (750 fish per week for 16 weeks). Produce a total maximum of 500 F_1 adults from each mutagenized male.[d] Keep track of the parental origin of each F_1 tank.

3 Once the first cohort of F_1 fish reach 4 months old, set up these (about 750) F_1 fish in pair matings against tester fish.[e]

4 On the following 2 days after setting up the fish, collect the layed eggs and label both eggs and the parental pair using serially numbered, duplicate tags. For each cross, sort 40 fertilized embryos, transfer to a new Petri dish and incubate at 27–29 °C. After the second day collection, return all tester fish to their tanks. Keep F_1 fish that mated in tagged, separate 1-litre boxes until the screen tests are finished (up to 10 days). Return F_1 fish that did not mate to tanks, pooling them without mixing them with respect their parental origin.

5 Score the collected test clutches for the corresponding mutant phenotype(s).

6 Discard all F_1 fish that produced wild-type clutches. If a phenotype is found, allow the parental F_1 fish to rest and feed for 2 weeks in the running water system and outcross to a wild-type stock. In the next generation, the mutation should appear in one-quarter of the embryos in one-quarter of random sibling crosses, and can be maintained as in Chapter 1.

Protocol 14 continued

7 Repeat weekly pair matings as in steps 3– 6, except including, after a 2-week rest, the F_1 fish that did not mate in their first trial. Discard all F_1 fish that do not mate after two trials. Alternate using groups of tester fish, so that they are used evenly and fish are allowed to rest for 1 week every 2–3 weeks.

a For each mutation, the number of genomes screened is equal to the number of test crosses multiplied by the observed allele frequency of the tester fish, as determined in step 1.

b All calculations are estimated as in *Protocol 5*, footnote b.

c For each mutation, the actual allele frequency is the square root of the fraction of crosses exhibiting its corresponding phenotype.

d As in *Protocol 5*, footnote c.

e Setting up 750 test crosses and collecting the resulting egg lays in one week requires two full-time workers.

It is also possible to perform an allele screen where the F_1 carriers have a mosaic germ line derived from mutagenesis of postmeiotic cells with ENU (*Figure 3b*) (8; D. Stainier, personal communication). This mutagenic method results in a mutational load estimated to be almost 10 times higher than that obtained using mutagenesis of premeiotic cells (23). Therefore, the numbers of screened clutches and tester fish needed are correspondingly reduced and tester fish can be feasibly preselected (for example, using PCR analysis to identify fish carrying DNA rearrangements, or performing random crosses to identify heterozygous carriers). In addition, because the phenotype to be screened for is already known, even a small mutant cluster is sufficient for the positive identification of a mosaic F_1 carrier.

Allele screens of mosaic germ lines after postmeiotic mutagenesis with ENU have identified point mutations (8). However, a majority of lesions induced using this treatment appear to be caused by chromosomal rearrangements or deletions (9). Therefore, although mosaic allele screens can be used to identify new point mutations, they may be most suitable to isolate overlapping deletions centred on a mutated gene, for example, to better define its genomic location.

Protocol 15

Allele screen using a mosaic mutagenized germ line and preselected tester fish

This protocol allows the identification of 2–4 newly induced alleles for a single mutation. The suggested genetic scheme (excluding mutagenesis, retests, and recovery) requires about 45 weeks (27 weeks for the screening phase).[a] Shorter screen duration times may be achieved by increasing the number of preselected tester fish and the number of test crosses per week. It is also possible to carry out screen tests using larger numbers of unselected fish carrying multiple mutations, as in *Protocol 14*. Germ lines mutagenized using other mutagens (e.g. gamma-rays) may also be screened using this procedure. Because mutations are recovered through haploid screening, mutations need to be induced and propagated in a genetic background that readily yields eggs (i.e. AB or golden-Tübingen) until the identification of non-mosaic carriers.

Protocol 15 continued

Equipment and reagents

- Five males treated with ENU mutagenesis of postmeiotic sperm, as in *Protocol 2*
- 90 identified carriers for the original mutation, which should become sexually mature and be selected for during the weeks just before the F_1 fish generated in step 1 become sexually mature. If heterozygous carriers are identified using random crosses, grow about 500 progeny fish from identified carriers and perform random sibling crosses in the next generation. Once heterozygous carriers are identified, they can be used to identify other carriers more effectively. Alternatively, if the mutation can be identified using a PCR approach, grow 200 fish and perform PCR assays on fin clips in the next generation (Chapter 5). In addition, in order to clearly distinguish fish carrying new alleles from those carrying the original mutation, tester fish should be marked with visible markers (e.g. *leopard-Long Fin* or wild type, if the mutagenized fish are AB or golden-Tübingen, respectively)

- Fish facility with minimum nursery and adult tank capacities of 40 each
- 70 pair mating set-ups (Chapter 1)
- 50 single boxes with a lid (Chapter 1)
- For the recovery phase only: materials for haploid production (as in *Protocol 13*)

Method

1 Outcross mutagenized males during the first 2 weeks after ENU mutagenesis (*Protocol 2*) to females of the same (egg-releasing) background to produce 1000 F_1 adult progeny.

2 Once the F_1 progeny reach sexual maturity, set up on the first week 40 pair matings between F_1 progeny and identified tester fish.

3 Collect successful clutches during the following 2 days. Label both embryos and fish with duplicate tags containing a serial test number. Sort up to 2×80 fertilized embryos per clutch and keep the corresponding F_1 fish separate outside of the running water system until the end of the screen test (up to 10 days). After the second-day collection, return the tester fish and the F_1 fish that did not mate to tanks.

4 Screen for clusters of embryos exhibiting the known phenotype.

5 If a clutch does not exhibit the mutant phenotype, discard the F_1 carrier.

6 If a clutch contains a cluster of embryos exhibiting the mutant phenotype, return the F_1 fish to the running water system in a separate tank and allow it to rest and feed for 2 weeks.

7 During the following weeks, set up each week 40 different F_1 fish, and the F_1 fish that did not lay 2 weeks before, against tester fish, and repeat steps 3–6, except that those fish that did not lay after being set up twice are discarded. Rotate tester fish so that they rest for 1 week every 2–3 weeks.

8 Cross the F_1 fish containing the putative mutation against fish of the same egg-releasing background to produce 4×65 F_2 embryos (see *Protocol 13*, footnote f).

9 Two weeks after the original test, reset the putative carrier F_1 female against tester males to verify the presence of the phenotype. If the pair mates and a cluster of embryos exhibiting the original phenotype is not observed, discard the F_1 female and corresponding growing F_2 clutches. If a cluster of embryos exhibiting the original phenotype is observed, or if the pair does not mate, continue growing the corresponding F_2 clutches.

10 Identify heterozygous carriers in the next generation: once the F_2 generation has reached at least 4 months of age, produce and screen haploid clutches from as many F_2 females as possible, or until 4–5 heterozygous carriers are identified (as in *Protocol 13*, step 9 for mosaic haploids).[b]

Protocol 15 continued

11 Outcross one of the identified nonmosaic heterozygous F_2 females and maintain the mutation as in Chapter 1.

[a] All calculations are estimated as in *Protocol 5*, footnote b.

[b] In this case the fraction of F_2 females heterozygous for the mutation should be approximately double the fraction of mutant embryos in the allele test screen in step 4 of this protocol.

5 Screen tests

5.1 Morphological screens

The morphological landmarks of the 5 days of development have been described in ref. 42 and the screening procedure and a model scoresheet can be found in ref. 41. Embryos are observed at regular intervals in general screens or at the appropriate stage in more specific screens. The morphology of the zebrafish embryo can be scored directly using a dissecting scope. After 24h, detailed observation requires that embryos be immobilized using MESAB (about a 1:10 dilution of the stock in *Protocol 6*), which needs to be replaced by several changes of fresh E3 medium if further observation of the same embryos is desired.

5.2 Screens using molecular markers and tissue stains

Using molecular markers and tissue stains allows the detection of more subtle developmental defects. A number of screens have used antibodies (34, 35, 43) and RNA probes (32, 35). The use of mixtures of probes allows to screen simultaneously for defects in a number of cell types, as well as provides internal standards for the staining procedure. Specific dyes can also be used to look at particular tissues, for example Alcian Blue to stain for cartilage in fixed embryos (44, 45), 2-(4-dimethyl-aminostyryl)-*N*-ethyl pyridinium iodide (DASPEI) to stain hair cells in lateral-line neuromasts in live embryos (46), and Acridine Orange to detect apoptotic cells in live embryos (47). The injection of fluorescent dyes, although more laborious, has also been used to label particular cells and their processes, as in the identification of mutations affecting retinotectal projections (48). In the future, transgenic lines expressing GFP in tissue-specific patterns will allow the rapid and simple screening of particular cell types in developing embryos (Chapter 5). Especially in those procedures requiring multiple steps, streamlining of the protocols and devices to simultaneously handle multiple samples is essential for large-scale screening approaches.

5.3 Locomotion and behavioural screens

Locomotion behaviour of developing larvae, such as rhythmic tail movements, the escape response and equilibrium control, have been identified visually or with simple 'touching' assays (49). More sophisticated assays involving monitoring of free-moving as well as immobilized embryos have also been used, such as in the identification of mutants defective in optokinetic and phototactic behaviours (50, 51).

6 Conclusions

The zebrafish is an unusually malleable genetic system that provides a wide range of genetic screening strategies. Rather than having exhausted the subject, the genetic screens that have already been performed provide the foundation for many screens to come. These will allow us to revisit development using more subtle techniques, such as molecular markers and GFP-expressing tissues, and more directed approaches, such as screens for genetically interacting components. Future screens will also explore entirely new subjects, such as internal organs, adult traits and new behavioural assays. Allele screens will remain instrumental for further genetic analysis and molecular cloning of already isolated mutations. Technical improvements, such as batteries of deletions scanning the genome, the refinement of zebrafish genomics, and the development of directed recombination in the germ line will open even more genetic possibilities. Through these developments, we will see zebrafish genetics continue to grow and, with it, its contribution to vertebrate biology.

Acknowledgements

We are grateful to Adam Amsterdam, Nancy Hopkins, and Didier Stainier for generously contributing unpublished data to protocols in this chapter.

References

1. Kimmel, C. B. (1989). *TIG*, **5**, 283.
2. Haffter, P., Granato, M., Brand, M., Mullins, M. C., Hamerschmidt, M., Kane, D. A., Odenthal, J., van Eeden, F. J. M., Jiang, Y.-J., Heisenberg, C.-P., Kelsh, R. N., Furutani-Seiki, M., Vogelsang, E., Beuchle, D., Schach, U., Fabian, C., and Nüsslein-Volhard, C. (1996). *Development*, **123**, 1.
3. Driever, W., Solnica-Krezel, L., Schier, A. F., Neuhauss, S. C. F., Malicki, J., Stemple, D. L., Stainier, D. Y. R., Zwartkruis, F., Abdelilah, S., Rangini, Z., Belak, J., and Boggs, C. (1996). *Development*, **123**, 37.
4. Grunwald, D. J. and Streissinger, G. (1992). *Genet. Res.*, **59**, 103.
5. Solnica-Krezel, L., Schier, A. F., and Driever, W. (1994). *Genetics*, **136**, 1401.
6. Mullins, M. C., Hammerschmidt, M., Haffter, P., and Nüsslein-Volhard, C. (1994). *Curr. Biol.*, **4**, 189.
7. Drost, J. B. and Lee, W. R. (1998). *Genetica*, **102/103**, 421.
8. Appel, B., Fritz, A., Westerfield, M., Grunwald, D. J., Eisen, J. S., and Riley, B. B. (1999). *Curr. Biol.*, **9**, 247.
9. Imai, Y., Feldman, B., Schier, A. F., and Talbot, W. S. (2000). *Genetics*, **155**, 261.
10. Fritz, A., Rozowski, M., Walker, C., and Westerfield, M. (1996). *Genetics*, **144**, 1735.
11. Walker, C. (1999). In *Methods in cell biology* (ed. W.H. Detrich, M. Westerfield, and L.I. Zon), Vol. 60, p. 43. Academic Press, San Diego.
12. Chakrabarti, S., Streissinger, G., Singer, F., and Walker, C. (1983). *Genetics*, **103**, 109.
13. Ando, H. (1998). *Neurosci. Lett.*, **244**, 81.
14. Levken, A. C., Helde, K. A., Thorpe, C. J., Rooke, R., and Moon, R. T. (2000). *Physiol. Genomics*, **2**, 37.
15. Amsterdam, A., Yoon, C., Allende, M., Becker, T., Kawakami, K., Burgess, S., Gaiano, N., and Hopkins, N. (1997). *Cold Spring Harbor Symp. Quant. Biol.*, **LXII**, 437.
16. Amsterdam, A. and Hopkins, N. (1999). In *Methods in cell biology* (ed. W.H. Detrich, M. Westerfield, and L.I. Zon), Vol. 60, p. 87. Academic Press, San Diego.
17. Amsterdam, A., Burgess, S., Golling, G., Chen, W., Sun, Z., Towsend, K., Farrington, S., Haldi, M., and Hopkins, N. (1999). *Genes Devel.*, **13**, 2713.
18. Gaiano, N., Allende, M., Amsterdam, A., Kawakami, K., and Hopkins, N. (1996). *Proc. Natl Acad. Sci. USA*, **93**, 7777.
19. Raz, E., van Luenen, H. G. A. M., Schaerringer, B., Plasterk, R. H. A., and Driever, W. (1997). *Curr. Biol.*, **8**, 82.
20. Fadool, J. M., Hartl, D. L., and Dowling, J. E. (1998). *Proc. Natl Acad. Sci. USA*, **95**, 5182.
21. Streisinger, G., Walker, C., Dower, N., Knauber, D., and Singer, F. (1981). *Nature*, **291**, 293.
22. Pelegri, F. and Schulte-Merker, S. (1999). In *Methods in cell biology* (ed. W.H. Detrich, M. Westerfield, and L.I. Zon), Vol. 60, p. 1. Academic Press, San Diego.
23. Riley, B. B. and Grunwald, D. J. (1995). *Proc. Natl Acad. Sci. USA*, **92**, 5997.
24. Corley-Smith, G. E., Lim, C. J., and Brandhorst, B. P. (1996). *Genetics*, **142**, 1265.

25. Ungar, A. R., Helde, K. A., and Moon, R. T. (1998). *Mol. Mar. Biotechnol.,* **7**, 320.

26. Corley-Smith, G. E., Lim, C. J., and Brandhorst, B. P. In *The zebrafish book,* p. 7.52. University of Oregon Press, Eugene (online version at http://zfin.org/zf_info/zfbook/zfbk.html).

27. Sakai, N., Burgess, S., and Hopkins, N. (1997). *Mol. Mar. Biotechnol.,* **6**, 84.

28. Gestl, E. E., Kauffman, E. J., Moore, J. L., and Cheng, K. C. (1997). *J. Heredity,* **88**, 76.

29. Halpern, M. E., Ho, R. K., Walker, C., and Kimmel, C. B. (1993). *Cell,* **75**, 99.

30. Westerfield, M., Liu, D. W., Kimmel, C. B., and Walker, C. (1990). *Neuron,* **4**, 867.

31. Felsendfeld, A. L., Walker, C., Westerfield, M., Kimmel, C., and Streisinger, G. (1990). *Development,* **108**, 443.

32. Moens, C. B., Yan, Y.-L., Appel, B., Force, A. G., and Kimmel, C. B. (1996). *Development,* **122**, 3981.

33. Westerfield, M. (1993). In *The zebrafish book,* p. 7.3. University of Oregon Press, Eugene (online version at http://zfin.org/zf_info/zfbook/zfbk.html).

34. Henion, P. D., Raible, D. W., Beattie, C. E., Stoesser, K. L., Weston, J. A., and Eisen, J. S. (1996). *Dev. Genet.,* **18**, 11.

35. Alexander, J., Stainier, D. Y. R., and Yelon, D. (1998). *Dev. Genet.,* **22**, 288.

36. Hatta, K., Kimmel, C. B., Ho, R. K., and Walker, C. (1991). *Nature,* **350**, 339.

37. Grunwald, D. J., Kimmel, C. B., Westerfield, M., Walker, C., and Streissinger, G. (1988). *Dev. Biol.,* **126**, 115.

38. Streisinger, G., Singer, F., Walker, C., Knauber, D., and Dower, N. (1986). *Genetics,* **112**, 311.

39. Neuhauss, S. (1996). Craniofacial development in zebrafish (*Danio rerio*): mutational analysis, genetic characterization, and genomic mapping. In *Fakultät für Biologie.* Eberhard-Karl-Universität Tübingen, Tübingen .

40. Hatta, K. and Kimmel, C. B. (1993). *Pers. Dev. Neurobiol.,* **1**, 257.

41. van Eeden, F. J. M., Granato, M., Odenthal, J., and Haffter, P. (1999). In *Methods in cell biology* (ed. W.H. Detrich, M. Westerfield, and L.I. Zon), Vol. 60, p. 21. Academic Press, San Diego.

42. Kimmel, C., Ballard, W. W., Kimmel, S. R., Ullmann, B., and Schilling, T. F. (1995). *Dev. Dyn.,* **203**, 253.

43. Beattie, C. E., Raible, D. W., Henion, P. D., and Eisen, J. S. (1999). In *Methods in cell biology,* (ed. W.H. Detrich, M. Westerfield, and L.I. Zon), Vol. 60, p. 71. Academic Press, San Diego.

44. Schilling, T. F., Piotrowski, T., Grandel, H., Brand, M., Heisenberg, C.-P., Jiang, Y.-J., Beuchle, D., Hammerschmidt, M., Kane, D. A., Mullins, M. C., van Eeden, F. J. M., Kelsh, R. N., Furutani-Seiki, M., Granato, M., Haffter, P., Odenthal, J., Warga, R. M., Trowe, T., and Nüsslein-Volhard, C. (1996). *Development,* **123**, 329.

45. Piotrowski, T., Schilling, T., Brand, M., Jiang, Y.-J., Heisenberg, C.-P., Beuchle, D., Grandel, H., van Eeden, F. J. M., Furutani-Seiki, M., Granato, M., Haffter, P., Hammerschmidt, M., Kane, D. A., Kelsh, R. N., Mullins, M. C., Odenthal, J., Warga, R. M., and Nüsslein-Volhard, C. (1996). *Development,* **123**, 345.

46. Whitfield, T. T., Granato, M., van Eeden, F. J. M., Schach, U., Brand, M., Furutani-Seiki, M., Haffter, P., Hammerschmidt, M., Heisenberg, C.-P., Jiang, Y.-J., Kane, D. A., Kelsh, R. N., Mullins, M. C., Odenthal, J., and Nüsslein-Volhard, C. (1996). *Development,* **123**, 241.

47. Furutani-Seiki, M., Jiang, Y.-J., Brand, M., Heisenberg, C.-P., Houart, C., Beuchle, D., van Eeden, F. J. M., Granato, M., Haffter, P., Hammerschmidt, M., Kane, D. A., Kelsh, R. N., Mullins, M. C., Odenthal, J., and Nüsslein-Volhard, C. (1996). *Development,* **123**, 229.

48. Baier, H., Klostermann, S., Trowe, T., Karlstrom, R. O., Nüsslein-Volhard, C., and Bonhoeffer, F. (1996). *Development,* **123**, 415.

49. Granato, M., van Eeden, F. J. M., Schach, U., Trowe, T., Brand, M., Furutani-Seiki, M., Haffter, P., Hammerschmidt, M., Heisenberg, C.-P., Jiang, Y.-J., Kane, D. A., Kelsh, R. N., Mullins, M. C., Odenthal, J., and Nüsslein-Volhard, C. (1996). *Development,* **123**, 399.

50. Brockerhoff, S. E., Hurley, J. B., Janssen-Bienhold, U., Neuhauss, S. C. F., Driever, W., and Dowling, J. E. (1995). *Proc. Natl Acad. Sci. USA,* **92**, 10545.

51. Brockerhoff, S. E., Hurley, J. B., Niemi, G. A., and Dowling, J. E. (1997). *J. Neurosci.,* **17**, 4236.

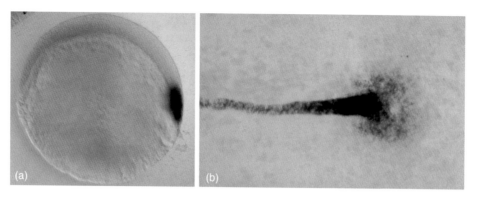

Plate 1 *In situ* hybridization using a single probe. Embryos were processed for *in situ* hybridization as described in *Protocol 1* of Chapter 2. (a) 60% epiboly embryo, dorsal to the right. The embro was hybridized with a digoxigenin-labelled probe against the *chordino* gene (9). Detection of the probe was carried out using alkaline phosphatase and BM Purple as a substrate. The embryo was mounted in a viewing chamber (Chapter 2, Section 4.1) using glycerol as a mounting medium. (b) Tailbud embryo, flat-mount. Anterior to the left, dorsal view. The embryo was hybridized with a digoxigenin-labelled probe against the *no tail* gene (10) and the probe detected as described in (a). Mounting was in benzyl benzoate/benzyl alcohol. Please note that while the stained cells of the notochord and the tailbud are very apparent (even the 'negative' nuclei in the anterior part of the notochord are visible), the remainder of the embryo cannot be seen, due to the strong clearing effect of benzyl benzoate/benzyl alcohol.

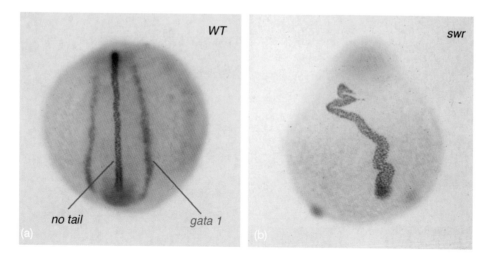

Plate 2 *In situ* hybridization combined with antibody staining. Embryos were processed according to *Protocol 3* of Chapter 2. (a) Five-somite-stage embryo (dorsal view) hybridized with the a digoxigenin-labelled probe against the *gata1* gene, depicted in blue. Subsequently, staining for detection of the No-tail protein was carried out, using a biotinylated secondary antibody and the avidin/biotinylated horseradish peroxidase system as the detection method. DAB was used as a substrate. (b) Mutant *swirl* embryo, processed as in (a). Note the complete lack of background staining, achieved through titration of all components involved in the antibody procedure.

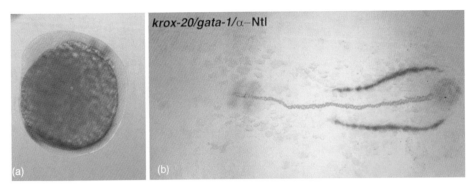

Plate 3 Detection of three different gene products within the same embryo (11). A 5-somite-stage embryo was processed according to *Protocol 5* of Chapter 2. Following hybridization with antisense probes that were labelled with digoxigenin (myoD, depicted in blue) and fluorescein, respectively (Krox–20, depicted in red), the staining reactions were carried out in that order. Subsequently, the protein distribution of the No-tail protein product was visualized by using an anti-No-tail antibody that was detected by a horseradish peroxidase reaction. The embryo was mounted for photography as described in *Protocol 8* of Chapter 2, using araldite as a mounting medium.

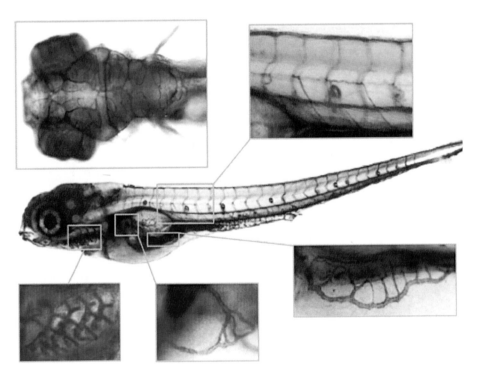

Plate 4 Alkaline phosphatase staining outlining the vasculature of a 4-day-old zebrafish larva. The larva was processed exactly as described in *Protocol 14* of Chapter 2. Clockwise from upper left: head vessels, dorsal view; intersomitic vessels, lateral view; subintestinal vessel, lateral view; vessel of the pectoral fin, lateral view; gill vessels, lateral view.

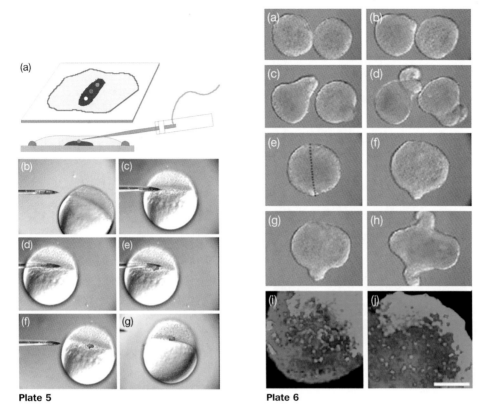

Plate 5

Plate 6

Plate 5 Transplantation technique. (a) Schematic of holding preparation. On a 25 × 25 mm coverslip, embryos are held in a strip of methyl cellulose overlaid with working medium. A ring of Vasoline® traps the droplet on the coverslip. The transplantation pipette approaches at about a 10° angle from the horizontal. (b) Pick-up of donor blastomeres. In this case, not all the collected blastomeres were labelled, stressing the need to check donor blastomeres for actual label. (c) The pipette is positioned to enter host. Normally this is the last movement of the pipette. (d) Using the stage control, the host embryo is moved on to the pipette, being careful to not pierce the yolk cell. To check if the pipette is in proper region of the embryo, a few blastomeres of the host can be drawn back into the pipette. (e) Expelling donor blastomeres. By moving the stage back and forth as the blastomeres are expelled, they can be mixed. (f) Movement of the host embryo off the pipette using the stage control. (g) Example of a transplantation in an early blastula, at the 2K cell stage.

Plate 6 Development of isolated blastoderms. At the 128-cell stage, blastoderms were dissected off yolk cells using cactus needles, allowed to develop in Danieau's medium, and time-lapse recorded in 0.1% agarose. (a–d) A pair of embryoids dissected at 2 h. By the beginning of the experiment, the blastoderm, which after dissection resembles a hemisphere, has fused at the vegetal margin; the little 'tail' grows from the vegetal side. Recordings are at 2.5 h (a), 4 h (b), 7 h (c), and 10 h (d). (e–h) A pair of blastoderms dissected at 2 h and fused vegetal side to vegetal side. The dotted line on the embryoid indicates the border between the two blastoderms. Recordings are at 3 h (e), 4 h (f), 7 h (g), and 10 h (h). (i and j) Examples of the cell mixing that occurs in embryoids between the fused blastoderms. In this experiment, blastoderms were labelled with rhodomine or fluoroscein, and dissected and fused between 2 and 3 h. The embryoids were fixed at 8 h, and the images were produced using a Confocal microscope at 40×. Scale bar: 300 μm for a–h, and 100 μm for i and j.(i and j were kindly contributed by R. M. Warga.)

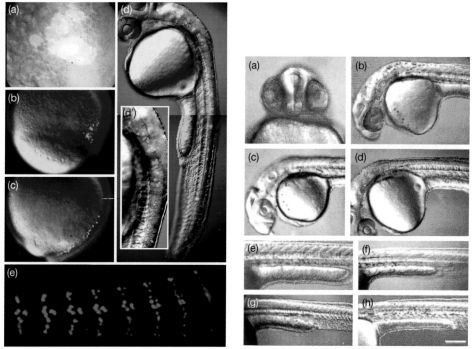

Plate 7   **Plate 8**

Plate 7 Behaviour of donor blastomeres after transplantation. (a) Transplantation of marginal (red) and animal-pole (green) blastomeres to host margin. At 70% epiboly (b) and tailbud (c), marginal donors have involuted and are moving past animal-pole blastomeres. (d) These marginal donors later form muscle at 24 hours. Animal-pole blastomeres make ectodermal derivatives. (e) From a different experiment; spreading and intercalation of transplanted blastomeres as they move into the shield.

Plate 8 Examples of histology of transplantations double labelled with rhodamine (red) and fluorescein (green). (a) Nasal view where one group of transplanted blastomeres formed cells in both neural retinas with a few cells in the telencephalon (red); the co-transplanted blastomeres formed cells in the telencephalon, the neural retina and the lens (green). (b) Side view of head, with muscle and hatching gland (red) and notochord (green). (c) Side view of head, with blood (on yolk sac), endothelium (in head), fin mesenchyme, pronephic duct, and muscle (red) and muscle and mesenchyme (green). (d) Side view of head, with CNS cells, including floorplate (green) and muscle (red). (e) Side view of trunk, a thin row of floor plate and, ventralwards, larger notochord cells (green). (f) Side view of trunk, with pronephric duct and a single endothelium cell on the yolk extension (red) and muscle (green). (g) Side view of trunk, with muscle (red) and mesenchyme (green). (h) Side view of trunk, with muscle and notochord sheath (red) and cells of the spinal cord (green). Clones for a, b, c, and d are sketched in *Plate 11*. Scale bar: 200 μm.
(Figure contributed by R. M. Warga and D.A.K.)

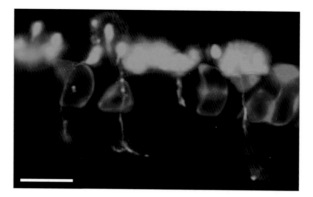

Plate 9 High-power view of donor cells in the ventral spinal cord (green) and in the notochord (red). Motoneurons are extending axons into adjacent somites. Scale bar: 50 μm. (Figure contributed by R. M. Warga.)

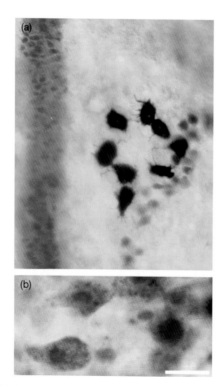

Plate 10 Examples of double-labelled material, using high-power views of endodermal progenitors lateral to the notochord. (a) Notochord progenitors labelled with antibody to Ntl and visualized with biotin/avidin. The biotin/avidin stain also detected a clone of endodermal progenitors that was labelled in the blastula at the 2K cell stage with biotin and rhodamine. Note the light and the dark portions of the clone, caused by leakage of label across the midbody into the sister of the injected blastomere. (b) Endodermal progenitors double labelled with biotin lineage tracer (large purple-blue cytoplasmic staining of blue peroxidase substrate reaction from Vecta stain) and antibody to Forkhead2 (brown nuclear staining of DAB). Note that the nuclei can be detected faintly in the blue intensified cells. Scale bar: 50 μm for a and 30 μm for b. (Figure contributed by R. M. Warga.

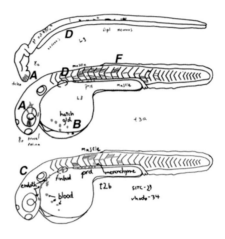

Plate 11 Examples of laboratory notes on a cutaway template. The top portion is a drawing of the medial regions of the 30-h fish, showing the central nervous system and the notochord; the bottom halves are drawings of the outer portions of the fish. The capital letters indicate the associated photograph from *Plate 8*. The cell count was done on the microscope, where more cells could be seen than were actually drawn. (Figure contributed by R. M. Warga and D.A.K.)

(a)   (b)   (c)

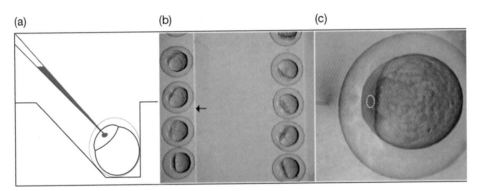

Plate 12 Microinjection of zebrafish embryos. (a) Side-view representation of a embryo sitting in an injection plate made using the mould described in *Figure 1* of Chapter 5. The injection amount and position shown is approximately what we recommend. (b) A series of two-cell stage embryos aligned within such an agarose microinjection plate. The walls of the injection plate trenches, against which the embryos are pressed, can be seen (arrow). (c) An embryo which has just been injected with a DNA solution. The injection tracer, which in this case is Phenol Red, has diffused but the original injected bolus size is indicated by a hatched circle. The injection needle can be seen at the left edge of the picture.

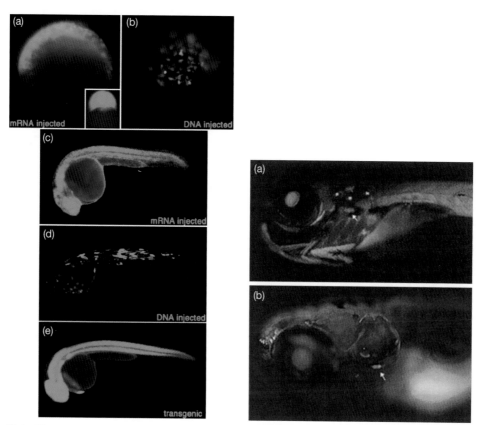

Plate 13   **Plate 14**

Plate 13 Comparing expression from injecting mRNA and DNA encoding GFP. (a) An embryo injected with 40 pg of GFP mRNA shows strong widespread expression by the shield stage. In fact, strong GFP fluorescence can be observed by 2–2.5 h (128–256-cell stage) after injection (inset). (b) In comparison, expression from DNA injection is less widespread and by shield stage only scattered cells are fluorescent. (c) By 24 h mRNA-injected larvae show strong GFP expression throughout the entire bod, whereas DNA injection gives a highly mosiac expression pattern and individual expressing cells can be recognized, in particular muscle cells (d). (e) Shows a transgenic embryo that carries a transgene driving GFP under the control of a ubiquitously expressed promoter (EF1-α).

Plate 14 GFP expression in germ line transgenic zebrafish. (a, b) Lateral views with anterior to the left and dorsal to the top. (a) GFP expression in lymphoid (arrow indicates thymus) and several non-lymphoid tissues of a 7-day embryo containing a small *rag1* promoter construct transgene. (b) In contrast, germ line transgenic embryos containing the GFP-modified *rag1* artificial chromosome express GFP in a pattern representing the endogenous gene.

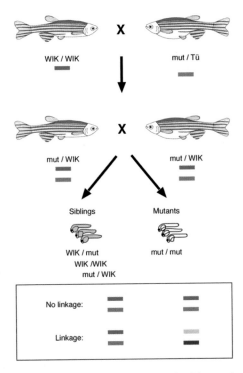

Plate 15 Map cross and bulked segregant analysis of a recessive zebrafish mutation. WIK/WIK, reference fish; mut/Tü, fish carrying the mutation in Tü background. Band sizes and intensities of a representative SSLP marker are indicated schematically. In case of no linkage between the mutation and the marker, the intensities in the mutant and sibling pool are the same. In case of a linkage, the Tü band is stronger in the mutant pool and the WIK band in the sibling pool. Only a quarter of F_1 crosses consist of two mutant carriers and yield mutant F_2 as shown, the others are discarded.

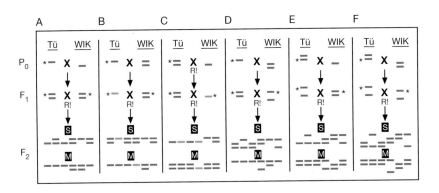

Plate 16 Possible allele systems of an SSLP. S, siblings; M, mutants; RI, recombination; red, Tü bands; blue, WIK bands; the asterisk indicates linkage with the mutation. The band sizes are arbitrary; Tü bands may occur above, below, or in between WIK bands. In each example, two out of six mutants are recombinant (17% recombination). (A) Two bands. (B) Homozygous F_1 fish, in the mutants only half of the recombinations are visible. (C) Homozygous F_1 fish with a recombination in the P_0, again only half of the recombinations are visible. (D) Three bands. The lower bands should be scored as one allele. (E) Three bands with a recombination in the P_0. The upper bands should be scored as one allele. (F) Four bands. The upper and lower bands should be scored as one allele. (Modified from ref. 26.)

Chapter 7
Mapping and cloning

Robert Geisler

Max-Planck-Institut für Entwicklungsbiologie, Abteilung Genetik, Spemannstr. 35, 72076 Tübingen, Germany

1 Maps of the zebrafish genome

In order to identify zebrafish genes affected by mutations, both mutations and candidate genes are commonly placed on genome maps (*Figure 1*). Two types of maps are relevant for this approach: genetic maps, in which the probability of meiotic recombination between markers is used to determine their distance and construct the map; and radiation hybrid maps, in which breakage induced by radiation takes the place of recombination.

Mutations can be placed only on genetic maps. This type of map is produced by scoring a large number of polymorphic markers on a panel of F_2 fish from a reference cross. Two markers are close to each other if they give a similar pattern of alleles across the panel. Mutations are placed on the map indirectly, by scoring previously mapped markers on a panel of mutant

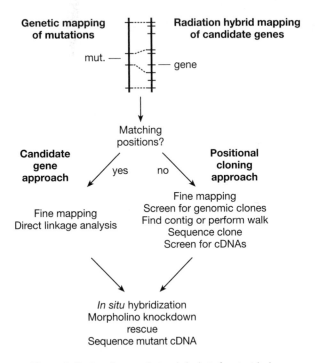

Figure 1 Strategy for mapping and cloning of mutant loci.

embryos and estimating the map position of the mutation from the observed linkages (for details see Section 2). The most useful type of markers for this purpose are currently SSLPs (simple sequence length polymorphisms, also called microsatellites). These markers consist of a $(CA)_n$ repeat of variable length flanked by two primer sequences (1, 2). SSLPs are frequent, scorable on standard agarose gels, robust and co-dominant, i.e. both alleles can be scored separately in a diploid individual, providing more information than a dominant marker. A contiguous map of the zebrafish genome has been produced by scoring SSLPs on the MGH panel (3–5); currently this map contains over 3000 SSLPs. A distance of 1 cM (centimorgan) on the MGH panel is equivalent to approximately 660 kb, and the resolution (determined by the panel size) is 1.1 cM, or approximately 690 kb. Another map of the zebrafish genome was produced using primarily RAPD (random amplified polymorphic DNA) markers. These markers consist of primers flanking arbitrary genomic sequences, and display either a polymorphic length, or amplify only in one line due to a polymorphism in one of the primer sequences. In the latter case they behave as dominant markers, i.e. only presence or absence can be scored. Therefore this map was produced on a panel of haploid fish, the HS panel (6). Smaller numbers of markers were scored on the MOP panel (7) and the GAT panel (W. Talbot, unpublished; Table 1).

Candidate genes have been mapped genetically as well as on radiation hybrid panels. Genetic mapping is a relatively slow and uncertain procedure, because it requires identification of a polymorphism associated with the gene which is then scored on a mapping panel, and it will not be further discussed here. Radiation hybrid panels consist of lines of somatic cell hybrids produced by fusion of lethally irradiated cells of the species of interest with a rodent cell line. Random fragments of the chromosomes are integrated into hamster chromosomes or retained as separate minichromosomes (8, 9). Markers are scored for presence or absence in DNA from the hybrid lines, and markers with similar patterns of positives are placed close to each other on the map. In contrast to genetic mapping, which relies on the fixed meiotic recombination rate, the breakage frequency can be controlled through the radi-

Table 1 Mapping panels

Panel	Type	Mapped markers[a]	Laboratory	Website
Goodfellow T51 (T51)	Radiation Hybrid	7722	R. Geisler, Y. Zhou	http://wwwmap.tuebingen.mpg.de http://zfrhmaps.tch.harvard.edu/ZonRHmapper [b]
Loeb/NIH/ 5000/4000 (LN54)	Radiation Hybrid	4050	I. Dawid	http://eclipse.nichd.nih.gov/ nichd/lmg/lmgdevb.htm
Boston MGH Cross (MGH)	Meiotic	3462	M. Fishman	http://zebrafish.mgh.harvard.edu
Heat shock (HS)	Meiotic	2812	J. Postlethwait, W. Talbot	http://www.neuro.uoregon.edu/postle/ mydoc.html http://zebrafish.stanford.edu/ genome/Frontpage.html [c]
Mother of pearl (MOP)	Meiotic	466	J. Postlethwait	http://www.neuro.uoregon.edu/postle/ mydoc.html
Gates et al. (GAT)	Meiotic	398	W. Talbot	http://zebrafish.stanford.edu/ genome/Frontpage.html

[a] Source: ZFIN database as of 20 May 2001.
[b] Websites are updated alternately.
[c] Websites point to the same maps.

ation dose, allowing higher resolution, and map distances more closely reflect actual physical distances. Because donor and host cells are derived from different species, almost any amplifiable marker can be scored as a dominant marker, without a need for polymorphism. Two radiation hybrid panels are available for the zebrafish: the Goodfellow T51 panel (10–12) and the Ekker LN54 panel (13, 14). On the Goodfellow T51 panel, 1 cR (centiray) is equivalent to approximately 60 kb, and the potential resolution is approximately 6 cR or 350 kb. The potential resolution of the LN54 panel is approximately 1 Mb. Additional markers are continuously being added to both the T51 and the LN54 map, primarily genes represented by ESTs (expressed sequence tags); the T51 map contains the largest number of markers of any zebrafish map.

Data obtained by radiation hybrid mapping can be submitted to a public mapping service for evaluation (for details, see Section 3). However, some markers (currently about 20%), cannot be successfully mapped on the T51 panel because they fall in gaps of the map. Since a lower resolution translates into significant linkage over larger physical distances, the LN54 panel allows positioning of most of these markers.

2 Mapping mutations using SSLP markers

2.1 Mapping approaches

For placing a zebrafish mutation on a genetic map, two methods are available: half-tetrad analysis and genome scanning.

In the first method, gynogenetic embryos are produced from a map cross by activating eggs with irradiated sperm and applying pressure to inhibit the second meiotic division ('early pressure', EP). In these embryos, a mutation shows strong linkage to a centromeric marker on the same chromosome independent of its map position, because a single recombination between the centromere and the chromosome would remove the mutant phenotype and multiple recombinations are rare (strong chiasmatic interference). Thus, a mutation can be localized roughly using only one SSLP per chromosome (15–18).

For genome scanning, a map cross between a line carrying the mutation (e.g. in Tübingen (Tü) background) and a reference line (e.g. WIK) is set up, brother–sister matings are performed in the F_1 generation and F_2 embryos are sorted by phenotype. The mutant embryos and their wild-type siblings are then scored for a set of SSLPs that covers all chromosomes at approximately equal distances. If an SSLP shows an increased frequency of the Tü allele in the mutant F_2 embryos, it is linked to the mutation. We will restrict our further discussion to this method.

A direct count of the allele frequency of each marker in the mutant embryos would be very time consuming, as well as impractical, with early zebrafish embryos which would not yield enough DNA to complete a genome scan. A convenient shortcut is pooled PCR, also known as 'bulked segregant analysis' (19, 20): DNA of a number of mutant embryos and wild-type siblings is pooled, only one PCR is performed with the mutant and sibling pool for each marker, and the intensities of the resulting bands are evaluated. An increased intensity of the Tü-derived band in the mutant embryos in comparison to the siblings indicates a potential linkage to the SSLP in question (*Figure 2*, see also *Plate 15*). It is necessary to use the siblings as a control because marker alleles often have intrinsic differences in intensity.

While map distances can be calculated from the band intensities, the bulked segregant approach on its own tends to give false positives (especially when considering weak linkages) because of slight differences in DNA preparation, background fluorescence, band irregularities and quantitation errors. Improvements such as electrophoresis on a capillary sequencer and a higher number of markers may allow a confident detection of linkages based on the bulked segregant approach alone. However, for the time being, it is prudent to confirm all potential linkages found by this approach by genotyping single embryos. If required, the map positions are later refined by typing more embryos and using additional SSLPs from the same genetic region.

2.2 Selection of zebrafish lines and SSLPs

Several inbred zebrafish lines are available for map crosses, but none of them is completely isogenic, and all share marker alleles to some extent. Most mutations have been induced in the Tübingen (Tü) line, for which the WIK line (21) is commonly used as a reference line. On average, 68% of the SSLPs from the set described below display a scorable polymorphism in a Tü × WIK cross. An alternative is the partially isogenized SJD line (22), which displays 75%

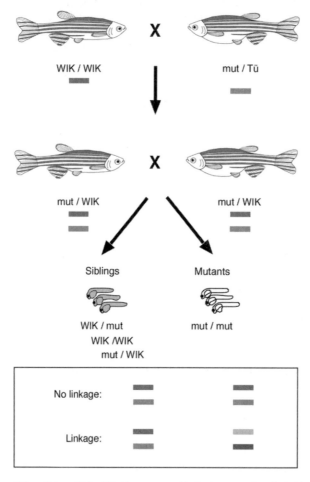

Figure 2 (see Plate 15) Map cross and bulked segregant analysis of a recessive zebrafish mutation. WIK/WIK, reference fish; mut/Tü, fish carrying the mutation in Tü background. Band sizes and intensities of a representative SSLP marker are indicated schematically. In case of no linkage between the mutation and the marker, the intensities in the mutant and sibling pool are the same. In case of a linkage, the Tü band is stronger in the mutant pool and the WIK band in the sibling pool. Only a quarter of F_1 crosses consist of two mutant carriers and yield mutant F_2 as shown, the others are discarded.

polymorphism in crosses with Tü. However, SJD is more difficult to maintain because of a skewed sex ratio (a common problem in strongly inbred zebrafish lines) and should be considered primarily if a mutation fails to be mapped with Tü × WIK. Two other zebrafish lines, AB (23) and TL (24), have a lower rate of polymorphism relative to Tü than WIK. The India (IN) line, which was used together with AB for the generation of the MGH panel, is probably not free of mutations affecting larval viability (4), as well as being harder to breed than WIK (21). The KC line from King's College, London has only been used as a source for cosmids. Wild-type lines can be obtained from the Zebrafish International Resource Center in Eugene, Oregon: <http://zfin.org/zf_info/stckctr/stckctr.html> or from the Tübingen Zebrafish Stockcenter <http://www.eb.tuebingen.mpg.de/abt.3/stockcenter/zebraf_stockcenter.html>.

To scan the genome for linkage to a mutation, we have selected a set of 192 SSLP markers from the MGH map that give strong and easily distinguishable PCR products (as determined on agarose gels) when amplified from the Tü line, and from the reference line WIK (*Table 2*). Each marker in the current set has been used successfully in the mapping of at least one Tübingen mutation. Since the genome size of the zebrafish is approximately 2500 cM (25), the average marker spacing is 13 cM. The limit for detection of linkages as described below is approximately 36 cM, so our SSLP set provides three times genome coverage. A different SSLP

Table 2 Tübingen marker set for genome scans (ver. 4). These SSLPs are recommended for genome scanning in Tü × WIK and electrophoresis on agarose gels. Primer sequences are available from the MGH website: <http://zebrafish.mgh.harvard.edu>

LG	SSLPs
1	Z4593, Z9394, Z5508, Z1705, Z1351, Z9704, Z11464, Z6802, Z1781
2	Z7634, Z4662, Z3430, Z1406, Z6617, Z1703, Z20550
3	Z872, Z8208, Z15457, Z9964, Z11227, Z3725, Z20058, Z6019
4	Z1525, Z9920, Z21636, Z7490, Z984
5	Z15414, Z11496, Z6727, Z10456, Z1390, Z3804, Z14143, Z4299, Z1202
6	Z740, Z13275, Z880, Z6624, Z10183, Z5294, Z13614, Z7666, Z4297, Z1680
7	Z3273, Z10785, Z1206, Z4706, Z1182, Z1059, Z8156, Z1239, Z13880, Z13936, Z5563
8	Z1634, Z1068, Z4323, Z13412, Z21115, Z789, Z10929, Z3526
9	Z1777, Z6268, Z4673, Z5080, Z1805, Z20031, Z10789, Z4577
10	Z9199, Z6410, Z8146, Z13632, Z1145, Z9701, Z3260
11	Z10919, Z3362, Z13411, Z1393, Z3527, Z1590
12	Z1778, Z21911, Z1473, Z4188, Z1358
13	Z1531, Z5643, Z6104, Z13611, Z5395, Z1627, Z7102, Z6657, Z1826, Z6007
14	Z1523, Z5436, Z1536, Z5435, Z4203, Z22107, Z1226, Z3984, Z1801
15	Z6312, Z6712, Z21982, Z4396, Z11320, Z13230, Z13822, Z7381, Z5223
16	Z3741, Z21155, Z6365, Z10036, Z1215, Z4670
17	Z4268, Z1490, Z22083, Z22674, Z9847, Z1408, Z4053
18	Z1136, Z1144, Z13329, Z8488, Z10008, Z3558, Z9154, Z5321
19	Z4009, Z160, Z3782, Z3816, Z11403, Z6661, Z7926, Z1803
20	Z9334, Z10056, Z11841, Z3964, Z7158, Z3954, Z22041, Z8554, Z4329
21	Z3476, Z1274, Z4492, Z10960, Z4425, Z1497, Z4074
22	Z1148, Z10673, Z9402, Z230, Z10321, Z21243
23	Z8945, Z4003, Z15422, Z4421, Z3157, Z176, Z1773
24	Z5075, Z1584, Z5413, Z23011, Z3399, Z22375, Z5657, Z3901
25	GOF15, Z1378, Z3490, Z5669, Z1462

set optimized for AB ×IN map crosses is available from the MGH website: <http://zebrafish.mgh.harvard.edu/cgi-bin/ssr_map/bulkseg_list.cgi>.

2.3 Mapping strategy

To map a mutation, set up map crosses and collect embryos according to the scheme given in *Figure 2* (see also *Plate 15*). Set up several F_1 crosses and keep only those that yield mutant progeny. F_2 embryos should be collected from at least two F_1 crosses and, if possible, from two map crosses to increase the chance that markers close to the mutation have a scorable polymorphism. The protocols for setting up crosses and collecting embryos have been described elsewhere in this book.

Because a large number of F_2 embryos need to be lysed for mapping, it is convenient to perform the lysis on 96-well plates using a very simple method (*Protocol 1*). Adult fins can by lysed in a similar way (*Protocol 2*). For bulked segregant analysis, embryos can either be lysed as a pool in an Eppendorf tube, or aliquots of single-embryo lysates can be pooled. The second approach has the advantage that the single embryos remain available for further genotyping. The unpurified lysates are used as PCR templates (*Protocol 3*). Alternatively, they may be purified, e.g. on spin columns, but we found that this results in little improvement for our SSLPs. Electrophoresis of PCR products is performed on agarose gels with ethidium bromide.

Protocol 1

Quick lysis of embryos

Equipment and reagents

- Safelock Eppendorf tubes
- Glass Pasteur pipettes
- 40 mm Petri dishes
- 500 ml dispenser bottle
- PCR plate
- Disposable plastic tray
- Sealing film for PCR plates (e.g. ThermoSeal PP, Sigma)

- Thermocycler or incubator
- Multipipette
- Map cross F_2 embryos
- 100% methanol (MeOH)
- 17 mg/ml proteinase K in double-distilled water (ddH$_2$O)
- 1 × TE (10 ml Tris–HCl, pH 8.0, 1 mM EDTA)

Method

1 Set up crosses and sort F_2 embryos by phenotype as described in the screening chapter. Transfer to Eppendorf tubes with 100% MeOH and store at −70 °C until use.[a]

2 Shake storage tubes with embryos and MeOH and pour them into a 40 mm Petri dish. Rinse tubes with MeOH (using a dispenser bottle). Add more MeOH to the dish if necessary so that the embryos remain covered.[b]

3 Pipette single embryos on a PCR plate with a glass Pasteur pipette or truncated 200 μl tip. Make sure that each embryo sinks to the bottom of the well. Return leftover embryos to one of the tubes and return the tube to the freezer. Use a fresh dish and pipette for the next mutant.[c]

4 Let MeOH evaporate on a PCR block at 70 °C while preparing the next plate (approximately 10 min).

5 In a plastic tray mix 250 μl 17 mg/ml proteinase K and 2.25 ml 1 × TE per plate. With a multipipette, add 25 μl to each well. Cover with sealing film and heat in a thermocycler or incubator:[d]

55 °C for 240 min

94 °C for 10 min.

Protocol 1 continued

6   With a pipetting robot or multipipette, pool 10 µl each of the sibling and mutant lysates, and add 45 µl sterile ddH$_2$O to the remainder. If no pooled DNA is required, just add 75 µl sterile ddH$_2$O.[e]

7   Cover plate with sealing film, and freeze at -20 °C.

[a] We label tubes with allele, F$_1$ cross, 'mut' or 'sibs' and the number of embryos. The same information is entered in the laboratory database. It is essential to keep F$_1$ crosses separate, as they may have different polymorphisms.

[b] For good results embryos must be stored in MeOH; this probably helps to disrupt the tissue. The embryos can be kept for several years, and remain usable for *in situ* hybridization and RNA preparation. This protocol does not work reliably for adult tissue.

[c] We generally place 48 siblings in row A–D of a plate and 48 mutant embryos from the same F$_1$ cross in row E–H. Rows are labelled accordingly on the rim of the plate. Each plate is also assigned a number, which is indicated twice on the rim of the plate, as well as entered in the database.

[d] Rub over sealing film with a cloth to ensure that wells are sealed.

[e] Dilution with ddH2O is necessary because 1 × TE inhibits the *Taq* polymerase. Each embryo will allow at least 20 PCR reactions. For stages after hatching, lysates may be aliquoted and further diluted if desired. In this case, PCR performance should be tested with a few markers to determine the appropriate dilution.

Protocol 2

Quick lysis of adult fins

Equipment and reagents

- Scalpel
- PCR plates
- Disposable plastic tray
- Sealing film for PCR plates (e.g. ThermoSeal PP, Sigma)
- Shaker
- Centrifuge for microtitre plates
- Multipipette

- Vacuum chamber
- Map cross P$_0$, F$_1$, or F$_2$ fish
- DNAzol (Molecular Research Center)
- 17 mg/ml proteinase K in ddH$_2$O (aliquots stored at −20 °C)
- 80% and 100% ethanol (EtOH)
- Sterile ddH$_2$O

Method

1   Set up crosses and sort fish by phenotype as described in the screening chapter. Freeze single fish in microtubes at−70 °C until use; or briefly anaesthetize live fish with 0.04% MESAB.

2   Cut one-half of each tail fin with a scalpel and place in a separate well of a PCR plate. After each use, dip the blade in EtOH and let it burn off. Return fish to freezer or to fish tank immediately.[a]

3   In a plastic tray, mix 240 µl 17 mg/ml proteinase K and 3.76 ml 1 × DNAzol per plate. With a multipipette, add 40 µl to each well. Cover plate with sealing film and shake overnight at room temperature, then spin for 10 min at 4000 rpm, and transfer the supernatant to a fresh plate.[b]

4   With a multipipette, add 20 µl 100% EtOH to each well, and spin for 15 min at 4000 rpm.[c]

5   Remove supernatant, add 60 µl 80% EtOH, and spin for 5 min at 4000 rpm. Repeat once.

6 Remove supernatant, dry for approximately 5 min in a vacuum chamber, and add 25 μl sterile ddH$_2$O.[d]

7 With a pipetting robot or multipipette, pool 10 μl each of the sibling and mutant lysates, and add 45 μl ddH$_2$O to the remainder. If no pooled DNA is required, just add 75 μl ddH$_2$O.[e]

8 Cover plate with sealing film, and freeze at −20 °C.

[a] This protocol also works for small amounts of other adult tissues.

[b] Rub over sealing film with a cloth to ensure that wells are sealed.

[c] Because DNAzol promotes precipitation of DNA, ½ volume of EtOH is sufficient.

[d] Watch the plate to prevent overdrying.

[e] Each preparation will allow at least 20 PCR reactions.

Protocol 3

Mutant PCR and electrophoresis

Equipment and reagents

- 96-well PCR plate
- Sealing film
- Thermocycler
- Agarose electrophoresis system
- 10 × PCR buffer (100 mM Tris–HCl (pH 8.3), 500 mM KCl, 15 mM MgCl$_2$ and 0.1% (w/v) gelatin, stored at 4 °C)
- 100 mM dATP, dCTP, dGTP, dTTP stocks (prepared from solid substance, stored at −20 °C)
- 20 μM forward and reverse primer (stored at 4 °C)
- 5 U/μl *Taq* polymerase

- 20 ng/μl pooled DNA or embryo lysates (from *Protocol 1*)
- 3 × Ficoll loading buffer (7.5% Ficoll, 0.125% Bromophenol Blue) with approximately 10 ng/μl undigested plasmid DNA as a loading control
- 10 ng/μl of 100 bp ladder (Pharmacia) in 1 × Ficoll loading buffer
- 1 × TBE (90 mM Tris-borate, 90 mM boric acid, 2 mM EDTA, pH 8.0)
- Metaphor agarose (Promega)
- Qualex Gold agarose (AGS)
- 10 mg/ml ethidium bromide

Method

1 On a PCR plate set up 20 μl reactions:

  (a) 14.28 μl reaction mix (2 μl of 10 × PCR buffer, 0.04 μl each of 100 mM dATP, dCTP, dGTP and dTTP, 12.12 μl water).[a]

  (b) 0.16 μl each of 20 μM forward and reverse primer.[b]

  (c) 0.4 μl of 5 U/μl *Taq* polymerase.

  (d) 5 μl of template DNA (either 20 ng/μl pooled DNA or ¹/₂₀ of an embryo).

2 Cycling:

  (a) Initial denaturing at 94 °C for 2 min.

  (b) 35 cycles of denaturing at 94 °C for 30 s, annealing at 60 °C for 30 s and extension at 73 °C for 1 min.

  (c) Final extension at 73 °C for 5 min.

3 Electrophoresis:

(a) Pour 1% Metaphor + 1% Qualex Gold agarose gels containing 1 × TBE and 0.8% ethidium bromide.[c]

(b) Add 5 μl of 3 × loading buffer to each sample. For each marker, run one lane with 80 ng of 100 bp ladder as a size standard.[d]

(c) Run at 200 V for 100 min in 1 × TBE buffer with 0.4% ethidium bromide.[e]

[a] Reaction mix is stored in 1.5 ml aliquots at −20 °C. Each aliquot is sufficient for two plates. Before distributing to the plates, the robot adds Taq polymerase, divides each aliquot in half, and adds the primers.

[b] Over extended periods, primer DNA can be stored at −20 °C rather than 4 °C, but it should not be thawed too often.

[c] We heat 3.6 litres of agarose in a steel bucket on a hotplate set to 225 °C while stirring with a magnetic stirrer until it has cleared (approximately 2 h), let it cool for 30 min, and add ethidium bromide while continuing to stir. The ends of the gel trays are sealed with sticky tape, combs are inserted and the trays placed in a rack one by one starting from the bottom, pouring 200 ml gel in each with a beaker. Special care is taken to apply sufficient gel to the top and bottom of each tray. Large bubbles are pushed aside with a pipette tip. Racks are covered with a plastic bag after the gels have set, and can be stored at room temperature for several days.

[d] We use gels with 4 × 30 lanes each and apply samples with a 12-channel CAPP multipipette. The lanes are half as wide as the distance between microtitre wells, so that PCR products from two plate rows are interspersed on the gel.

[e] If the same gel box was used before, reduce the run time to 90 or 80 min, depending on the temperature of the TBE. Electrophoresis is faster at a higher temperature. TBE can be re-used up to five times.

After testing mutant and sibling pools of a mutation with the entire set of 192 markers, evaluate the band intensities to find potentially linked markers. Attempt to confirm the most promising potential linkages by PCR of the 48 siblings (used only as a control) and 48 mutant embryos (from which a map distance is calculated). We initially try to confirm the best three potential linkages, and if this fails, another three. Once a linkage has been found, try all nearby markers that have not been shown to be homozygous, in order to find the two closest markers, and calculate a map position from the linkage to these markers, as described below. If no linkage can be confirmed after exhausting all potential linkages, it may be worth performing another genome scan with all 192 markers and a different map cross.

2.4 Set-up for high-throughput mapping

Pipetting on 96-well plates can be performed with a 12-channel multipipette. However, if more than a few mutations are to be mapped, a pipetting robot is very helpful to reduce strain and the probability of the pipetting errors. In Tübingen, a Biomek 2000 (Beckman Coulter) is used. Cooled labware adapters (from Beckman or home-built to a similar design), attached to a thermostat, are used to keep reagent tubes and microtitre plates on the worksurface below 10 °C. The cooling is not absolutely required, but increases the reliability of the PCR and the lifetime of reagents. We run 12 plates in parallel on TouchDown (Hybaid) cyclers.

In Tübingen home-built gel boxes, gel trays, combs and storage racks are used for mutant and radiation hybrid mapping gels (Figure 3). The trays are made of acrylic, have a size of 400

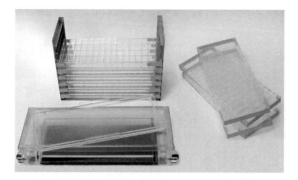

Figure 3 Electrophoresis equipment. Background, gel trays and storage rack. Foreground, gel box and UV-transparent tray for imaging.

× 135 mm and 16 grooves for inserting combs. For mutant mapping, four combs with 30 teeth each, made of heat-stable polycarbonate (Tecanat, Ensinger), are inserted. Samples are applied only to the middle 24 wells (plus one for a molecular weight standard) to reduce edge effects. Metal racks holding up to nine gel trays are used for space-saving gel pouring and storage. After running the electrophoresis, gels are transferred to a clean tray of slightly larger dimensions made of UV-transparent acrylic, with handles and closed ends as well as a slightly raised base, to avoid picking up contaminations from the bench. This tray is placed on the transilluminator against the guide rail and photographed in different positions, sliding it along the rail. To reduce the volume of toxic waste, gels can be dried after use. For this purpose we transfer them to perforated metal trays after photography, place them in a rack of the same type used for storage, and dry them in a fume hood. Metal trays and drying racks are carried in a large metal pan to avoid contaminating the lab with toxic dust. A silicon lubricant is occasionally sprayed on the metal trays to prevent gels from sticking.

Digital gel imaging systems are commercially available, but a home-built set-up may offer greater flexibility and better integration with an existing computer network. We capture images with NIH Image 1.61 (developed at the US National Institutes of Health) on a Power Macintosh G3 computer equipped with a Cohu video camera and a Scion LG digitizer board, using on-chip integration to increase sensitivity. This set-up is described in the NIH Image manual. A high-quality bandpass filter (e.g. Thermo Corion) is essential; we found that red or orange Wratten filters, commonly used in photography, do not reduce background sufficiently when used with some UV transilluminators. To simplify handling of long gels, the camera is mounted such that 'left' on the transilluminator is 'up' on the image. A guide rail made of acrylic is glued on the back of the transilluminator surface to help positioning the gel trays. We have created NIH Image macros for capturing gel images and semi-automatic scoring of the images. Consecutive images are numbered automatically and saved with a single mouse-click. For this purpose a trackball is placed in the 'dirty' part of the gel room, so that images can be captured without contaminating the computer, keyboard or mouse.

For high-throughput mapping, a laboratory database is necessary. We found that FileMaker Pro offers a high degree of flexibility and ease of use, but the choice of database management software should depend on the familiarity of researchers with a particular program and on locally available bioinformatics support. In Tübingen, FileMaker Pro database files are used to keep track of mutants (including P_0, F_1 and F_2 fish and template DNA), markers, pooled PCRs and single-embryo PCRs, respectively. All template and PCR plates are entered into the database and assigned identification numbers before they are set up. For PCR plates, sticky labels (Avery) are printed from the database and applied to the plates. Digital gel images are saved with filenames matching the plate numbers, and quantification and scoring results obtained with NIH Image are imported into the corresponding database records. In the pooled PCR database file, one record represents two wells on a PCR plate, namely a sibling pool and the corresponding mutant pool, tested with one marker. This allows us to identify the most promising potential linkages by a simple find and sort operation. For single-embryo PCR, one record represents two adjacent rows on a PCR plate. Map distances are calculated by summarizing recombinations over all rows of one mutant tested with a particular marker.

When mapping several mutations, the pooled PCR is best performed in parallel, so that one gel image shows sibling and mutant pools of several different mutations, all tested with the same marker. This makes it easier to distinguish marker alleles from artefacts. In Tübingen, sets of 24 or 48 mutations are tested in parallel, filling either half a microtitre plate or a full plate with sibling and mutant pools. *Figure 4* illustrates the pipetting scheme for a set of 24 mutations. PCR, electrophoresis, and imaging of nine pooled or single-embryo plates can be performed by a researcher in parallel in approximately 3 h.

2.5 Evaluating pooled PCR gels

In pooled PCR experiments, a marker is potentially linked if the ratio of the band intensities in the mutant pool is different from that in the sibling pool. Intensities can be quantified in NIH Image using a method described in the chapter 'Analyzing electrophoretic gels' of the NIH Image Manual, which consists of plotting intensities along a lane, drawing a baseline, and integrating the area between a peak (i.e. a band) and its baseline. If two bands are not

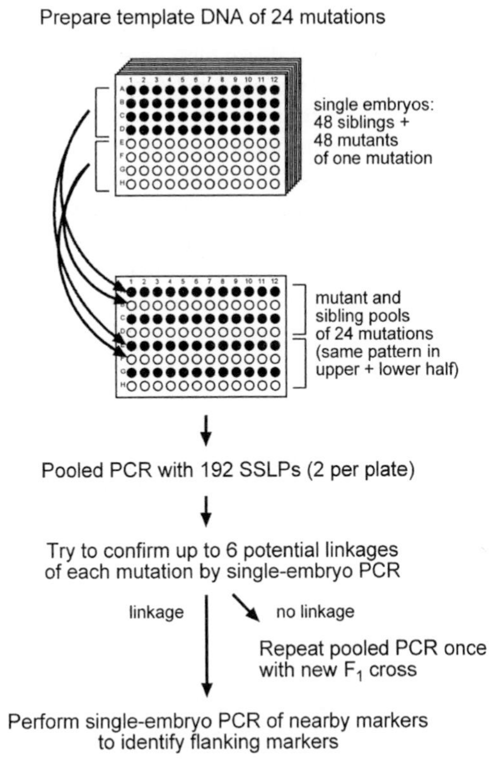

Figure 4 Pipetting scheme and PCR strategy.

completely separated, a common baseline is drawn, and they are separated by a vertical line drawn at the lowest point of the plot between them.

Because this method is too laborious for routine use, we have developed an NIH Image macro that further automates it. After making a selection with the mouse to indicate the approximate position of the bands, the macro automatically finds the peaks, calculates the baselines, integrates the intensities and writes the results to a text file with a label that matches the record number in our database. In contrast to the original method, we calculate a separate baseline for each peak even if the intensity between the peaks drops only slightly. This accentuates intensity differences in small peaks located on the shoulder of bigger ones. Also, we do not calibrate each gel, but assume that our set-up is approximately linear (using the calibration setting 'Uncalibrated OD'). We believe that quantification errors introduced by non-linearity of the imaging set-up are small compared to ones arising from factors such as non-linear amplification and irregularities of the gels. Other software for quantifying bands is commercially available.

To calculate a recombination rate, θ, from pooled PCR results of a recessive mutation, the ratio of the intensities of the two sibling bands, r_s, as well as that of the two mutant bands, r_m, is determined. Then the following formula is applied, which takes into account how both ratios change in the presence of linkage:

If $r_m < r_s$:

$$\theta = (1.5r_m + 0.5r_s)/(r_m - r_s) = \{[(1.5r_m + 0.5r_s)/(r_m - r_s)]^2 - [2r_m/(r_m - r_s)]\}^{1/2}$$

If $r_m > r_s$:

$$\theta = (1.5/r_m + 0.5/r_s)/(1/r_m - 1/r_s) = \{[(1.5/r_m + 0.5/r_s)/(1/r_m - 1/r_s)]^2 - [2/r_m/(1/r_m - 1/r_s)]\}^{1/2}$$

After all the pooled PCR results have been quantified, potential linkages with a recombination rate up to 3% are visually re-evaluated, and classified as good if one band is stronger in the mutants, while the other is stronger in the siblings, and amplification and image quality are good; as marginal if the band intensities are only slightly different, amplification is weak, or the gel shows smear or other artefacts; or as bad if the supposed recombination was a false positive caused by dirt or gel background. *Figure 5* shows an example of a good potential linkage.

In addition, any potential linkage that is seen but cannot be quantified, e.g. because the bands are not separated or because of gel problems, is immediately entered in the database as good or as marginal, using the same criteria as described above. For a small mutant mapping project it is probably sufficient to evaluate the bands only visually rather than quantifying them, although some marginal potential linkages may be difficult to find.

2.6 Evaluating single-embryo PCR gels

Pooled PCR is subject to several kinds of error. Therefore in Tübingen all pooled PCR results are confirmed by genotyping single individuals, and map distances and LOD scores are calculated only from these genotypes. The downside of this approach is that some experience and understanding of the allele systems is required to interpret the band patterns correctly and avoid false positives.

Genotyping results of mutant F_2 embryos should be verified either by genotyping the P_0 and F_1 fish, or by comparing a number of F_2 siblings to the mutants. P_0 and F_1 DNA can be prepared from fins (*Protocol 2*) or whole adults (*Protocol 4*). However, genotyping of F_2 siblings is preferred in Tübingen, because DNA preparation from adults requires additional effort and the fish may occasionally be lost before freezing. Also, genotyping the siblings together with the mutants allows a direct comparison of the band patterns on the gels and gives more confidence that the patterns are interpreted correctly.

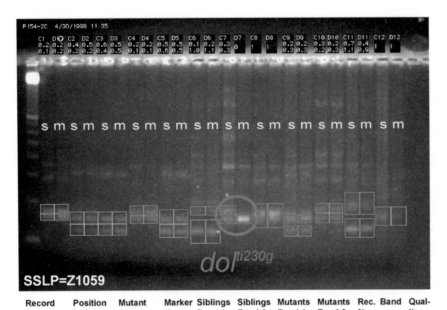

Record	Position in Set	Mutant Allele	Marker Name	Siblings Band 1	Siblings Band 2	Mutants Band 1	Mutants Band 2	Rec. %	Band	Quality
P154-2C7	M17C7	AI30G	z1059	0.271	0.353	0.000	1.160	0	2	xxx

Figure 5 Pooled PCR gel. Top, agarose gel with pooled PCR products of 12 mutants (48 mutant embryos and siblings each) for the SSLP Z1059. The mutation *dol*[ti230g] has a strong potential linkage to Z1059, because the sibling pool (s) shows two bands and the mutant pool (m) only one band. Bottom, database record with the measured band intensities. The potential linkage was classified as good (three stars) and was subsequently confirmed by single-embryo PCR (26).

Protocol 4

DNA isolation from whole adult fish (after ref. 29)

Equipment and reagents

- Mortar and pestle[a]
- 50 ml Falcon tubes
- Centrifuge for Falcon tubes
- Pasteur pipette with heat-sealed tip
- Liquid nitrogen
- Lysis buffer (10 mM Tris–HCl, pH 8, 100 mM EDTA, 1% SDS)
- 17 mg/ml proteinase K in ddH$_2$O (aliquots stored at -20 °C)

- 3 M NaCl
- Phenol
- Phenol/chloroform/IAA (isoamyl alcohol) 25:25:1
- Chloroform/IAA 25:1
- 70% and 100% EtOH
- 1 × TE (10 ml Tris–HCl, pH 8.0, 1 mM EDTA)
- 10 mg/ml DNase-free RNase[b]

Method

1 Freeze single adult fish in liquid nitrogen and either use immediately, or store at -70 °C.

2 Grind in a pre-chilled mortar with liquid nitrogen.

ROBERT GEISLER

3 Transfer powder to a Falcon tube with 10 ml lysis buffer, add proteinase K to a final concentration of 100 μg/ml, and incubate overnight at 37 °C on a rotating shaker.

4 Extract with 10 ml phenol, phenol/chloroform/IAA and chloroform/IAA, by vortexing for 1 min, spinning for 5 min, and transferring the aqueous phase into a fresh tube.

5 Precipitate with 1 ml of NaCl solution and 20 ml EtOH, spin for 15 min, wash with 70% EtOH, spin again, and transfer to a fresh tube with a heat-sealed Pasteur pipette.

6 Resuspend in 2 ml TE, add DNase-free RNase to a final concentration of 100 μg/ml, and incubate at 37 °C for 2 h.

7 Repeat extraction and EtOH precipitation.

8 Resuspend in 1–2 ml TE.[c]

[a] Clean thoroughly before use.

[b] DNase-free RNase can be prepared by boiling RNase stock solution for several minutes (30).

[c] This protocol yields several milligrams of DNA. Such an amount is required, for example, for testing large numbers of markers for polymorphism. Otherwise consider using fin clips (*Protocol 2*).

In Tübingen the same template plates from which pooled DNA was obtained are used for single-embryo PCR. The upper half of the plate contains siblings and is only used as a control. The mutant embryos in the lower half are manually scored as homozygous for one marker allele, homozygous for the other allele, or heterozygous, and the scores exported to the database using NIH Image macros (*Figure 6*). The recombination rate is then obtained simply by counting the haploid marker alleles (heterozygotes are counted once, homozygotes twice) and dividing the count of the less-frequent allele, n_1 or n_2, by the total number of haploid genotypes, n (twice the total number of diploid embryos scored): if $n_1 < n_2$: $\theta = n_1/n$; if $n_1 > n_2$: $\theta = n_2/n$.

The zebrafish lines currently available for map crosses are not completely isogenic and share some alleles, so that F_1 fish may be homozygous for a given marker (approximately one-third in Tü × WIK F_1 crosses). If a homozygous F_1 is not recognized, it will give rise to false linkages. Therefore, whenever only two genotypes are present in the mutants (homozygotes for one allele and heterozygotes), check whether homozygotes for the other allele are found among the siblings. If this is not the case, a homozygous F_1 must be assumed (*Figure 7b, c*, see also *Plate 16b, c*). To obtain the correct recombination rate in the presence of a homozygous F_1, divide the number of recombinants by the number of mutant embryos, rather than twice that number, as each embryo contributes only one informative meiosis (this is similar to performing a backcross instead of an outcross). Unfortunately the recombination rate in male zebrafish is as much as 15-fold lower than that in females, in particular around the centromeres (A. Singer and J. Postlethwait, personal communication). Because the ratio is region-specific, map distances obtained using only male or female recombinations cannot be easily combined with sex-averaged distances to calculate map positions.

Another possibility is the presence of three or four bands in an F_1 cross. Once it is known which of these bands originate from the same P_0 fish, they can be counted as one allele, and the recombination rate can be calculated without any modifications. In the case of three bands, look for homozygotes in both the mutants and siblings. Only one of the bands should occur in a homozygous state. This band therefore represents one allele, and the two other bands the other allele (*Figure 7d, e*, see also *Plate 16d, e*). In the case of four bands, the origin of

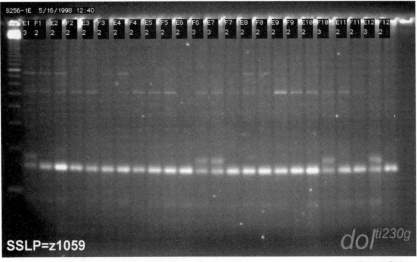

Record	Templ. Row	Mutant Allele	Marker Name	Genotypes	Hom. F1	Status
S256-1E	T268A	AI30G	z1059	3222222222223322222232232	☐	OK
S256-1G	T268C	AI30G	z1059	2222222223232322223232232	☐	OK

Figure 6 Single-embryo PCR gel. Top, agarose gel with PCR products of 24 dol^{ti230g} embryos for the SSLP Z1059. Embryos are scored as: 1, homozygous for the upper band (not seen on this gel); 2, homozygous for the lower band; 3, heterozygous. The gel confirms strong linkage of the *dolphin* locus with Z1059. Only five recombination events are seen. Bottom, database records showing 10 recombination events in a total of 48 analysed dol^{ti230g} embryos, corresponding to a map distance of 10.6 cM and a LOD score of 15.0. Siblings are not shown. (From ref. 26.)

the bands can only be guessed from the F_2 results: if there is linkage, the two bands that occur more frequently than the others should represent the linked allele (*Figure 7f*, see also *Plate 16f*).

Problematic band patterns can also arise for several other reasons, including alleles that do not amplify at all (due to a deletion or simple nucleotide polymorphism), alleles suppressing other alleles (behaving as dominant), additional bands correlated with an allele (due to annealing of a primer at a second site), and electrophoresis artefacts. As a general rule, make sure that the sibling band pattern is significantly different from that of the mutants if a linkage is assumed, and test additional markers if the scoring is questionable.

2.7 Calculation of a map position

Calculation of map distances between mutations and markers is straightforward and can be carried out with a calculator or spreadsheet program (or, as in Tübingen, within the laboratory database). To obtain a map distance comparable to the published map from the recombination rate, apply the Kosambi mapping function:

$$m = \frac{1}{4} \ln[(1 + 2\theta)/(1 - 2\theta)] \text{ Kosambi cM}$$

This is one of several mapping functions that aim to correct large map distances such that they are additive and 50% recombination corresponds to an infinite distance. It was used for the published SSLP map (4, 5). For small distances, 1 Kosambi cM is approximately equal to 1% recombination.

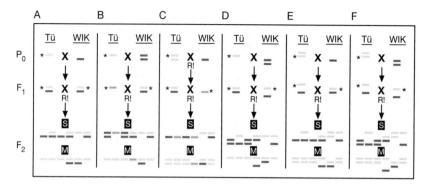

Figure 7 (see Plate 16) Possible allele systems of an SSLP. S, siblings; M, mutants; RI, recombination; red, Tü bands; blue, WIK bands; the asterisk indicates linkage with the mutation. The band sizes are arbitrary; Tü bands may occur above, below, or in between WIK bands. In each example, two out of six mutants are recombinant (17% recombination). (A) Two bands. (B) Homozygous F_1 fish, in the mutants only half of the recombinations are visible. (C) Homozygous F_1 fish with a recombination in the P_0, again only half of the recombinations are visible. (D) Three bands. The lower bands should be scored as one allele. (E) Three bands with a recombination in the P_0. The upper bands should be scored as one allele. (F) Four bands. The upper and lower bands should be scored as one allele. (Modified from ref. 26.)

As a measure of the significance of a linkage, a LOD score can be calculated. The LOD score (logarithm of odds) is the decadic logarithm of the ratio of the likelihoods of two hypotheses, namely that there is linkage at the expected map distance, or that there is no linkage (27, 28). It can be calculated as:

$$\text{LOD} = n \left(\theta \lg\theta + (1 - \theta) \lg(1 - \theta) - \lg 0.5\right).$$

By convention, a LOD score of 3 is used as a threshold for significant linkage in genetic mapping, meaning that linkage at the expected map distance is 1000 times more likely than no linkage. With 48 tested embryos, such a LOD score is obtained for map distances up to approximately 36 Kosambi cM or 31% recombination. However, false positives with LOD scores above the threshold are probable when several thousand mapping experiments are performed, such as in a genome scan for several mutations. Furthermore, the LOD score does not take into account the possibility of human error and the effect of polymorphisms in the genetic background, which may enhance the visibility of a phenotype and therefore appear weakly linked. For these reasons a LOD score of 3 should be taken as evidence for significant linkage only when it is confirmed by a second marker. If a linkage is slightly below the threshold, a decision can be made by testing additional embryos: if the linkage is real, the LOD score should increase proportionally to the number of genotypes.

A mutation is in the interval between two markers if the recombinants for these two markers are not correlated. A method of testing for this is to consider only the individuals that are heterozygous for the closest linked marker (i.e. have a single recombination) and determine the fraction of these that are also heterozygous for a second linked marker. If this fraction is less than 50%, the recombinants are not correlated, and the mutation is likely to be in the interval between the two markers. If it is more than 50%, both markers are likely to be on the same side of the mutation. The occurrence of recombinations along the chromosome can be visualized by sorting the genotype vectors appropriately (an example is shown in *Figure 8*).

Once markers flanking the mutation have been determined, it is possible to omit individuals that are recombinant for both of the markers from further mapping experiments, and recalculate the map distances without them, on the grounds that they are probably missorted

Marker	cM	Genotypes
Z4706	36.7	2222222223222222222222223322232322222222222222222
Z4999	36.7	1111111113111111111111113111311111111111111111111
Z1182	45.0	02222222222332222222223222222222232222222222222232
Z1059	52.3	32222222222332222223223222222222223223222232322232

Figure 8 Recombinations in *dol^{ti230g}* mutant embryos. Recombination sites are visualized by sorting the genotypes according to marker position. 1, Homozygous for the upper band; 2, homozygous for the lower band; 3, heterozygous (compare Figure 6). Shaded = recombinant individual; dotted line = inferred position of the *dolphin* locus between 36.7 and 45.0 cM from the top of LG7, recombinants above and below are not correlated.

or contaminated. However, some true recombinants may be lost by this procedure if the markers are only weakly linked, since two recombinations can occasionally happen on the same chromatid (chiasmatic interference in the zebrafish being incomplete).

Map distances determined on a panel of mutant embryos are often different from published distances. In order to determine a map position, d_m, for a mutation flanked by two markers, scale the observed map distances, d_{m1} and d_{m2}, such that they fit in the interval between the published marker positions, d_1 and d_2 (*Figure 9*):

$$d_m = d_1 + d_{m1}[(d_2 - d_1)/(d_{m1} + d_{m2})].$$

In Tübingen the following criteria are applied for successful rough mapping:

- Linkage with two markers out of the 192-marker set, if possible flanking markers, otherwise the two closest markers on one side.
- A LOD score greater or equal 3 for both linked markers.
- At least 24 mutant embryos tested with both linked markers, including at least one recombinant for each marker, if any recombinants exist.

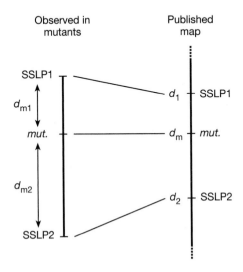

Figure 9 Placing a mutant on a published map. Map distances determined on a panel of mutant embryos (d_{m1}, d_{m2}) are scaled to fit into the interval between two published marker positions (d_1, d_2) in order to obtain a map position for the mutation (d_m). For *dolphin* (shown in *Figure 8*) d_{m1} = 3.1 cM, d_{m2} = 5.3 cM, d_1 = 36.7 cM. d_2 = 45.0 cM, resulting in a map position d_m = 39.8 cM from the top of LG7.

2.8 Fine mapping

Once a map position has been established, additional SSLPs in the same region can be tested in order to find one that is closer to the mutation. Out of the several thousand SSLPs available, approximately 20–30% show a usable polymorphism in Tü × WIK map crosses. However, SSLPs appear to be clustered in particular regions, therefore a linkage closer than 1 cM, necessary for positional cloning, can currently be found only for a minority of mutations.

More F_2 embryos can be used to improve the resolution and accuracy of the map distances. In positional cloning projects, typically 1500–3000 mutant embryos are collected and typed for the closest markers. With 3000 embryos (6000 meioses) one recombination is equivalent to 0.02 cM or 13 kb. In order to reduce the number of PCRs that have to be performed, it is possible to pool the mutant embryos, and test single embryos only from those pools that contain a recombinant. The possible pool size depends on the marker. Only a small number of siblings need to be tested as a control for each F_1 cross.

3 Radiation hybrid mapping of candidate genes

Any stretch of zebrafish sequence can be scored on a radiation hybrid panel, as long as it is unique. Primer pairs for radiation hybrid mapping can be designed, e.g. with Primer3 (available online at the Whitehead Institute <http://www-genome.wi.mit.edu/cgi-bin/primer/primer3_www.cgi>). The optimal size range for the PCR product is 100–400 bp; products of more than 800 bp do not amplify reliably under our conditions. We perform PCR and gel electrophoresis in a manner very similar to mutant mapping (Protocol 5). The only differences are the smaller reaction volume and template DNA concentration, the use of standard agarose gels with 16 combs rather than high-resolution agarose gels with four combs, and shorter electrophoresis, since very little size resolution is required (Figure 10).

Protocol 5

Radiation hybrid PCR and electrophoresis

Equipment and reagents

- 96-well PCR plate
- Sealing film
- Thermocycler
- Agarose electrophoresis system
- 10 × PCR buffer (100 mM Tris–HCl (pH 8.3), 500 mM KCl, 15 mM $MgCl_2$, and 0.1% (w/v) gelatin, stored at 4 °C)
- 100 mM dATP, dCTP, dGTP, dTTP (prepared from solid substance, stored at −20 °C)
- 20 μM forward and reverse primer (stored at 4 °C)
- 5 U/μl Taq polymerase

- 10 ng/μl radiation hybrid and control DNA (stored at −20 °C)
- 3 × Ficoll loading buffer (7.5% Ficoll, 0.125% Bromophenol Blue) with approximately 10 ng/μl undigested plasmid DNA as a loading control
- 10 ng/μl of 100 bp ladder (Pharmacia) in 1 × Ficoll loading buffer
- 1 × TBE (90 mM Tris-borate, 90 mM boric acid, 2 mM EDTA, pH 8.0)
- Qualex Gold agarose (AGS)
- 10 mg/ml ethidium bromide

Method

1 On a PCR plate set up 10 μl reactions:

   (a) 7.14 μl reaction mix (1 μl of 10 × PCR buffer, 0.02 μl each of 100 mM dATP, dCTP, dGTP and dTTP, 6.06 μl water).[a]

Protocol 5 continued

(b) 0.08 μl each of 20 μM forward and reverse primer.[b]

(c) 0.2 μl of 5 U/μl *Taq* polymerase.

(d) 2.5 μl of 10 ng/μl radiation hybrid or control DNA. Double the concentration for duplicates of markers that have initially given only faint bands.[c]

2 Cycling:

(a) Initial denaturing at 94 °C for 2 min.

(b) 35 cycles of denaturing at 94 °C for 30 s, annealing at 60 °C for 30 s and extension at 73 °C for 1 min. Use 2 min extension time for markers that give a PCR product of more than 800 bp. Adjust the annealing temperature if necessary.

(c) Final extension at 73 °C for 5 min.

3 Electrophoresis:

(a) Pour 2% Qualex Gold agarose gels containing 1 × TBE and 0.8% ethidium bromide.

(b) Add 5 μl of 6 × loading buffer to each sample.[d] For each marker, run one lane with 80 ng of 100 bp ladder as a size standard.

(c) Run at 200 V for 20 min in 1 × TBE buffer with 0.4% ethidium bromide.[e]

[a] Reaction mix is freshly prepared, but can also be stored at −20 °C (see *Protocol 2*).

[b] Over extended periods, primer DNA can be stored at −20 °C rather than 4 °C, but it should not be thawed too often.

[c] DNA of the T51 radiation hybrid panel, including positive (zebrafish) and negative (Wg3H) controls, is commercially available from ResGen <http://www.resgen.com>, as a 25 ng/μl solution. We aliquot and dilute 1 ml of each stock solution robotically on three deep well plates (Beckman). DNA of the LN54 panel is available on request from the lab of Mark Ekker.

[d] We use gels with 16 × 30 lanes each (for details see *Protocol 2*) and apply samples with a 12-channel CAPP multipipette. The lanes are half as wide as the distance between microtitre wells, so that PCR products from two plate rows are interspersed on the gel.

[e] TBE can be re-used at least 15 times.

The Tübingen set-up for high-throughput radiation hybrid mapping is a Beckman Core System with a Biomek 2000 for pipetting, an ORCA robot for plate handling, an ALPS 100 plate sealer and Tetrad (MJ Research) cyclers with motorized bonnets (12 blocks in total), allowing automatic overnight operation. The same gel and imaging equipment is used as for mutant mapping. Products of 36 PCR plates can be run and imaged in parallel by one researcher in about 3 h.

PCR products of the hybrid lines are scored as positive, negative, or undecided, according to the following rules:

• Run each marker in duplicate. Score even very weak bands as positive if they are clearly distinguishable from the background. Score lanes that have gel problems or contain only smear as undecided.

• In case of any discordancy (i.e. a hybrid scored as positive on one plate and as negative on the other), re-score the bands.

• If the discordancy persists, run a third plate and use the radiation hybrid (RH) types supported by two of the plates. Because some PCR reactions may fail to give scorable bands, this procedure may leave a small fraction of the RH types undecided.

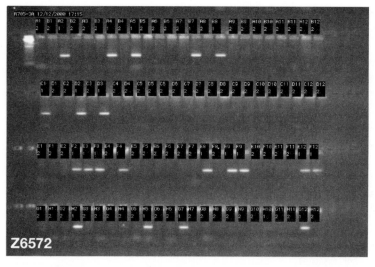

Marker	RH Types
Z7330	221222122222111122222221221222222222222222212222222111212222222211212222222222222222221222222122
Z10729	221222122222111122222221221222222222222222212222222111212222222211222222222221222222221222222122
fj47h06.x1	221222122222112122222221221222222222222222212222222111212222222211222222222221222222221222222122
Z6572	221222121222121222222221221222222222222222212222222111212222222212112221122212222221221222222122
Z3892	221222121222121222222221221222222222222222212222222111222222222212112221122212222221221222222122
Z13220	221222121222121222222221221222222222222222212222222111222222222212112221122212222221221222222122

Figure 10 Radiation hybrid gel. Top, agarose gel with PCR products of the T51 radiation hybrid panel for the marker Z6572 (LG18). RH lines 1–96 appear in numerical order and are scored as: 1, positive; 2, negative; 0, undecided. The weak band in lane H10 (RH line 92) was scored as positive and was successfully reproduced on a second gel. Lanes G12 and H12 (RH lines 95 and 96) contain positive and negative controls, and are not used for map calculation. Bottom, RH types (score vectors) of markers around Z6572, sorted by radiation hybrid map order. Highlighting of the positives (shaded) allows us to visualize the breakpoints of zebrafish DNA fragments present in the RH lines. Markers are placed on the map such that the number of breakpoints is minimal. However, the panel provides only a limited amount of information, so that some ambiguities remain, e.g. placing Z13220 between fj47h06.x1 and Z6572 would result in the same number of breakpoints. Duplicate markers are not shown.

When in doubt which band should be scored, check the size of the positive control (line 95 of the T51 panel). A marker may still be scorable even if the negative control (line 96 of the T51 panel) shows a weak band of the same size; in this case score bands as positive only if they are clearly stronger than in the negative control. Because the hybrids were produced from diploid cells, a marker may amplify two bands, occurring at a similar frequency (especially if it is an SSLP); in this case both must be scored. The average retention (i.e. fraction of positive hybrids) is 18% for the T51 panel and 28% for the LN54 panel, and the lowest retention of a region in the T51 panel is approximately 6%. A smaller fraction of positive hybrids may therefore indicate a problem with the marker being tested. On the other hand, the retention reaches 100% at the position of the selectable HPRT marker that was used for construction of the T51 panel (on LG14), so high retention alone is not a sufficient reason to reject a marker. High retention in combination with unreproducible band intensities probably indicates a primer pair that recognizes a repeat sequence.

Mapping data can be submitted online to one of the following mapping services:

• The Geisler lab <http://wwwmap.tuebingen.mpg.de> integrates submitted markers into the next update of the T51 map. Updates occur approximately once a month. The map position relative to neighbouring SSLP markers is reported to the submitter, and the marker is added to the publicly available map and to the ZFIN database, with appropriate credits. Genes may be mapped as anonymous markers, but submitters are requested to reveal their identity after publishing them.

- The Zhou lab server <http://zfrhmaps.tch.harvard.edu/ZonRHmapper> instantly calculates an approximate map position on the T51 panel, but does not integrate submitted data into the map. Alternatively, primer sequences or primers can be submitted for mapping, and the resulting data will be added to the map and to ZFIN.

- The Dawid lab <http://mgchd1.nichd.nih.gov:8000/zfrh/beta.cgi> integrates submitted markers into the LN54 map. Genes may be mapped as anonymous markers.

In principle, it is possible to calculate a map on your own, based on the published scoring data, but this requires considerable effort. The T51 map is calculated with SAMapper 1.0, created at the Stanford Human Genome Center, with slight modifications (currently not publicly available); the LN54 map is calculated with RHMapper, available from the Whitehead Institute <http://www-genome.wi.mit.edu>.

4 Matching mutations with candidate genes

4.1 Searching maps and databases

After establishing a map position for a mutation, a search for potential candidate genes can be conducted. Genes within several centimorgans of the mutant locus should be considered (note that the accuracy of genetic and radiation hybrid maps is always lower than their resolution). This can be a problem for loci near a centromere, where genetic recombination appears to be strongly reduced, resulting in a large number of candidate genes within a few centimorgans. In these cases either very accurate mapping information for the mutation or a very convincing candidate gene is required. A good starting point for a candidate gene search is the ZFIN website <http://zfin.org/ZFIN>. It shows results for the various mapping panels side by side, and also as an integrated map that uses the MGH map as a framework (ZMAP; A. Day et al., unpublished). It also shows published map positions of mutations. The most recent results for a particular panel can be obtained from the website of the respective mapping group (see Table 1). Contacts with the submitters of anonymous T51 markers can be established by sending a request to the Geisler or Zhou lab (for 'unp' and 'chunp' markers, respectively).

Another approach to finding candidate genes is to identify and map new or previously unmapped members of known gene families. Helpful resources in this respect are the LocusLink service at the National Center of Biotechnology Information (NCBI) <http://www.ncbi.nlm.nih.gov/LocusLink> as well as ZFIN <http://zfin.org/ZFIN>, which provide an interface to curated information on known genes. The TIGR Zebrafish Gene Index <http://www.tigr.org/tdb/zgi/zgi.html> and the UniGene division of GenBank <http://www.ncbi.nlm.nih.gov/UniGene/Dr.Home.html> list unique genes represented by clusters of overlapping ESTs. The WashU Zebrafish Genome Resources Project <http://zfish.wustl.edu> has generated and clustered the majority of all zebrafish ESTs and provides information on homology obtained by BLAST searching. The corresponding cDNA clones can be obtained from the American Type Culture Collection <http://www.atcc.org/>, Incyte Genomics <www.incyte.com>, or the Resource Center of the German Human Genome Project (RZPD) <http://www.rzpd.de>; however, each clone should be resequenced or checked by PCR with appropriate primers in order to confirm its identity, since a significant proportion may be mislabelled.

To identify homologues of known genes on your own, perform BLAST searches of GenBank <http://www.ncbi.nlm.nih.gov/BLAST>, the TIGR database <http://tigrblast.tigr.org/tgi/>, and the zebrafish genome trace archive <http://www.ncbi.nlm.nih.gov/genome/

guide/D_rerio.html>, which contains shotgun sequences covering most of the zebrafish genome. Traces can also be searched at the Sanger Centre <http://www.sanger.ac.uk/ Projects/D_rerio/blast_server.shtml> or the Ensembl project <http://trace.ensembl.org/ perl/ssahaview>. Genes identified in this way may have to be cloned by PCR. As an alternative to the computational approach, homologues can be identified by screening filters of cDNA libraries or by PCR with degenerate primers. Suitable methods are described, for example in ref. 30 and on the websites of the RZPD <http://www.rzpd.de> and ResGen <www.resgen. com>, who also provide filters, pools, and screening services for several libraries.

Candidate genes can also be found by utilizing the extensive conservation of synteny (gene order) between the human and zebrafish genome. Eighty per cent of genes and ESTs analysed belong to conserved synteny groups (31). Comparative maps of zebrafish and human are available from the WashU website <http://zfish.wustl.edu> (31, 32) as well as from the Talbot lab website <http://zebrafish.stanford.edu> (33, 34). The latter has also some information on conservation of synteny between zebrafish and mouse. The Jackson Laboratory website <http://www.jax.org> compares mouse, rat, and human. Comparative maps provide a general idea of the size of the syntenic regions, and the correspondence of human chromosomes to zebrafish chromosomes. To quickly identify a specific region of the human genome that may correspond to the vicinity of a mutation, perform a LocusLink search for human homologues of zebrafish genes that occur near the mutation. Click on the map position and zoom outward to browse the region. If a promising human candidate gene is found in the region, possible zebrafish homologues can in turn be identified by BLAST searches.

4.2 Evidence for the identity of a mutant locus

Once a candidate gene has been identified, a range of experiments can be performed in order to prove the identity of the mutant locus:

- Fine mapping of the mutation and mapping of the gene on additional panels can exclude some potential matches.
- Genotyping of single-nucleotide polymorphisms associated with the gene can demonstrate a direct linkage between the mutation and the gene (see below).
- In situ hybridization can show that the gene is expressed in the affected tissues and that its expression is altered in the mutant.
- Knock down of the candidate gene by injections of morpholinos (modified antisense oligonucleotides) can partially or completely phenocopy the mutation.
- Rescue of the mutant phenotype by cDNA injections can be attempted.
- Sequencing of the mutant transcript (from multiple alleles, if available) is required to provide final proof for the identity of the mutation.

Only some of these experiments are typically performed, depending on the cost and effort of sequencing and on the faith put in the candidate gene. Methods for in situ hybridization, morpholino and cDNA injections are outside the scope of this chapter, and are discussed elsewhere in this volume. For each step of cDNA cloning and sequencing, a variety of kits is available, e.g. the RNAClean kit (Thermo Hybaid) for total RNA preparation, the OligoTex mRNA Purification System (Qiagen) for poly(A)-RNA isolation, and the SuperScript kit (GIBCO BRL) for reverse-transcriptase PCR. Further cloning and sequencing methods can be found, for example, in ref. 30.

4.3 Demonstrating a direct linkage by SNP genotyping

A direct linkage between a mutation and a candidate gene can be demonstrated by identifying a single nucleotide polymorphism (SNP) associated with the gene and genotyping mutant embryos for it; there should be no recombinants or at least very few (intragenic recombination cannot be completely ruled out). SNPs are less convenient for making maps than SSLP markers because their detection is more difficult with current technology, but they are more frequent and therefore better suited to identifying a direct linkage. They even occur in coding sequences, in particular in the third position of each codon where tRNAs often recognize alternative bases. SNPs can be detected by direct sequencing, mass spectrometry, or hybridization on custom-made microarrays. However, we will restrict our discussion here to single-strand conformational polymorphism (SSCP) and denaturing gradient gel electrophoresis (DGGE) analysis, which are particularly useful for detecting SNPs on a small scale, such as in a cloning project.

The majority of SNPs are detectable as single-strand conformational polymorphisms (SSCPs) that can produce different secondary structures (stem-loops). To detect these polymorphisms a double-stranded PCR product is first denatured by heating under alkaline conditions, then chilled rapidly and separated on a non-denaturing polyacrylamide gel at a low temperature. This allows the single strands to anneal to themselves, but prevents renaturation of double strands (35) (*Protocol 6*). The optimal size for the PCR product is around 150 bp (36). Approximately 50% of primer pairs from non-coding regions show an SSCP in zebrafish map crosses (37). Different primer pairs derived from the candidate gene sequence are tested until a polymorphic band is detected (*Figure 11*). Additional bands that appear in all lanes either represent secondary structures that do not include the polymorphic base, or renatured PCR products.

Protocol 6

Detection of single-strand conformational polymorphisms (SSCPs) (slightly modified from ref. 41)

Equipment and reagents

- Flatbed electrophoresis system (Multiphor II, Pharmacia)
- Staining trays
- PCR products of SNP markers[a]
- Denaturing solution (10 mM EDTA, 500 mM NaOH)
- Loading buffer (formamide, 0.1% (v/v) of saturated Bromophenol Blue solution and 0.1% (v/v) of saturated xylene cyanol solution)
- Precast acrylamide SSCP gel with matching electrode wicks and buffers (CleanGel SSCP, ETC Elektrophorese-Technik)
- Silver staining kit for DNA (Pharmacia)

Method

1   Mix 5.4 μl PCR product with 0.6 μl denaturing solution and 2.4 μl loading buffer. Incubate for 10 min at 95 °C, and immediately put on ice.

2   Load total samples on a precast acrylamide SSCP gel.

3   Run at 4 °C for 10 min at 300 V, and for 3–4 h at 200 V.[b]

4   Stain gel with a silver staining kit according to manufacturer's instructions.

[a] PCR is performed as described in *Protocol 1*.

[b] This corresponds to voltage gradients of 20 V/cm and 13.3 V/cm, respectively.

An alternative method of detecting SNPs on gels is denaturing gradient gel electrophoresis (DGGE) (*Protocol 7*). In this method melting-point differences are detected by running PCR

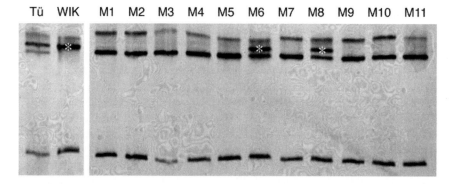

Tü WIK   M1   M2   M3   M4   M5   M6   M7   M8   M9   M10   M11

Figure 11 Linkage analysis of the mutation *blc^ah04b* and the candidate gene *wnt10br* using a single-strand conformational polymorphism (SSCP). Tü and WIK, P$_0$ fish. M1–M11, mutant F$_2$ embryos. M6 and M8 are recombinant and show the WIK band (asterisks). Forward primer: ATGCTGTACAGGCAGTGTGC. Reverse primer: TCTGCAAAAGTCAAAAACCTGA.

products on a gradient gel with increasingly denaturing properties (which is equivalent to applying a temperature gradient). The strands separate at a distance corresponding to their melting point. The denatured strands are held together in a Y-shaped conformation by a 5′ (GC)-rich sequence or 'GC-clamp' attached to one of the primers, preventing further migration (38–40). In a preliminary experiment, the same heterozygous DNA sample is run across the entire width of the gel, and the denaturing gradient is at a right angle to the voltage, producing an image that reflects the shape of the melting curve (*Figure 12a*). A double band in the region of the curve where the DNA begins to melt indicates that an SNP is detectable. For genotyping, a shallow gradient corresponding to this region of the curve is applied in parallel to the voltage. DNA of different genotypes stops migrating when the melting point is reached, resulting in different band sizes (*Figure 12b*). While DGGE requires pouring of gradient gels and an electrophoresis system with precise temperature control, it allows the detection of nearly all SNPs in the amplified sequence.

Protocol 7

Denaturing gradient gel electrophoresis (DDGE) (Bio-Rad, slightly modified)

Equipment and reagents

- Vertical electrophoresis system with gradient mixer (DCode Universal Mutation Detection System, Bio-Rad)
- Capillary pipette (Hamilton)
- Staining tray
- Fluorescence scanner (Molecular Dynamics)
- 40% acrylamide solution (rotiphorese 29:1, Roth)
- PCR products of SNP markers[a]

- Urea
- 10% (w/v) ammonium persulphate (APS; stored at −20 °C)
- 10% (w/v) N,N,N,N,-tetramethylene diamine (TEMED; stored at 4 °C)
- 1 × TAE buffer (40 mM Tris-acetate, 2 mM EDTA, pH 8.0)
- SYBR-Green I stock solution (Molecular Probes)

Protocol 7 continued

Method

1 Pour 6.5% polyacrylamide gels (33 ml, 18 × 18 cm) with denaturing conditions decreasing from the bottom to the top, e.g. from 100% to 0%:[b]

   (a) fill the distal cylinder of the gradient mixer with 16.5 ml of non-denaturing solution (1 × TAE, 16.2% (v/v) 40% acrylamide solution);

   (b) fill the proximal cylinder with 16.5 ml of denaturing solution (1 × TAE, 16.2% (v/v) 40% acrylamide solution, 42% (w/v) urea);

   (c) add 150 µl 10% (w/v) ammonium persulphate and 8 µl 10% (w/v) N,N,N,N,-tetramethylene diamine, and pour immediately.

2 Load PCR products with a capillary pipette and run electrophoresis according to manufacturer's instructions.

3 Stain the gel for 10 min with 1:10 000 (v/v) SYBR-Green I solution in 1 × TAE.

4 Scan on a fluorescence scanner.

[a] PCR is performed as described in *Protocol 1*. MacMelt software (Bio-Rad) can be used to design suitable primers with 5' GC-clamps and to predict melting curves.

[b] For preliminary experiments, pour gels with a 0–100 % denaturing gradient at a right angle to the electrodes. For genotyping, pour gels with a shallow gradient parallel to the electrodes, corresponding to the region near the previously identified melting point.

5 Positional cloning of mutations

5.1 Genomic libraries

For the majority of zebrafish mutations there are currently no obvious candidate genes. In these cases a positional cloning approach may be taken, in which the mutation is first localized on a single genomic clone, and then a search for transcripts of the clone is performed (see also ref. 42, and the Zon Lab Guide to Positional Cloning in the Zebrafish <http://134.174.23.167/zonrhmapper/positionCloningGuide.htm>). Only a few positional clonings have been published to date, including *one-eyed pinhead* (43), *sauternes/alas2* (44), *weis-sherbst/ferroportin1* (45), *foggy* (46) and *miles apart* (47), but several more are under way. Genomic libraries based on different vectors and lines are available for the zebrafish (*Table 3*). All these libraries have been arrayed on microtitre plates, so that individual clones can be obtained. In general, a large insert size is desirable for a library because this makes it is easier to reach the mutant locus from a flanking marker, and a high genome coverage, because it increases the likelihood that a positive clone is found.

    Yeast artificial chromosomes (YACs) have particularly large inserts (48). Two YAC libraries (Zebrafish YAC and Zebrafish YAC II) were made from the AB zebrafish line, with 6.1-fold and 4.7-fold coverage and an average insert size of 300 kb and 420 kb, respectively (49, 50). However, YACs are prone to recombination and difficult to handle, which makes them less suitable for positional cloning than shorter genomic clones.

    P1 artificial chromosomes (PACs) use a phagemid vector that allows them to be maintained in bacteria at a low copy number, resulting in a low recombination rate, while a phage replicon can be induced to generate larger amounts of DNA (51). Bacterial artificial chromosome (BACs) use a low-copy-number vector derived from the F factor (52). PACs and BACs have similar, intermediate insert sizes. A PAC library with 6.6-fold coverage and 120 kb insert size was generated from the AB zebrafish line (53). All of the published positional cloning experi-

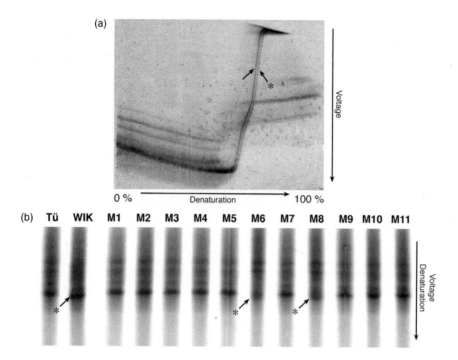

Figure 12 Linkage analysis of the mutation blcah04b and the candidate gene tarama by denaturing gradient gel electrophoresis (DGGE). (a) Preliminary experiment with heterozygous DNA, the denaturation gradient is at a right angle to the voltage. A single nucleotide polymorphism is visible as a double band at approximately 60% denaturation (asterisk). Therefore a gradient of 40–80% denaturing conditions was chosen for genotyping. (b) Genotyping experiment, the denaturation gradient is parallel to the voltage. Tü and WIK, P0 fish. M1–M11, mutant F2 embryos. M6 and M8 are recombinant and show the WIK band (asterisks). Forward primer (with GC-clamp): CGGGCGGGGGCGGCGGGACGGGCGCGGGGCGCGGCGGGCGGGATGTTGAGGATCCATCCTGTGAT. Reverse primer: CTGCAGTACAATTGTCCTGGCCA.

ments utilized this library. A BAC library with 4.7-fold coverage and the same insert size is commercially available from Incyte Genomics (formerly Genome Systems).

For the zebrafish genome project new BAC libraries were made from the Tü line, because this is the genetic background for the majority of the zebrafish mutations. The RPCI-71 and CHORI-211 libraries were generated by P. de Jong, Roswell Park Cancer Institute and Children's Hospital Oakland Research Institute (unpublished). They have 1.7-fold and 10.4-fold coverage and an average insert size of 85 kb and 165 kb, respectively. The former was generated from embryos and contains a fraction of clones with a small insert size, possibly due to an endogenous nuclease activity, while the latter was created from testes. Two additional libraries (DanioKey Pilot and DanioKey) were generated by T. Jesse of the company Keygene N.V., also from Tü testes. They have 1.0-fold and 10.2-fold coverage and an insert size of 150 kb and 175 kb, respectively. The four Tü-derived BAC libraries are being partially fingerprinted by restriction digests in order to produce contigs of overlapping clones (R. Geisler, R. Plasterk, and J. Rogers, unpublished). Genomic sequencing of the zebrafish will mostly be carried out on these contigs, and anchoring of the contigs on the existing maps will eventually produce a physical map of the zebrafish genome. However, the unanchored contigs are immediately useful for positional cloning as soon as a member of a contig is identified in a cloning experiment.

Table 3 Genomic libraries

Library[a]	Type	Line	Coverage	Clones arrayed	Insert size (kb)	Creator	Filters from	Pools from
Zebrafish YAC (HACHy914)	YAC	AB	6.1	34560	300	L.I. Zon	RZPD	RZPD
Zebrafish YAC II (MGH_y932)	YAC	AB	4.7	19008	420	T.P. Zhong	RZPD	RZPD, ResGen
Zebrafish PAC (dZ, BUSMP706)	PAC	AB	6.6[b]	104064	120	C.T. Amemiya	Incyte, RZPD	Incyte, RZPD
Incyte Zebrafish BAC	BAC	AB	4.7[b]	73728	120	Incyte	Incyte	Incyte
RPCI-71 (bZ, RPCIB728)	BAC	Tü	1.7	33408	85[c]	P. de Jong	RZPD, CHORI	RZPD
CHORI-211 (zC)	BAC	Tü	10.4	105907	165[c]	P. de Jong	RZPD, CHORI	RZPD
DanioKey Pilot (zKp)	BAC	Tü	1.0	11808	150[3]	T. Jesse	RZPD	RZPD
DanioKey (zK, HUKGB735)	BAC	Tü	10.2	99285	175[c]	T. Jesse	RZPD	RZPD
Zebrafish Cosmid (ICRFc68)	Cos-mid	KC	1.3	55296	40	C. Burgtorf	RZPD	RZPD
Zebrafish Cosmid (ICRFc70)	Cos-mid	AB	1.3	55296	40	C. Burgtorf	RZPD	RZPD
Zebrafish Cosmid (ICRFc71)	Cos-mid	Tü	1.3	55296	40	C. Burgtorf	RZPD	RZPD

[a] Prefixes used by the Sanger Centre (lower case) as well as prefixes and library numbers used by the RZPD (upper case) are given in brackets. Sanger Centre clone names consist of a prefix, plate and well, e.g. bZ2K3; RZPD clone names consist of a prefix, library number, well (two digits) and plate, e.g. RPCIB728K032.

[b] Coverage as indicated by Incyte Genomics.

[c] Size determined by restriction fingerprinting (R. Koch and R. Plasterk, unpublished).
RZPD = Resource Center of the German Human Genome Project <www.rzpd.de>, Incyte = Incyte Genomics <www.incyte.com>, CHORI = Children's Hospital Oakland Research Institute <http://www.chori.org/bacpac/>. Information on vectors etc. is available from the distributors.

Finally, three short-insert libraries are available, based on a cosmid vector (54). These libraries were generated by the laboratory of H. Lehrach from the lines AB, KC and Tü, respectively. Each of them has 1.3-fold coverage and an average insert size of 40 kb (55).

5.2 Screening for genomic clones

In order to start a positional cloning experiment, a mutation must be mapped close to a genetic marker. Genomic clones are then identified by screening genomic libraries either with the marker itself or with nearby cDNA clones, using one of the following methods:

1. Database searches for sequenced genomic clones can be performed at NCBI <http://www.ncbi.nlm.database nih.gov/BLAST> or the Sanger Centre <http://www.sanger.ac.uk/Projects/D_rerio/blast_server.shtml>. Only a minority of genomic clones are currently represented in these databases, but the number will increase rapidly as more of the zebrafish genome is sequenced. In addition, BAC sequencing is expected to yield a large number of SSLPs suitable for genetic mapping, the location of which on a specific BAC will be known from the outset.

2. Pools of genomic clones can be screened by PCR. For libraries distributed by the RZPD, each primary pool contains purified DNA from eight or nine library plates. When a primary pool is found to be positive, a set of secondary pools can be obtained. In the case of 384-well plates, each of these sets consists of eight plate pools, 16 row pools, and 24 column pools. Screening of the secondary pools should give at least one positive for a plate, row, and column, pinpointing an individual positive clone, which can then be ordered from the distributor (*Figure 13*). If there are multiple positive clones in the same primary pool, the identification may be ambiguous; in this case the true positives must be identified later. The same PCR and electrophoresis conditions can be used as for mutant mapping (cf. *Protocol 3*), except that the number of cycles may have to be increased and the small number of reactions does not require robotic pipetting.

3. Gridded library filters (produced by spotting of bacterial colonies) can be screened by hybridization (*Protocol 8*). On RZPD filters, coordinates of positive clones are obtained by counting from left to right and from top to bottom, starting with 1. To simplify the counting, the clones are spotted in blocks of 3 × 3, 4 × 4 or 5 × 5. Most filters have a duplicate pattern in which clones are spotted twice in one block, so for each true positive another positive should be present in a particular position of the same block, providing an internal control (*Figure 14*). Clones are ordered by specifying the coordinates. Additional protocols and instructions for calculating coordinates are available from the RZPD website: <https://www.rzpd.de/general/html/glossary/protocols/>).

(a)

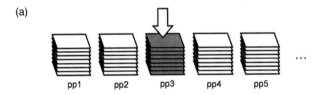

(b)

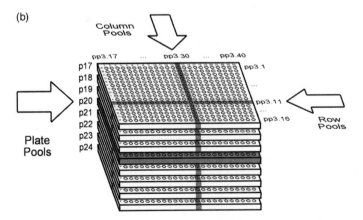

Figure 13 Screening of library pools. In this example, each primary pool was made from eight 384-well library plates. (a) One PCR is performed with each of the pools, resulting in amplification (shaded) for pool pp3. (b) PCR is performed with the corresponding 48 secondary pools. Positive results are obtained for plate pool p20 (= plate 20), row pool pp3-11 (= row K), and column pool pp3-30 (= column 14). The RZPD clone number of the positive clone (row, column, plate) is therefore K1420.

Protocol 8

Screening of library filters (after ref. 30 and RZPD protocols)

Equipment and reagents

- Hybridization box
- Incubator
- Shaker
- Radiation counter and safety equipment
- Water bath or heat block
- Centrifuge
- Ice bath
- Autoradiography scanner (Molecular Dynamics), or set-up for exposure and development of autoradiography films
- Library filters
- 20 × SSPE (3.6 M NaCl, 200 mM NaH₂PO₄, 20 mM EDTA, pH 7.4)

- 100 × Denhardt's solution (2% BSA, 2% Ficoll, 2% polyvinyl pyrrolidone)
- 10 mg/ml herring or salmon sperm DNA (sonicated, boiled, and chilled on ice)
- 10% SDS
- Probe DNA (PCR product or plasmid insert)
- [α-^{32}P]dCTP
- Labelling kit (e.g. Megaprime kit, Amersham)
- Screwcap microtubes
- 70% and 100% EtOH
- 3 M sodium acetate (NaAc), pH 5.6
- ddH₂O
- Saran wrap

Method

1 Add 25 ml prehybridization solution (5 × SSPE, 5 × Denhardt's solution, 0.5% SDS, 20 mg/ml herring sperm DNA) to library filters, prehybridize on a shaker for 1 h at 65 °C.[a]

2 Prepare radioactively labelled probe with [α-^{32}P]dCTP and a labelling kit, according to manufacturer's instructions. Precipitate by adding 2 volumes of EtOH, ¹⁄₁₀ volume of 3 M NaAc and 50 μg carrier DNA, spin for 15 min, wash with 70% EtOH, spin again and resuspend in 100 μl ddH₂O. Count radiation before and after precipitation to check for proper label incorporation.[b]

3 Boil probe for 5 min in a screwcap microtube, chill on ice, add to the prehybridization solution, and hybridize for at least 12 h at 65 °C.

4 Wash the filters twice with 2 × SSPE, 0.1% SDS at room temperature for 10 min.

5 Wash twice with 1 x SSPE, 0.1% SDS at 65 °C for 15 min.

6 Wrap filter in Saran wrap and perform autoradiography.

7 For removal of the probe (stripping) pour boiling 0.5% SDS on to the filters and let it cool to room temperature.[c]

[a] Several filters may be stacked in one hybridization box. Specificity may be improved by adding genomic zebrafish DNA or (CA)$_n$ oligonucleotides to the prehybridization solution to saturate repetitive sequences (42).

[b] Radioactive labelling is preferred in order to achieve a high sensitivity. The probe can also be purified on a spin column (Qiagen).

[c] Filters can be reused 6–8 times after stripping (except for YAC filters), or 10–12 times without stripping by allowing the radioactive label to decay.

When obtaining a clone from a library, streak it out on an agar plate, perform a miniprep, check its identity by PCR, and prepare a glycerol stock (*Protocol 9*). Because inserts may be partially or completely lost, it is important to determine the insert size. This can then be done by pulsed-field gel electrophoresis (PFGE), which resolves large DNA fragments in an electric field of changing orientation (*Protocol 10*).

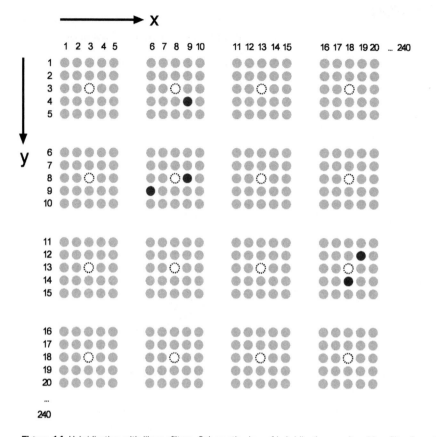

Figure 14 Hybridization with library filters. Schematic view of hybridization results with a filter from the RZPD. The coordinates of the positive spots (dark grey) are determined with the help of background spots (light grey) or positive control spots (see filter documentation). In order to see the background, an exposure may be performed before stringent washing. Each clone is spotted twice in one 5 × 5 block, therefore the spot at $x = 9$, $y = 4$ must be a false positive. Because the double spots can occur only in specific orientations ('vectors'), the orientation can be used to confirm the location of the spots within a block. The central spot of each block is free of DNA, but contains dye that can be used to align films or transparent printouts with the filters.

Protocol 9

Plasmid miniprep and preparation of glycerol stocks

Equipment and reagents

- Platinum wire loop
- Bunsen burner
- Microwave oven
- Microcentrifuge
- Ice bath
- Vacuum concentrator (SpeedVac, Savant)

- LB medium (1% Bacto tryptone, 0.5% Bacto yeast extract, 1.0% NaCl, 1.0% MgSO$_4$, adjusted to pH 7.2 with NaOH and autoclaved)
- LB agar (LB medium with 1.5 % agar)
- 10–100 mg/ml stock solutions of chloramphenicol (in EtOH), kanamycin or ampicillin (in H$_2$O), aliquots stored at −20°C

- 90 mm Petri dish
- BAC, PAC, or cDNA culture from a library
- 100% EtOH
- Autoclaved toothpicks
- 15 ml culture tubes
- Alkaline lysis solution I (5 mM sucrose, 10 mM EDTA, 25 mM Tris, pH 8.0)

- Alkaline lysis solution II (200 mM NaOH, 1% (w/v) SDS)
- Alkaline lysis solution III (3 M sodium acetate, pH 4.8)
- Autoclaved glycerol

Method

1 Plating:

   (a) Melt 15 ml LB agar in a microwave oven; after cooling add 20 µg/ml chloramphenicol (for BACs), 12.5 µg/ml kanamycin (for PACs) or 20 µg/ml ampicillin (for cDNA clones), and pour into a sterile Petri dish.[a]

   (b) Streak out culture with a wire loop, sterilized by dipping in EtOH and flaming. Incubate overnight at 37 °C.

   (c) Pick three colonies with a sterile toothpick, inoculate culture tubes with 5 ml of LB medium with 20 µg/ml chloramphenicol, 12.5 µg/ml kanamycin, or 20 µg/ml ampicillin. Incubate overnight at 37 °C on a rotating shaker.[b]

2 Alkaline lysis miniprep (after ref. 30):

   (a) Spin 1.5 ml of overnight culture for 10 min, and remove supernatant.[c]

   (b) Add 200 µl of solution I, mix, add 400 µl of solution II, mix thoroughly by inversion, and incubate on ice for 15 min.

   (c) Add 300 µl of solution III, mix thoroughly by inversion, incubate on ice for 15 min, spin for 5 min, and transfer supernatant into a fresh tube.

   (d) Add 500 µl isopropanol, spin for 20 min, wash with 70% EtOH, spin again, and dry in a Speed-Vac.

   (e) Resuspend DNA in 100–200 µl of TE.[d]

3 Preparation of glycerol stocks:

   (a) Check the identity and siz ture, add 250 µl of sterile glycerol. Store at −70 °C.

[a] Recommended concentrations of antibiotics vary widely; however, higher concentrations may promote loss of insert DNA.

[b] PACs can be grown just with kanamycin, or optionally induced at an OD_{590} of 0.1 by adding 1 mM IPTG.

[c] This protocol can be used for up to 30 ml of culture.

[d] Miniprep DNA is sufficiently pure for restriction digests and pulse-field electrophoresis. BAC and PAC DNA suitable for sequencing can be prepared with a Psi Clone Mini BAC or Big BAC kit (Princeton Separation).

[e] PCR is performed as described in *Protocol 1*. The PCR primers should be different from the primers or probe used for screening, as a safeguard against false positives.

Protocol 10

Determination of BAC or PAC insert size by pulsed-field gel electrophoresis (PFGE)

Equipment and reagents

- Incubator
- Pulsed field electrophoresis system (CHEF Mapper, Bio-Rad)
- Staining tray
- Shaker
- Fluorescence scanner (Molecular Dynamics) or gel imaging system (with green filter)
- BAC or PAC DNA (from a miniprep)
- 10 U/µl NotI or other restriction enzyme[a]
- 10 × restriction buffer (supplied by enzyme manufacturer)
- 10 mg/ml BSA
- ddH$_2$O
- 1 × TBE (90 mM Tris-borate, 90 mM boric acid, 2 mM EDTA, pH 8.0)

- Pulsed field certified agarose (Bio-Rad)
- 6 × Ficoll loading buffer (15% Ficoll, 0.25% Bromophenol Blue)
- PFGE marker in agarose (e.g. λ DNA PFGE Markers, Amersham Pharmacia Biotech, or MidRange II PFG Marker, New England BioLabs)
- 10 ng/µl 2–30 kb marker (e.g. Analytical Marker DNA Wide Range, Promega, or λ ×DNA HindIII digest, New England Biolabs) in 1 × Ficoll loading buffer
- 10 ng/µl 0.5–10 kb marker (e.g. GeneRuler 1 kb Ladder, Fermentas) in 1 × Ficoll loading buffer

Method

1 Mix 1 µg of purified DNA or several µgrams of miniprep DNA with 0.3 µl BSA, 3 µl 10 × restriction buffer, 1 µl NotI, and ddH$_2$O in a final volume of 30 µl, and digest for 1 h at 37 °C.

2 During digestion, pour a 100 ml gel containing 1% pulsed field certified agarose and 0.5 × TBE, using the CHEF Mapper casting stand.

3 Pour 5 litres 0.5 × TBE in the gel chamber and put the black platform in the middle. On the power supply, switch on power and pump switch. On the cooler, switch on power and set the temperature to 14 °C. Set the pump speed to 70.

4 Apply an aliquot of PFGE marker to a well of the gel, put the gel in the chamber, and apply 15 µl of 2–30 kb and 0.5–10 kb markers to additional wells. Add 5 µl of 6 × loading buffer to each digest and apply all to one well.

5 Select auto algorithm mode with a low molecular weight of 10 kb and a high molecular weight of 200 kb, start run.

6 Place gel in a staining tray with 200 ml 1:10 000 (v/v) SYBR-Green I solution in 0.5 × TBE, and shake for 30 min at room temperature in the dark.

7 Scan on a fluorescence scanner.

[a] NotI excises the inserts of all existing zebrafish BAC and PAC libraries.

If an identified clone is from one of the fingerprinted BAC libraries (RPCI-71, CHORI-211, DanioKey Pilot, or DanioKey), it may already be part of a contig; this can be determined by searching the WebFPC server at the Sanger Centre <http://www.sanger.ac.uk/Projects/D_rerio/> for the clone name. If the clone is not yet in a contig, it may be used to start a genomic walk. For this purpose, the ends of the clone are sequenced, PCR primers are designed from the sequences (e.g. with Primer3 <http://www-genome.wi.mit.edu/cgi-

bin/primer/primer3_www.cgi>), and used to identify further clones with one of the methods listed above. Several walking steps of this kind may have to be performed.

To locate a mutation on a specific genomic clone within a walk or contig, polymorphisms must be identified in the end sequences of the genomic clones and used for genotyping of mutant individuals. If SSLPs are not available, it may be necessary to identify SNPs, e.g. by SSCP analysis or by direct sequencing, and to continue sequencing until a polymorphism is found. As explained in Section 2.7, a correlation between the recombinants of two markers indicates that both markers are on the same side of the mutation. Therefore it is sufficient to genotype only those individuals for the new polymorphisms that were previously identified as recombinants for a flanking marker. As the mutant locus is approached, fewer of the recombinants for the flanking marker from which the walk was started should show a recombination. Recombinants for the flanking marker on the opposite side should show a recombination only when the walk is extended beyond the mutant locus, indicating the exact position of the mutation (*Figure 15*).

A limitation for positional cloning experiments is the suppression of genetic recombination in the centromeric regions. A comparison of radiation hybrid and genetic maps indicates that the physical distance corresponding to 1 cM is by at least an order of magnitude larger near the centromeres than near the telomeres, as reported for the human genome (56). Therefore positional cloning experiments may be impractical within several centimorgan of the centromeres. To determine whether this is a problem for a particular mutant locus, check whether the ratio of radiation hybrid and genetic map distances in the region of the mutation is far larger than the average of approximately 16 cR/cM. Alternatively, compare the positions of centromeric markers (available, for example, from the WashU website <http://www.genetics.wustl.edu/fish_lab/frank/cgi-bin/fish/meioticmap/meioticmaps.html>) to the map location of the mutation.

5.3 Analysing a genomic clone

When a mutation has been localized on a genomic clone, the clone may either be sequenced completely, or just used to search for cDNAs. In addition to local sequencing facilities and commercial services, sequencing of entire genomic walks is offered by the Sanger Centre <http://www.sanger.ac.uk/Projects/D_rerio/contig.shtml>. The sequence then needs to be

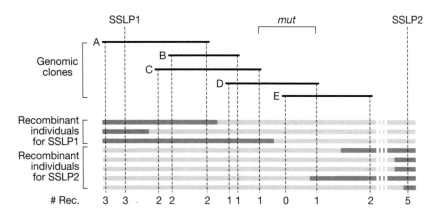

Figure 15 Schematic view of a genomic walk. The walk was started from SSLP1. The ends of the genome clones were tested for recombination in individuals previously identified as recombinant for SSLP1 or SSLP2. This places the mutant locus between the right end of clone C and the right end of clone D. Black = recombinant; grey = not recombinant.

annotated in order to find transcripts that may be hit by the mutation. Integrated software for genomic sequence annotation, such as the 'pipeline' that the Ensembl project <http://www.ensembl.org> employs for the human and mouse genome, is not yet publicly available. However, some of the programs that make up this pipeline are available as public web services:

- Repeats in the sequence can be masked with RepeatMasker <http://repeatmasker.genome.washington.edu/cgi-bin/RepeatMasker>
- Transcripts can be predicted, e.g. with GenScan <http://genes.mit.edu/GENSCAN.html>
- Information about homology and conserved domains can be obtained by searching Gene-Bank for the predicted peptides with BLASTP, by searching the NCBI Conserved Domain Database (CD-Search, included by default in BLASTP) or the SMART database of conserved domains <http://smart.embl-heidelberg.de/>
- Matches with ESTs can be obtained by searching the EST division of GenBank or the TIGR database on the nucleotide level with BLASTN, however, this type of search is slow and feasible only for short sequences. A much quicker, but less comprehensive search for ESTs and SSLPs can be conducted with Electronic PCR <http://www.ncbi.nlm.nih.gov/genome/sts/epcr.cgi> which uses primer sequences and expected product sizes contained in the UniSTS database to simulate PCR experiments.
- GeneWise <http://www.sanger.ac.uk/Software/Wise2/genewiseform.shtml> and Gene-Builder <http://l25.itba.mi.cnr.it/~webgene/genebuilder.html> can be used to refine alignments with peptides, but currently don't accept complete PAC or BAC sequences.

(Note that web forms don't accept more than 32000 characters per field; longer sequences must be uploaded as a file.)

Even if a genomic clone is being sequenced, a search for cDNAs should still be performed because predictions of transcripts have a limited reliability, and cDNA clones can be used for rescue experiments. The options available for cDNA screening are similar to those outlined for genomic libraries, i.e. database searching (assuming that a genomic sequence is available), PCR with pools, or hybridization with filters. More than 100000 zebrafish ESTs have been generated by partial sequencing of cDNA clones, by the WashU-Zebrafish Genome Resources Project (32) and to a smaller extent by the labs of C.L. Hew (57, 58) and C.C. Liew (59). ESTs matching a genomic sequence can be retrieved by searching GenBank, as described above, and the corresponding cDNA clones can then be ordered from the American Type Culture Collection, Incyte Genomics, or the RZPD. Several cDNA libraries are available for screening as pools (using primers derived from potential transcript sequences) or as filters (Table 4). As

Table 4 cDNA libraries

Library [a]	Stage or tissue	Clones arrayed [b]	Insert size [kb]	Creator by	Arrayed	Available [c]
Appel/Eisen 15–19-hour embryo (460)	15–19 h	n/d	1.7	B. Appel	IMAGE	F
Zebrafish SJD adult male (1763)	Adult male	n/d	n/d	J. Barnes	IMAGE	F, P
Zebrafish SJD day 8 fin regeneration (1764)	Fin, 8-day regeneration	n/d	n/d	J. Barnes	IMAGE	F
Zebrafish adult retina (1663, UWS_p760)	Retina	27648	3.0	S.E. Brockerhoff	IMAGE, RZPD	G, P
RZPD 609 (1733, MPMGp609)	Late somitogenesis + liver	38784	n/d	M.D. Clark	IMAGE, RZPD	G, F, P

Table 4 cDNA libraries

Library [a]	Stage or tissue	Clones arrayed [b]	Insert size [kb]	Creator by	Arrayed	Available [c]
Zebrafish embryo cDNA (ICRFp524)	Late somitogenesis	65 280	n/d	M.D. Clark	RZPD	G, F
Zebrafish liver cDNA (MPMGp532)	Liver	20 352	n/d	M.D. Clark	RZPD	P
Zebrafish shield cDNA (MPMGp567)	Shield	110 592	2.0	M.D. Clark	RZPD	G, F
Gong zebrafish ovary (1776)	Ovary	n/d	n/d	Z. Gong	IMAGE	F
Gong zebrafish testis (1776)	Testis	n/d	n/d	Z. Gong	IMAGE	F
Ekker maternal (452)	Oocytes + embryos	n/d	1.2	P. Hackett	IMAGE	F
Ekker embryo – early gastrulation (453)	6 h	n/d	1.7	P. Hackett	IMAGE	F
Ekker embryo – post-segmentation (454)	24 h	n/d	1.5	P. Hackett	IMAGE	F
Campbell zebrafish ovary (1808)	Ovary	n/d	2.0	Invitrogen	IMAGE	F
Zebrafish cDNA (DKFZp717)	15–17 h	27 648	1.1	B. Korn	RZPD	G, F
Zebrafish adult caudal fins cDNA (WUSMp623)	Fin, 0-day regeneration	27 648	1.6	R. Lee	RZPD	G, F
Zebrafish fin day 1 regeneration (1750, WUSMp624)	Fin, 1-day regeneration	27 648	1.9	R. Lee	IMAGE, RZPD	G, F, P
Zebrafish fin day 3 regeneration (1751, WUSMp625)	Fin, 3-day regeneration	27 648	1.3	R. Lee	IMAGE, RZPD	G, F, P
Zebrafish adult olfactory (1753, UCDMPp574)	Olfactory epithelium	27 648	1.7	J. Ngai	IMAGE, RZPD	G, F
Zebrafish adult brain (1754, MPMGp611)	Brain	55 296	2.6	J. Ngai	IMAGE, RZPD	G, F, P
Zebrafish neuronal (1752)	Brain	n/d	n/d	NN	IMAGE	F
Stainier zebrafish heart(626)	Heart	n/d	n/d	D. Stainier	IMAGE	F
SuganoSJD adult male (1812)	Adult male	n/d	n/d	S. Sugano	IMAGE	F
Sugano–Kawakami DRA (1333)	Adult (line AB)	n/d	n/d	S. Sugano [d]	IMAGE	F
Zebrafish C32 fin (1703)	Fin	n/d	n/d	N. Wu	IMAGE, ResGen	F
Zebrafish gridded kidney (1650, CHBOp575)	Kidney	110 592	2.0	L.I. Zon [e]	IMAGE, RZPD	G, F, P
Zebrafish kidney cDNA random primed (CHBOp576)	Kidney	110 592	2.0	L.I. Zon [e]	RZPD	G, F

Libraries were arrayed by the IMAGE consortium <http://image.llnl.gov/>, the RZPD <http://www.rzpd.de>, or ResGen <www.resgen.com>. Filters and pools of IMAGE libraries are available from the RZPD.

[a] IMAGE library names are given if possible. Library numbers used by IMAGE as well as prefixes and library numbers used by the RZPD are given in brackets.

[b] Number of arrayed clones for RZPD libraries only.

[c] G, GeneFinder pools (to preselect positive filters for screening); F, filters; P, pools.

[d] Ref. 60.

[e] Ref. 61.

an alternative to screening libraries, full-length cDNAs can be cloned from mRNA by rapid amplification of cDNA ends (RACE-PCR) (62). This term describes a variety of methods in which adapters are annealed or ligated to the transcript sequence, and the 5' and 3' ends are amplified with an internal, gene-specific primer and an adapter-specific primer. The fragments can then be ligated to recover the full-length cDNA. Several kits are available for RACE-PCR, e.g. the Marathon cDNA Amplification Kit (Clontech).

To demonstrate that a specific gene identified by positional cloning has been hit by a mutation, the same methods are available as in a candidate gene approach. In addition, it may be possible to achieve partial rescue of the mutant phenotype by PAC or BAC injection (63), using injection methods described elsewhere in this volume.

Acknowledgements

I would like to thank all present and former members of the Geisler and Haffter labs for their input. In particular I would like to thank Gerd-Jörg Rauch and Silke Rudolph-Geiger for selecting SSLPs and contributing to the embryo lysis and PCR protocols; Christian Busold for comparing the degree of polymorphism of the WIK and SJD lines relative to Tü; Jeremy Keenan for mapping *dolphin*; Manfred Klein for designing equipment; Axel Küchler for the SSCP and DDGE protocols and for linkage analysis of *bleached*; Günter Raddatz and Stefan Schuster for discussions on BAC annotation; Brit Rentzsch and Henry Roehl for the fin clip protocol; Eva Rief and Darren Gilmour for discussions on PAC and BAC screening; Gertrud Scheer for photography; Samuel Sidi for the PFGE protocol; Axel Küchler, Gerd-Jörg Rauch, and Christoph Seiler for critical reading of this manuscript; Ralf Dahm for editorial support and for his patience; Pascal Haffter for devising mutant mapping and radiation hybrid mapping strategies; and Christiane Nüsslein-Volhard, in whose department this work was carried out. I would like to acknowledge the support of the German Human Genome Project (DHGP Grant 01 KW 9919) and the US National Institutes of Health (NIH Grant 1 R01 DK55377–01A1).

References

1. Litt, M. and Luty, J.A. (1989). *Am. J. Hum. Genet.*, **44**, 397–401.
2. Tautz, D. (1989). *Nucleic Acids Res.*, **17**, 6463–71.
3. Goff, D.J., Galvin, K., Katz, H., Westerfield, M., Lander, E.S., and Tabin, C.J. (1992). *Genomics*, **14**, 200–2.
4. Knapik, E.W., Goodman, A., Atkinson, O.S., Roberts, C.T., Shiozawa, M., Sim, C.U., Weksler-Zangen, S., Trolliet, M.R., Futrell, C., Innes, B.A., Koike, G., McLaughlin, M.G., Pierre, L., Simon, J.S., Vilallonga, E., Roy, M., Chiang, P.W., Fishman, M.C., Driever, W., and Jacob, H.J. (1996). *Development*, **123**, 451–60.
5. Knapik, E.W., Goodman, A., Ekker, M., Chevrette, M., Delgado, J., Neuhauss, S., Shimoda, N., Driever, W., Fishman, M.C., and Jacob, H.J. (1998). *Nat. Genet.*, **18**, 338–43.
6. Kelly, P.D., Chu, F., Woods, I.G., Ngo-Hazelett, P., Cardozo, T., Huang, H., Kimm, F., Liao, L., Yan, Y.L., Zhou, Y., Johnson, S.L., Abagyan, R., Schier, A.F., Postlethwait, J.H., and Talbot, W.S. (2000). Genetic linkage mapping of zebrafish genes and ESTs. *Genome Res.*, **10**, 558–67.
7. Postlethwait, J.H., Yan, Y.L., Gates, M.A., Horne, S., Amores, A., Brownlie, A., Donovan, A., Egan, E.S., Force, A., Gong, Z., Goutel, C., Fritz, A., Kelsh, R., Knapik, E., Liao, E., Paw, B., Ransom, D., Singer, A., Thomson, M., Abduljabbar, T.S., Yelick, P., Beier, D., Joly, J.S., Larhammar, D., Rosa, F., Westerfield, M., Zon, L., Johnson, S., and Talbot, W.S. (1998). *Nat. Genet.*, **18**, 345–9.
8. Goss, S.J. and Harris, H. (1975). *Nature*, **255**, 680–4.
9. Walter, M.A., Spillett, D.J., Thomas, P., Weissenbach, J., and Goodfellow, P.N. (1994). *Nat. Genet.*, **7**, 22–8.
10. Kwok, C., Korn, R.M., Davis, M.E., Burt, D.W., Critcher, R., McCarthy, L., Paw, B.H., Zon, L.I., Goodfellow, P.N., and Schmitt, K. (1998). *Nucl. Acids Res.*, **26**, 3562–6.
11. Kwok, C., Critcher, R., and Schmitt, K. (1999). *Methods Cell Biol.*, **60**, 287–302.

12. Geisler, R., Rauch, G.-J., Baier, H., van Bebber, F., Broß, L., Davis, R.W., Dekens, M., Finger, K., Fricke, C., Gates, M.A., Geiger, H., Geiger-Rudolph, S., Gilmour, D., Glaser, S., Gnügge, L., Habeck, H., Hingst, K., Holley, S., Keenan, J., Kirn, A., Knaut, H., Lashkari, D., Maderspacher, F., Martyn, U., Neuhauss, S., Neumann, C., Nicolson, T., Pelegri, F., Ray, R., Rick, J., Roehl, H., Roeser, T., Schauerte, H. E., Schier, A. F., Schönberger, U., Schönthaler, H.-B., Schulte-Merker, S., Seydler, C., Talbot, W.S., Weiler, C., Nüsslein-Volhard, C. and Haffter, P. (1999). *Nat. Genet.*, **23**, 86–9.

13. Ekker, M., Ye, F., Joly, L., Tellis, P., and Chevrette, M. (1999). *Methods Cell Biol.*, **60**, 303–21.

14. Hukriede, N.A., Joly, L., Tsang, M., Miles, J., Tellis, P., Epstein, J.A., Barbazuk, W.B., Li, F.N., Paw, B., Postlethwait, J.H., Hudson, T.J., Zon, L.I., McPherson, J.D., Chevrette, M., Dawid, I.B., Johnson, S.L., and Ekker, M. (1999). *Proc. Natl Acad. Sci. USA*, **96**, 9745–50.

15. Johnson, S.L., Africa, D., Horne, S., and Postlethwait, J.H. (1995). *Genetics*, **139**, 1727–35.

16. Kauffman, E.J., Gestl, E.E., Kim, D.J., Walker, C., Hite, J.M., Yan, G., Rogan, P.K., Johnson, S.L., and Cheng, K.C. (1995). *Genomics*, **30**, 337–41.

17. Johnson, S.L., Gates, M.A., Johnson, M., Talbot, W.S., Horne, S., Baik, K., Rude, S., Wong, J.R., and Postlethwait, J.H. (1996). *Genetics*, **142**, 1277–88.

18. Mohideen, M.A., Moore, J.L., and Cheng, K.C. (2000). *Genomics*, **67**, 102–6.

19. Michelmore, R.W., Paran, I., and Kesseli, R.V. (1991). *Proc. Natl Acad. Sci USA*, **88**, 9828–32.

20. Neuhauss, S. (1996). Craniofacial development in zebrafish (*Danio rerio*): mutational analysis, genetic characterization, and genomic mapping. Dissertation, Universität Tübingen.

21. Rauch, G.-J., Granato, M., and Haffter, P. (1997). A polymorphic zebrafish line for genetic mapping using SSLPs on high-percentage agarose gels. Technical Tips Online T01208.

22. Johnson, S.L. and Zon, L.I. (1999). *Methods Cell Biol.*, **60**, 357–9.

23. Chakrabarti, S., Streisinger, G., Singer, F., and Walker, C. (1983). *Genetics*, **103**, 109–24.

24. Haffter, P., Odenthal, J., Mullins, M.C., Lin, S., Farrell, M.J., Vogelsang, E., Haas, F., Brand, M., van Eeden, F.J.M., Furutani-Seiki, M., Granato, M., Hammerschmidt, M., Heisenberg, C.-P., Jiang, Y.-J., Kane, D.A., Kelsh, R.N., Hopkins, N., and Nüsslein-Volhard, C. (1996). *Dev. Genes Evol.*, **206**, 260–76.

25. Postlethwait, J.H., Johnson, S.L., Midson, C.N., Talbot, W.S., Gates, M., Ballinger, E.W., Africa, D., Andrews, R., Carl, T., Eisen, J.S., *et al.* (1994). *Science*, **264**, 699–703.

26. Rauch, G.-J. (2000). Von der Mutation zum Gen im Zebrafisch (*Danio rerio*). Dissertation, Universität Tübingen.

27. Ott, J. (1992). *Am. J. Hum. Genet.*, **51**, 283–90.

28. Liu, B.H. (1998). *Statistical genomics: linkage, mapping, and QTL analysis.* CRC Press, Boca Raton, FL.

29. Schauerte, H. (2000). Die Sonic hedgehog-Signaltransduktion in der Mittellinie des Zebrafisches (*Danio rerio*). Dissertation, Universität Tübingen.

30. Sambrook, J., Fritsch, E.F., and Maniatis, T. (1989). *Molecular cloning: A laboratory manual,* (2nd edn). Cold Spring Harbor Laboratory Press, Cold Spring Harbor, New York.

31. Barbazuk, W.B., Korf, I., Kadavi, C., Heyen, J., Tate, S., Wun, E., Bedell, J.A., McPherson, J.D., and Johnson, S.L. (2000). *Genome Res.*, **10**, 1351–8.

32. Johnson, S.L., Waterman, R.E., Noriega-Fenton, L., Sheng, X., Hersey, C., Rauch, G.-J., Geisler, R., Fischer, D.R., Barbazuk, W.B., Cox, K., Ekker, M., Hukriede, N., Daniluk, D., Song, A., Arodi, S., Foote, H., Sugano, S., Kawakami, K., Wu, N., Brockerhoff, S., Lee, R., Ngai, J., Clark, M., Prange, C., Zon, L.I., Clifton, S., McPherson, J.D., and Zhou, Y. (2001). Submitted.

33. Gates, M.A., Kim, L., Egan, E.S., Cardozo, T., Sirotkin, H.I., Dougan, S.T., Lashkari, D., Abagyan, R., Schier, A.F., and Talbot, W.S. (1999). *Genome Res.*, **9**, 334–47.

34. Woods, I.G., Kelly, P.D., Chu, F., Ngo-Hazelett, P., Yan, Y.L., Huang, H., Postlethwait, J.H., and Talbot, W.S. (2000) *Genome Res.*, **10**, 1903–14.

35. Orita, M., Iwahana, H., Kanazawa, H., Hayashi, K., and Sekiya, T. (1989). *Proc. Natl Acad. Sci. USA*, **86**, 2766–70.

36. Sheffield, V.C., Beck, J.S., Kwitek, A.E., Sandstrom, D.W., and Stone, E.M. (1993). *Genomics*, **16**, 325–32.

37. Förnzler, D., Her, H., Knapik, E.W., Clark, M., Lehrach, H., Postlethwait, J.H., Zon, L.I., and Beier, D.R. (1998). *Genomics*, **51**, 216–22.

38. Fischer, S.G. and Lerman, L.S. (1983). *Proc. Natl Acad. Sci. USA*, **80**, 1579–83.

39. Myers, R.M., Fischer, S.G., Lerman, L.S., and Maniatis, T. (1985a). *Nucleic Acids Res.*, **13**, 3131–45.

40. Myers, R.M., Fischer, S.G., Maniatis, T., and Lerman, L.S. (1985b). *Nucleic Acids Res.*, **13**, 3111–29.
41. Kishimoto, Y., Lee, K.H., Zon, L., Hammerschmidt, M., and Schulte-Merker, S. (1997). *Development*, **124**, 4457–66.
42. Talbot, W.S. and Schier, A.F. (1999). *Methods Cell Biol.*, **60**, 259–86.
43. Zhang, J., Talbot, W.S., and Schier, A.F. (1998). *Cell*, **92**, 241–51.
44. Brownlie, A., Donovan, A., Pratt, S.J., Paw, B.H., Oates, A.C., Brugnara, C., Witkowska, H.E., Sassa, S., and Zon, L.I. (1998). *Nat. Genet.*, **20**, 244–50.
45. Donovan, A., Brownlie, A., Zhou, Y., Shepard, J., Pratt, S.J., Moynihan, J., Paw, B.H., Drejer, A., Barut, B., Zapata, A., Law, T.C., Brugnara, C., Lux, S.E., Pinkus, G.S., Pinkus, J.L., Kingsley, P.D., Palis, J., Fleming, M.D., Andrews, N.C., and Zon, L.I. (2000). *Nature,* **403**, 776–81.
46. Guo, S., Yamaguchi, Y., Schilbach, S., Wada, T., Lee, J., Goddard, A., French, D., Handa, H., and Rosenthal, A. (2000). *Nature,* **408**, 366–9.
47. Kupperman *et al.* 2000
48. Coulson, A., Waterston, R., Kiff, J., Sulston, J., and Kohara, Y. (1988). *Nature,* **335**, 184–6.
49. Zhong, T.P., Kaphingst, K., Akella, U., Haldi, M., Lander, E.S., and Fishman, M.C. (1998). *Genomics,* **48**, 136–8.
50. Amemiya, C.T., Zhong, T.P., Silverman, G.A., Fishman, M.C., and Zon, L.I. (1999). *Methods Cell Biol.*, **60**, 235–58.
51. Ioannou, P.A., Amemiya, C.T., Garnes, J., Kroisel, P.M., Shizuya, H., Chen, C., Batzer, M.A., and de Jong, P.J. (1994). *Nat. Genet.*, **6**, 84–9.
52. Shizuya, H., Birren, B., Kim, U.J., Mancino, V., Slepak, T., Tachiiri, Y., and Simon, M. (1992). *Proc. Natl Acad. Sci. USA*, **89**, 8794–7.
53. Amemiya, C.T. and Zon, L.I. (1999). *Genomics,* **58**, 211–13.
54. Collins, J., and Bruning, H.J. (1978). *Gene*, **4,** 85–107.
55. Burgtorf, C., Welzel, K., Hasenbank, R., Zehetner, G., Weis, S., and Lehrach, H. (1998). *Genomics,* **52**, 230–2.
56. Payseur, B.A. and Nachman, M.W. (2000). *Genetics,* **156**, 1285–98.
57. Gong, Z., Yan, T., Liao, J., Lee, S.E., He, J., and Hew, C.L. (1997). *Gene*, **201**, 87–98.
58. Gong, Z. (1999). *Methods Cell Biol.*, **60**, 213–33.
59. Ton, C., Hwang, D.M., Dempsey, A.A., Tang, H.C., Yoon, J., Lim, M., Mably, J.D., Fishman, M.C., and Liew, C.C. (2000). *Genome Res.*, **10**, 1915–27.
60. Suzuki, Y., Yoshitomo-Nakagawa, K., Maruyama, K., Suyama, A., and Sugano, S. (1997). *Gene,* **200**, 149–56.
61. Blake, T., Adya, N., Kim, C.H., Oates, A.C., Zon, L., Chitnis, A., Weinstein, B.M., and Liu, P.P. (2000). *Blood,* **96**, 4178–84.
62. Frohman, M.A., Dush, M.K., and Martin, G.R. (1988). *Proc. Natl Acad. Sci. USA*, **85**, 8998–9002.
63. Yan, Y.L., Talbot, W.S., Egan, E.S., and Postlethwait, J.H. (1998). *Genomics,* **50**, 287–9.

Appendix 1
List of suppliers

Agar Scientific Ltd., 66A Cambridge Road,
Stansted, Essex CM24 8DA
Tel: +44(0)1279 813519
Fax: +44(0)1279 815106
http://www.agarscientific.com

Ambion (Europe) Ltd., Spitfire Close,
Ermine Business Park, Huntingdon,
Cambridgeshire PE29 6XY, UK
Tel: 0800 181 3273
Tel: 0800 182 3578
Fax: 0800 181 3276
Fax: 0800 182 3579

Amersham Biosciences Europe GmbH,
Munzinger Str. 9,
D-79111Freiburg, Germany

Amersham Biosciences,
928 East Arques Avenue,
Sunnyvale, CA 94085-4520, USA
Toll free: 1-800-333-5703
Tel: 1-408-773-1222
Fax: 1-408-773-8343

Amersham Pharmacia Biotech UK Ltd, Amersham
Place, Little Chalfont, Buckinghamshire HP7 9NA,
UK (see also Nycomed Amersham Imaging UK;
Pharmacia)
Tel: 0800 515313
Fax: 0800 616927
URL: http//www.apbiotech.com/

Anderman and Co. Ltd, 145 London Road,
Kingston-upon-Thames, Surrey KT2 6NH,
UK
Tel: 0181 5410035
Fax: 0181 5410623

Aquaculture Supply, 33418 Old St Joe Rd,
Dade City, FL 33525, USA
Tel: 352-567-8540
Fax: 352-567-3742

Aquaculture Supply, Chemin de Aulx, CH-1228
Plan-les-Ouates, Geneva, Switzerland
Tel: (+41) 22-794-8000
Fax: (+41) 22-794-9143

ASI, Applied Science Intruments,
29391 W. Enid, Eugene, OR 97402,
USA

American Type Culture Collection (ATCC),
PO Box 1549, Manassas, VA 20108,
USA

Aventis S.A., F-67917 Strasbourg,
Germany
URL: http://www.hoechst.com

Avery Dennison Corporate Center,
150 North Orange Grove Boulevard,
Pasadena, CA 91103-3596, USA
Tel: (626) 304-2000
Fax: (626) 792-7312

BD Biosciences Clontech,
Tullastrasse 4, 69126 Heidelberg, Germany
Tel: 49 6221 3417-0
Fax: 49 6221 303 511

Beckman Coulter (UK) Ltd, Oakley Court,
Kingsmead Business Park, London Road,
High Wycombe, Buckinghamshire HP11 1JU,
UK
Tel: 01494 441181
Fax: 01494 447558
URL: http://www.beckman.com/

Beckman Coulter Inc., 4300 N. Harbor Boulevard, PO Box 3100, Fullerton, CA 92834–3100, USA
Tel: 001 714 8714848
Fax: 001 714 7738283
URL: http://www.beckman.com/

Becton Dickinson and Co., 21 Between Towns Road, Cowley, Oxford OX4 3LY, UK
Tel: 01865 748844
Fax: 01865 781627
URL: http://www.bd.com/

Becton Dickinson and Co., 1 Becton Drive, Franklin Lakes, NJ 07417–1883, USA
Tel: 001 201 8476800
URL: http://www.bd.com/

Bio 101 Inc., c/o Anachem Ltd, Anachem House, 20 Charles Street, Luton, Bedfordshire LU2 0EB, UK
Tel: 01582 456666
Fax: 01582 391768
URL: http://www.anachem.co.uk/
Bio 101 Inc., PO Box 2284, La Jolla, CA 92038–2284, USA
Tel: 001 760 5987299
Fax: 001 760 5980116
URL: http://www.bio101.com/

Bio-Rad Laboratories Ltd, Bio-Rad House, Maylands Avenue, Hemel Hempstead, Hertfordshire HP2 7TD, UK
Tel: 0181 3282000
Fax: 0181 3282550
URL: http://www.bio-rad.com/

Bio-Rad Laboratories Ltd, Division Headquarters, 1000 Alfred Noble Drive, Hercules, CA 94547, USA
Tel: 001 510 7247000
Fax: 001 510 7415817
URL: http://www.bio-rad.com/

Bio-Rad Laboratories GmbH, Heidemannstrasse 164, D-80939 München, Postfach 45 01 33, D-80901 München, Germany
Tel: 089-318840
Fax: 089-318-84123

BioTech Trade & Service GmbH, Molecular Research Center, St. Leon-Rot, Germany
Tel: 49-6227-51308
Fax: 49-6227-53694

Birkenstock, BIRKI Schuh GmbH, Rheinstr. 2-4, 53560 Vettelschoss, Germany
Tel: (+49) 2645-942-800
Fax: (+49) 2645-942-801

Boehringer-Mannheim (see Roche)

Carl Roth GmbH & Co., Schoemperlenstr. 1-5, 76185 Karlsruhe, Germany
Postfach 10 01 21, 76231 Karlsruhe, Germany

Carolina Biological Supply Co., 2700 York Road, Burlington, NC 27215-3398, USA
Tel: 800-334-5551 (US), 336-584-0381 (International)
URL: http://www.carolina.com/

CP Instrument Co. Ltd, PO Box 22, Bishop Stortford, Hertfordshire CM23 3DX, UK
Tel: 01279 757711
Fax: 01279 755785
URL: http//:www.cpinstrument.co.uk/

Dage-MTI, Inc., 701N Roeske Avenue, Michigan City, IN 46360, USA

DMZ Universal (see Zeitz)

Dupont (UK) Ltd, Industrial Products Division, Wedgwood Way, Stevenage, Hertfordshire SG1 4QN, UK
Tel: 01438 734000
Fax: 01438 734382
URL: http://www.dupont.com/

Dupont Co. (Biotechnology Systems Division), PO Box 80024, Wilmington, DE 19880–002, USA
Tel: 001 302 7741000
Fax: 001 302 7747321
URL: http://www.dupont.com/

Eastman Chemical Co., 100 North Eastman Road, PO Box 511, Kingsport, TN 37662–5075, USA
Tel: 001 423 2292000
URL: http//:www.eastman.com/

Ehret GmbH, Postfach 1230, 7830 Emmendingen 14, Germany
Tel: (+49)7641-92650
Fax: (+49)7641-47972

EMP Biotech GmbH,
Max Delbrüeck Center for Molecular Medicine,
Biomedical Research Campus,
Robert-Röessle-Str. 10,
D-13125 Berlin, Germany
Tel: +49 (30) 94 89 2201
Fax: +49 (30) 94 89 3201
URL: http://www.empbiotech.com

Ensinger GmbH,
Rudolf-Diesel-Strasse 8,
71154 Nufringen, Germany
Tel: +49 (0) 7032/819-0

Eppendorf AG, Barkhausenweg 1, 22331
Hamburg, Germany

ETC Elektrophorese-Technik,
Gewerbepark Carl Schirm, Bahnhofstr.26,
D-72138 Kirchentellinsfurt, Germany
Tel: +49/7121/60253,
Fax: +49/7121/60252

FiMö Aquaristik GmbH, Hückerstr. 113, D-32257
Bünde, Germany
Tel: (+49) 5223-1888-40
Fax: (+49) 5223-1888-49

Fisher Scientific UK Ltd, Bishop Meadow Road,
Loughborough, Leicestershire LE11 5RG, UK
Tel: 01509 231166
Fax: 01509 231893
URL: http://www.fisher.co.uk/

Fisher Scientific, Fisher Research, 2761 Walnut
Avenue, Tustin, CA 92780, USA
Tel: 001 714 6694600
Fax: 001 714 6691613
URL: http://www.fishersci.com/

Fluka, PO Box 2060, Milwaukee, WI 53201, USA
Tel: 001 414 2735013
Fax: 001 414 2734979
URL: http://www.sigma-aldrich.com/

Fluka Chemical Co. Ltd, PO Box 260, CH-9471,
Buchs, Switzerland
Tel: 0041 81 7452828
Fax: 0041 81 7565449
URL: http://www.sigma-aldrich.com/

Genome Systems,
Incyte Genomics, Inc., 3160 Porter Drive,
Palo Alto, CA 94304, USA
Tel: +1 650-855-0555
Fax: +1 650-855-0572

Gibco-BRL (see Life Technologies)

Greiner Labortechnik GmbH, Postfach 1162,
72632 Frickenhausen, Germany
Tel.: (+49) 7022-948506,
Fax: (+49) 7022-948 514

Hamilton Bonaduz AG,
PO Box 26,
CH-7402 Bonaduz,
Switzerland
Tel: 41-81-660-60-60

Hamilton Deutschland GmbH,
PO Box 110565,
D-64220,
Darmstadt/FRG, Germany
Tel: +49-6151-98-02-0
Fax: +49-6151-89-17-33

Hoechst (see Aventis)

Hybaid Ltd, Action Court, Ashford Road, Ashford,
Middlesex TW15 1XB, UK
Tel: 01784 425000
Fax: 01784 248085
URL: http://www.hybaid.com/

Hybaid US, 8 East Forge Parkway, Franklin,
MA 02038, USA
Tel: 001 508 5416918
Fax: 001 508 5413041
URL: http://www.hybaid.com/

HyClone Laboratories, 1725 South HyClone Road,
Logan, UT 84321, USA
Tel: 001 435 7534584
Fax: 001 435 7534589
URL: http//:www.hyclone.com/

Inve Aquaculture, Hoogveld 91, B-9200
Dendermonde, Belgium
Tel: +32.52.259070
Fax: +32.52.259080

Invitrogen Corp., 1600 Faraday Avenue,
Carlsbad, CA 92008,
USA
Tel: 001 760 6037200
Fax: 001 760 6037201
URL: http://www.invitrogen.com/

Invitrogen BV, PO Box 2312, 9704 CH Groningen,
The Netherlands
Tel: 00800 53455345
Fax: 00800 78907890
URL: http://www.invitrogen.com/

Invitrogen GmbH, Technologiepark Karlsruhe, Emmy-Noether Strasse 10, 76131 Karlsruhe, Germany
Tel: 0800-0 83 09 02
Fax: 0800-0 83 34 35

Leitz GmbH & Co KG, Siemensstraße 64, D-70469 Stuttgart, Germany
Tel: +49 711 8103-0
Fax: +49 711 8103-486
URL: http://www.leitz.de

Life Technologies Ltd, PO Box 35, 3 Free Fountain Drive, Inchinnan Business Park, Paisley PA4 9RF, UK
Tel: 0800 269210
Fax: 0800 243485
URL: http://www.lifetech.com/

Life Technologies Inc., 9800 Medical Center Drive, Rockville, MD 20850, USA
Tel: 001 301 6108000
URL: http://www.lifetech.com/

Marine Biotech, Inc., 54 West Dane Street, Unit A, Beverly, MA 01915, USA
Tel: (978) 927-8720
Fax: (978) 921-0231
URL: http://www.marinebiotech.com;

Merck and Co. Inc., Whitehouse Station, NJ, USA

Merck Sharp & Dohme Research Laboratories, Neuroscience Research Centre, Terlings Park, Harlow, Essex CM20 2QR, UK
URL: http://www.msd-nrc.co.uk/

MSD Sharp and Dohme GmbH, Lindenplatz 1, D-85540, Haar, Germany
URL: http://www.msd-deutschland.com/

Millipore (UK) Ltd, The Boulevard, Blackmoor Lane, Watford, Hertfordshire WD1 8YW, UK
Tel: 01923 816375
Fax: 01923 818297
URL: http://www.millipore.com/local/UKhtm/

Millipore Corp., 80 Ashby Road, Bedford, MA 01730, USA
Tel: 001 800 6455476
Fax: 001 800 6455439
URL: http://www.millipore.com/

Miyako Kagaku Co. Ltd, New Kokusai Bldg. 4-1. Marunouchi 3 Chome Chiyoda-Ku, Tokyo 100-0005, Japan

MJ Research, Inc., Waltham, MA 02451 USA
Tel: 888-735-8437
Fax: 617-923-8080

Molecular Dynamics, Formerly known as: Amersham Biosciences, 928 East Arques Avenue, Sunnyvale, CA 94085-4520, USA
Toll free: 1-800-333-5703
Tel: 1-408-773-1222
Fax: 1-408-773-8343
URL: http://www.mdyn.com

Molecular Probes, PO Box 22010, Eugene, Oregon97402-0469, USA

Molecular Probes Europe BV, PoortGebouw, Rijnsburgerweg 10, 2333 AA Leiden, The Netherlands
Tel: +31-71-5233378
Fax: +31-71-5233419

Molecular Research Center, Inc., 5645 Montgomery Road, Cincinnati, OH 45212, USA
URL: http://www.mrcgene.com/

Müller & Pfleger GmbH Aquarien, Industriegebiet Kreuzwiese 13, D-67806 Rockenhausen, Germany
Tel: (+49) 6361 92160
Fax: (+49) 6361 7644

Narshige, 1710 Hempstead Turnpike, East Meadow, NY 11554, USA

Narishige International Ltd., Unit 7, Willow Business Park, Willow Way, London SE26 4QP, UK
Tel: 44 (0) 20 8 699 9696
Fax: 44 (0) 20 8 291 9678

New England Biolabs, 32 Tozer Road, Beverley, MA 01915–5510, USA
Tel: 001 978 9275054

New England Biolabs GmbH, Bruningstrasse 50, Geb.G 810, 65926 Frankfurt am Main, Germany
Tel: (0)69/305 23140
Free Call: 0800/246 5227 (0800 BIOLABS)
Fax: (0)69/305 23149
Free Fax: 0800/246 5229 (0800 BIOLABX)

Nikon Inc., 1300 Walt Whitman Road,
Melville, NY 11747–3064, USA
Tel: 001 516 5474200
Fax: 001 516 5470299
URL: http://www.nikonusa.com/
Nikon Corp., Fuji Building, 2–3, 3-chome,
Marunouchi, Chiyoda-ku, Tokyo 100, Japan
Tel: 00813 32145311
Fax: 00813 32015856
URL: http://www.nikon.co.jp/main/index_e.htm/

Nycomed Amersham Imaging, Amersham Labs,
White Lion Rd, Amersham,
Buckinghamshire HP7 9LL, UK
Tel: 0800 558822 (or 01494 544000)
Fax: 0800 669933 (or 01494 542266)
URL: http//:www.amersham.co.uk/
Nycomed Amersham, 101 Carnegie Center,
Princeton, NJ 08540, USA
Tel: 001 609 5146000
URL: http://www.amersham.co.uk/

Perkin Elmer Ltd, Post Office Lane, Beaconsfield,
Buckinghamshire HP9 1QA, UK
Tel: 01494 676161
URL: http//:www.perkin-elmer.com/

Pharmacal Research Labs, 33 Great Hill Rd.
Naugatuck, CT 06770, USA
Tel: 1-800-243-5350
URL: http://www.pharmacal.com/

Pharmacia, Davy Avenue, Knowlhill,
Milton Keynes, Buckinghamshire MK5 8PH,
UK
(also see Amersham Pharmacia Biotech)
Tel: 01908 661101
Fax: 01908 690091
URL: http//www.eu.pnu.com/

Pharmacia GmbH,
Am Wolfsmantel 46,
D-91058 Erlangen,
Germany
Tel: 49 9131 62-0
Fax: 49 9131 62-1202

Princeton Separations
PO Box 300,
Adelphia, NJ 07710, USA
Toll Free: 800 223 0902
Tel: 732 431 3338
Fax: 732 431 3768

Promega UK Ltd, Delta House, Chilworth
Research Centre, Southampton SO16 7NS,
UK
Tel: 0800 378994
Fax: 0800 181037
URL: http://www.promega.com/
Promega Corp., 2800 Woods Hollow Road,
Madison, WI 53711–5399,
USA
Tel: 001 608 2744330
Fax: 001 608 2772516
URL: http://www.promega.com/

Promega GmbH,
High-Tech-Park,
Schildkrötstraße 15,
Mannheim D-68199, Germany
Tel: (49) 621 85010
Fax: (49) 621 8501 222
URL: http://www.promega.com/de/

Qiagen UK Ltd, Boundary Court,
Gatwick Road, Crawley, West Sussex RH10 2AX,
UK
Tel: 01293 422911
Fax: 01293 422922
URL: http://www.qiagen.com/
Qiagen Inc., 28159 Avenue Stanford,
Valencia, CA 91355,
USA
Tel: 001 800 4268157
Fax: 001 800 7182056
URL: http://www.qiagen.com/

QIAGEN GmbH,
Max-Volmer-Straße 4,
40724 Hilden,
Germany
Tel: 02103-29-12000
Fax: 02103-29-22000

Roche Diagnostics Ltd, Bell Lane, Lewes,
East Sussex BN7 1LG, UK
Tel: 0808 1009998 (or 01273 480044)
Fax: 0808 1001920 (01273 480266)
URL: http://www.roche.com/
Roche Diagnostics Corp., 9115 Hague Road,
PO Box 50457, Indianapolis, IN 46256,
USA
Tel: 001 317 8452358
Fax: 001 317 5762126
URL: http://www.roche.com/

Roche Diagnostics GmbH, Sandhoferstrasse 116,
68305 Mannheim, Germany
Tel: 0049 621 7594747
Fax: 0049 621 7594002
URL: http://www.roche.com/
Roche Molecular Biology
URL: http://biochem.roche.com.

Schwarz Germany, Aquarienbau Schwarz,
Maselmühlweg 40-42, 37081 Göttingen,
Germany
Tel: (+49) 551-3850780
Fax: (+49) 551-3850788
Email: Aqua_Schwarz@t-online.de

Schleicher and Schuell Inc., Keene, NH 03431A,
USA
Tel: 001 603 3572398

Schülke&Mayr GmbH Germany, Robert-Koch-Str.
2, 22851 Norderstedt, Germany
Tel: (+49)40-521 00-0
Fax: (+49)40-521 00-318

Semadeni, Tägetlistr. 35-39, Industriezone Obere
Zollgasse, Postfach, CH-3072 Ostermundingen,
Switzerland
Tel: (+31) 931 3531
Fax (+31) 9311625.

Shandon Scientific Ltd, 93–96 Chadwick Road,
Astmoor, Runcorn, Cheshire WA7 1PR,
UK
Tel: 01928 566611
URL: http//www.shandon.com/

J. L. Shepherd and Associates,
1010 Arroyo Street, San Fernando,
CA 91340, USA

Sigma–Aldrich Co. Ltd, The Old Brickyard,
New Road, Gillingham, Dorset SP8 4XT, UK
Tel: 0800 717181 (or 01747 822211)
Fax: 0800 378538 (or 01747 823779)
URL: http://www.sigma-aldrich.com/
Sigma Chemical Co., PO Box 14508, St Louis,
MO 63178, USA
Tel: 001 314 7715765
Fax: 001 314 7715757
URL: http://www.sigma-aldrich.com/

Stratagene Inc., 11011 North Torrey Pines Road,
La Jolla, CA 92037, USA
Tel: 001 858 5355400
URL: http://www.stratagene.com/

Stratagene Europe, Gebouw California,
Hogehilweg 15, 1101 CB Amsterdam Zuidoost,
The Netherlands
Tel: 00800 91009100
URL: http://www.stratagene.com/

Tetra Germany: For Tetra-AZ, contact Dr Schmidt
or Dr Kürzinger, Tetra Werke, Herrenteich 78,
49304 Melle, Germany
Tel: (+49) 5422-105273
Fax: (+49) 5422-42985.

That Fish Place, 237 Centerville Road Lancaster,
PA 17603, USA
Tel: (+1)-800-733-3829
Fax: (+1)-800-786-3829

Thermo Corion, 8E Forge Parkway, Franklin,
MA 02038, USA
Tel: 508-528-4411
Fax: 508-520-7583

Thermo Hybaid, Head office, Action Court,
Ashford Road, Ashford, Middlesex TW15 1XB, UK

Thermo Spïectronic, 820 Linden Avenue,
Rochester, NY 14625, USA

United States Biochemical (**USB**), PO Box 22400,
Cleveland, OH 44122, USA
Tel: 001 216 4649277

Vector Laboratory, 30 Ingold Road, Buslingame,
CA 94010, USA

Victoreen, Inovision Radiation Measurements,
6045 Cochran Road, Cleveland, OH 44139,
USA

Wheaton Science Products, 1501 N. 10th Street,
Millville, NJ 08332-2093, USA

World Precision Instruments, Liegnitzer Strasse 15,
D-10999 Berlin, Germany
Tel: 030-6188845
Fax: 030-6188670

World Precision Instruments, International Trade
Center, 175 Sarasta Center Road, Sarasota,
FL 34240, USA

Zeitz-Instruments, Vertriebs GmbH,
Peter-Schlemihl-Srabe 19,
81377 München, Germany

Zoo Fachring e. G., Postfach 9, 7127 Pleidelsheim,
Germany
Tel: (+49) 7144 - 8119-0.

Atlas of embryonic stages of development in the zebrafish

Ralf Dahm

Max-Planck-Institut für Entwicklungsbiologie, Abteilung Genetik, Spemannstr. 35, 72076 Tübingen, Germany

Introduction

This stage atlas is intended as a quick reference guide to staging zebrafish during embryogenesis. It covers the normal development from fertilization up until the embryos hatch from the chorion 3 days after fertilization. The text is subdivided into characteristic developmental periods. A brief description summarizing the main events and characteristics is given for each developmental time point. This is followed by descriptions of exemplary stages, which are illustrated with images, and the figure in brackets after each stage denotes when the stage begins. All images were taken using a Zeiss Axiophot compound microscope. However, the description of the features characteristic for the developmental stages described in this appendix generally refer to what is visible in a stereomicroscope. If a special microscopy technique is needed in order to visualize certain features, this is indicated in the text.

Please note that all embryos up until 48 hours post-fertilization (hpf) were dechorionated prior to taking the images. The stages were termed as described in ref. 1. All reference to the period of time elapsed after fertilization is for embryos that developed at a constant temperature of 28.5°C. Moderate increases or decreases in temperature will lead to a faster or slower development, respectively.

Day 1

Zygote period (0–¾ hpf)

See *Figure 1*.

The zygote period begins with the fertilization of the oocyte and ends with the first cleavage at approximately ¾ hpf. Five to seven minutes post-fertilization (mpf), the second polar body is excluded. Between 16 and 18 mpf, the pronuclei fuse and 24–26 mpf, the first mitotic spindle forms. Cleavage begins 35 minutes after fertilization and is complete by 45 minutes.

1 cell (¼ hpf)

When the oocyte is laid, the chorion still closely surrounds the cell and the cytoplasm has not yet separated from the yolk. Contact with water induces the chorion to swell and lift off the cell. Additionally, it induces *cytoplasmic streaming*, the process by which the non-yolk cytoplasm of the zygote streams towards the animal pole of the cell, resulting in a clear cytoplasmic part at the animal pole and a vegetal part rich in yolk granules. This segregation of yolk and non-yolk cytoplasm continues throughout the early cleavages.

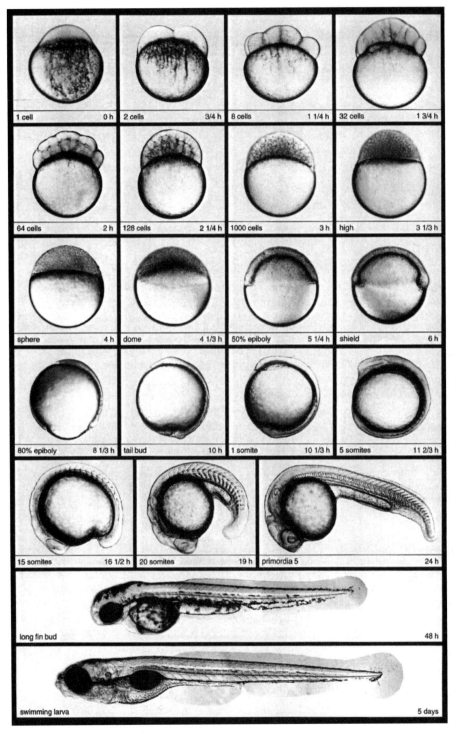

Figure 1 DIC images of live zebrafish at different developmental stages. The time after fertilization is given in hours (h) when raised at 28.5°C.

Cleavage period (¾–2¼hpf)

See *Figure 1*.

The cleavage period begins with the first cell division and includes embryos up until the 64-cell stage. After the first cleavage has occurred, the *blastomeres* divide synchronously at regular intervals of approximately 15 minutes. The early cleavages occur at regular orientations perpendicular to each other. The cleavages are incomplete, leaving the cells connected with each other and the underlying yolk cell.

Within each stage it is possible, using DIC optics, to determine the position in the cell cycle from the shape of the blastomere nuclei. During early interphase the nuclei have a globular appearance, whereas they are spherical in late interphase blastomeres. As the cells enter into mitosis, the nuclei become ellipsoidal. The orientation of the long axis of the ellipsoidal nuclei indicates the orientation of the future cleavage furrow. Finally the nuclei disappear during prophase.

Staging pointer

During the cleavage period, cell number (and later the arrangement of cells) is the principal way to stage the embryos.

Stages

2 cells (¾hpf)

The first cleavage furrow develops at the animal pole and passes down towards the vegetal pole. As are all furrows up until the 32-cell stage, it is oriented vertically. The first cleavage is not complete and the resulting cells still form a continuous cytoplasm with the yolk cell.

The following five cleavages (i.e. including those giving rise to the 64-cell stage embryo) are strictly oriented with respect to the first.

4 cells (1hpf)

The second cleavage occurs at a right angle to the first one, resulting in a 2 × 2 arrangement of the first four cells. The second cleavage furrow also arises at the animal pole and progresses downwards towards the yolk cell.

8 cells (1¼hpf)

The third round of cleavages produces two furrows that form simultaneously and are oriented parallel to the first furrow. This results in a 2 × 4 array of blastomeres.

16 cells (1½hpf)

The furrows of the fourth cycle of cleavages are oriented parallel to the second cleavage furrow, with one being located on either side. The result of the fourth cleavage is a 4 × 4 arrangement of cells that is slightly longer in one axis. The fourth cleavage is the first when blastomeres (the central four cells) become completely separated from each other. The 12 *marginal blastomeres* remain connected to the yolk cell by cytoplasmic bridges. The cleavages that follow the fourth cleavage completely separate the cells located in the centre. The marginal blastomeres on the other hand, remain connected to the yolk cell until mid-blastula (around the 512-cell stage).

32 cells (1¾hpf)

The fifth set of cleavage furrows generally run in four parallel planes that are located between those of the first and third cleavages. It is, however, not unusual that the cleavage furrows at this stage of development deviate from the strictly rectangular pattern observed earlier, and start running in oblique orientations. This results in a more or less irregular array of 4 × 8 cells.

64 cells (2 hpf)

The cleavages of the sixth round of cellular divisions are the first to run horizontally. As a result, 64-cell stage embryos can easily be confused with 32-cell stages. However, with a little experience cell size can be used in animal pole views to distinguish rapidly between the two, as the cells in the 64-cell stage embryo are smaller than the ones in the 32-cell ones. When viewed laterally, the cell mound in a 64-cell stage embryo looks significantly higher than that in a 32-cell stage embryo.

Blastula period (2¼–5¼ hpf)

See *Figure 1*.

The blastula begins with the 128-cell stage embryo and ends when gastrulation begins, at around the fourteenth round of cleavages. The orientation of the cleavage planes in this period is less precise than during the cleavage period. As a consequence, the arrangement of the blastomeres is less regular and is therefore no longer a reliable method for staging of embryos. Instead, the relative size of the cells, as compared to earlier stages, is a reasonably accurate means for staging. Additionally, as the number of cells increases, the cell mount becomes increasingly refractile, giving it a greyish appearance.

During the early part of the blastula period the cell divisions still occur more or less synchronously and at the same frequency as before (approximately once every 15 min). In the zebrafish embryo the *blastula* is a solid mound of cells without a blastocoele.

The blastula period includes the *mid-blastula transition* (MBT) at around the 512-cell stage, the formation of the *yolk syncytial layer* (YSL) at 1000 cells and the onset of epiboly at 4⅔ hpf. The MBT is characterized by compaction of the cell mound, an increase in the length of the cell cycle, the loss of the synchronicity of the divisions, and the onset of zygotic transcription. It should, however, be noted that these events do not occur simultaneously.

Staging pointer

During the blastula period the size of the cells as well as the shape of the cell mound are used for staging.

Stages

128 cells (2¼ hpf)

When viewed from the side, the cell mound now has a smooth and regular appearance, no cells bulge out. In contrast to earlier cleavages, the orientations of the furrows in the seventh round of cellular divisions are irregular. As a consequence, it is now no longer possible to determine the lineage of a cell based on its position in the embryo.

1000-cell stage (3 hpf)

The YSL is formed beginning with the tenth round of cleavages. At this stage, the marginal blastomeres disintegrate and release their cytoplasm (including their nuclei) into that of the adjacent yolk cell. As a consequence, the 1000-cell stage is the first time that the blastomeres are no longer connected with each other or with the yolk cell.

The YSL contains initially approximately 20 nuclei as a ring at the margin of the blastodisc. The nuclei of the YSL continue dividing for another three cycles, before they stop, immediately prior to the onset of epiboly, and instead the nuclei enlarge. During the divisions, the nuclei spread to eventually cover the entire yolk surface underlying the blastodisc. The YSL persists throughout embryogenesis.

The eleventh set of divisions passes through the blastodisc as a wave of cleavages. However, already many cells can be seen to divide asynchronously from their neighbours. Subsequent divisions in the developing embryo are no longer synchronous.

High (3⅓hpf)

At the high stage, the cell mound bulges out noticeably from the yolk cell. In side views, considerably more than 11 tiers of cells are visible in the outermost layer. The outermost layer is called the *enveloping layer* and the cells within the embryo proper are referred to as *deep cells*. The divisions are now highly asynchronous in all regions of the embryo.

Sphere (4hpf)

At this stage the cell mound has considerably flattened in the animal–vegetal direction, when compared to a high-stage embryo. However, the interface between the yolk cell and the blastomere cell mound still is a flat, horizontal line. In the sphere-stage, rearrangements of cells take place with increased speed when compared to later stages of development. This is the stage of the last nuclear divisions in the YSL.

Dome (4⅓hpf)

The shape changes of the blastula are accompanied by significant changes in the shape of the yolk cell. One of the most distinct signs of the onset of epiboly is that the yolk cell bulges up towards the animal pole, creating a dome-like shape. This movement flattens the cell mass out, thus increasing the surface area between the YSL and the blastomeres. The upward 'doming' of the yolk cell coincides with a radial outward movement of the deep blastomeres towards the periphery. At this stage, the YSL nuclei begin to enlarge and are easily visualized.

Epiboly (4⅔–10hpf)

See *Figure 1*.

Epiboly begins in the late blastula period. It is characterized by the spreading and a concurrent thinning of both the blastodisc cell mound and dispersal of the nuclei of the YSL over the yolk cell. The thickness of the blastoderm covering the yolk is relatively uniform. The enveloping layer is now a monolayer, whereas the deep cells are arranged in multiple layers. Epiboly continues up until the end of gastrulation, when the entire yolk is covered.

Staging pointer

The extent to which the blastoderm has covered the yolk cell (described as percent epiboly) is a robust way to determine an embryo's developmental stage. There are two phases of linear progression of epiboly, interrupted by a phase during which epiboly stalls. Between the onset of epiboly at around 4⅔ hpf and the beginning of gastrulation at 5¼hpf, the yolk is covered at a rate of about 15% every 30min. During the first hour of gastrulation (up until the shield has formed) epiboly is halted. At 6½hpf it progresses at a rate of approximately 15% every hour until it has fully covered the yolk at 10hpf.

Gastrula period (5¼–10 hpf)

See *Figure 1*.

In parallel to epiboly, the developing embryo undergoes gastrulation. This process is characterized by the morphogenetic cell movements of involution, convergence, and extension, generating the three *primary germ layers* and the *embryonic axes*. In the zebrafish, there is no blastopore. Instead cells of the deep cell layer at the margin of the blastoderm involute. As the blastoderm folds back on itself, it gives rise to the *germ ring*—a thickening of the blastoderm that can best be seen in lateral views of the embryo. The germ ring is thus composed of two layers of cells: the outer *epiblast* and the inner *hypoblast*. The epiblast contributes cells to the hypoblast throughout gastrulation.

The first step in *brain development* in the zebrafish embryo is the induction of a neural fate in the dorsal ectoderm. The underlying axial mesoderm (including the notochord) signals to the adjacent ectoderm and thus induces the formation of the neural plate on the dorsal side of the embryo. The neural plate is already visible at the end of gastrulation as a prominent thickening at the dorsal side with the anterior-most region—the region of the presumptive brain—being particularly thick (10 hpf). The thickness of the neural plate is not due to the presence of several layers of cells. Instead, its cells adopt a columnar shape and organize into a pseudostratified epithelium. By contrast, the ventral and lateral ectodermal cells retain their cuboidal shape.

Stages

50% epiboly (5¼ hpf)

Gastrulation begins with cell involution at around 50% epiboly. At this stage, radial intercalations of the blastomeres have resulted in a blastoderm of very uniform thickness. When viewed from the animal pole, the rim of the blastoderm is not yet thickened.

As epiboly is temporarily halted during the first hour of gastrulation (shield formation), staging during this time is done by determining the thickness of the marginal blastoderm. A few minutes after 50% epiboly has been reached, a thickened germ ring appears in the marginal region around the blastoderm rim (germ-ring stage, not shown). Initially, this ring is of a uniform thickness all around the circumference of the blastoderm rim. Subsequently, however, convergence movements of cells in the germ ring generate a local thickening—the *embryonic shield*.

Shield (6 hpf)

The shield is most easily visualized in animal pole views of the embryos. In side views it is apparent that both the epiblast and the hypoblast are thickened in the shield region. The shield marks the location of the future dorsal side of the embryo. As the cells located at the animal pole will give rise to the head structures, it is possible at this stage to determine both the *antero-posterior (A–P) axis* as well as the *dorso-ventral (D–V) axis* for the first time.

As soon as the shield has formed, epiboly proceeds at a relatively constant rate (approximately 15% per hour) until the blastoderm completely covers the yolk.

80% epiboly (8⅓ hpf)

As epiboly continues, the shield becomes progressively less distinctive. In ventral views the anterior axial hypoblast (the precursor of the prechordal plate) can be seen at the animal pole. When viewed dorsally, the boundaries in the hypoblast between the axial mesoderm and the paraxial segmental plate mesoderm become discernible, i.e. the *notochord* forms.

The dorsal blastoderm is now considerably thicker than that in the rest of the embryo. Lateral views show that the blastoderm above the margin is thinner than elsewhere in the embryo. Cells from this evacuation zone leave ventrally by epiboly and dorsally by convergence movements.

The dorsal epiblast thickens at the anterior pole giving rise to the *brain anlagen*. At this stage, the first postmitotic cells are present, e.g. those that will form axial somite-derived muscles, the notochord, and specific neurons in the hindbrain.

The uncovered yolk during the late stages of epiboly is often referred to as the *yolk plug*.

Tail bud (10 hpf)

Approx. 10–15 min after epiboly is complete, the embryo develops a distinct thickening at its caudal (posterior) end, the *tail bud*. The cells in the tail bud will contribute to the developing tail and, in a smaller proportion, to the trunk. In the zebrafish the first visible landmark of neural development is the *neural plate*, a distinct thickening of the dorsal ectoderm at around 9 hpf. The neural plate forms along the entire axis of the embryo. The thickening is most prominent in the anterior region (near the animal pole), the location of the prospective head. In this region the neural plate has a shallow mid-sagittal groove. The cells located in the anterior part of the neural plate will give rise to the brain, whereas those in the posterior part will contribute to the trunk spinal chord.

At the anterior-most edge of the neural plate, postmitotic prechordal plate hypoblast cells accumulate in a prominent bulge of hatching gland cells.

Segmentation period (10–24 hpf)

See *Figures 1* and *2*.

The segmentation period is characterized by the formation of the somites. Moreover, the embryo elongates along the A–P axis, the tail bud develops larger and the *rudiments of the primary organs* become apparent. On the cellular level, the first cells differentiate morphologically. Towards the end of this period, the first cells terminally differentiate and, for the first time, *body movements* appear.

Somites are mesodermal segments that form in bilateral pairs every approximately 30 min as the developing embryo extends posteriorly. Formation of each somite occurs when clusters of cells in the mesenchymal presomitic mesoderm form a segment border, and subsequently organize into an epithelial sphere. The majority of the cells in a somite belong to the *myotome*, which will give rise to a muscle segment, the myomere. As the muscle fibres in the myotome elongate, the myotome adopts a V-shape, with the apex pointing anteriorly. Each myotome is separated into a dorsal and a ventral half by a horizontal band of connective tissue, the *myoseptum*, running along the A–P axis. The first cells in a somite that differentiate into muscle fibres are the adaxial cells, i.e. those closest to the developing notochord and the location of the presumptive myoseptum. Apart from the myotome, each somite also gives rise to the *sclerotome*, which forms the majority of the axial skeleton. In contrast to the somites in some other vertebrate species, the somites in the zebrafish are not transient. While epithelial somites are transient structures, present only in the embryonic zebrafish, the segmental muscle arrangement persists throughout adulthood.

As the somites, the *notochord* differentiates from anterior to posterior. During differentiation, some cells swell and thus give the notochord its structure. Others later form an epithelial monolayer that covers the notochord as a sheath.

Early during the segmentation period, the *brain* rudiment is discernible from the spinal chord rudiment, based on its larger size. At this stage, however, the brain is still unstructured. Brain morphogenesis starts during the first half of the segmentation period when the presumptive CNS is still a neural keel and prior to the formation of the neurocoel. In contrast to other vertebrates, the neural tube does not invaginate as a hollow vesicle (primary neurula-

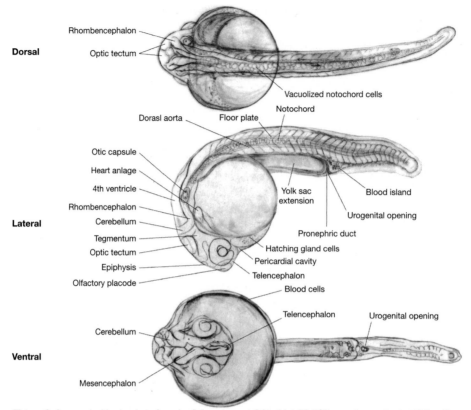

Figure 2 Camera lucida drawing of a zebrafish embryo at 24 hpf (at 28.5 °C), as observed using DIC optics.

tion). Instead, the formation of the neural tube begins with the formation of a neural keel by epithelial infolding at the midline (*c.* 13 hpf). The neural keel subsequently rounds to form a neural rod by 16 hpf. The lumen of the neural tube (neurocoel) then forms secondarily by cavitation of an initially solid rod (secondary neurulation) between 18 and 19.5 hpf. The first visible sign of brain morphogenesis is the development of 10 distinctive bulges, the neuromeres, along the A–P axis. The anterior-most three are particularly large and well developed. They correspond to the two forebrain parts (telencephalon and diencephalon) and the midbrain (mesencephalon). During the second half of the segmentation period, two expansions of the diencephalon give rise to the hypothalamus primordium (ventrally) and that of the epiphysis (dorsally). In parallel, the midbrain is subdivided horizontally into a ventral midbrain tegmentum and a dorsal midbrain (optic) tectum.

Staging pointer

During the segmentation period the number of somites provides an easy and reliable way to stage developing embryos. The somites develop sequentially in an anterior to posterior direction. The rate of somitogenesis is relatively constant. However, the first six somites develop faster (approximately 1 every 20 min) than the later ones (approximately 1 every 30 min).

The developing somites are best visualized using DIC optics. Only fully segmented somites should be counted.

The growth of the tail is so marked during this period that the embryos can also readily be staged based on their overall shape and length.

Stages

1 somite (10⅓ hpf)

The first somite border that develops is the posterior border of the first somite. The anterior border of the first somite appears shortly afterwards, more or less at the time when the second somite forms.

At this stage the *polster* is easily visualized when embryos are viewed along their A–P axis from the anterior pole.

5 somites (11⅔ hpf)

The presumptive brain appears significantly thicker and the *optic vesicles* begin to grow out laterally of the diencephalon. In the anterior trunk, the neural plate is beginning to develop into the *neural keel*. Additionally, the *Kupffer's vesicle*—a transient structure that contributes to tail mesodermal structures—can now be seen deep in the tail bud.

15 somites (16½ hpf)

At this stage the optic vesicle is a prominent structure in the anterior head of the embryo. In addition, the *otic placode* has formed about half way between the presumptive eye and the first somite (at the location of rhombomere r5). Also the *trigeminal placode*—the precursor of sensory neurons of the trigeminal ganglion—is visible. Its position is dorsolateral to the hindbrain, between the optic and otic primordia.

The *brain* shows four distinctive subdivisions (from anterior to posterior): the telencephalon, the diencephalon, the midbrain, and the hindbrain. When viewed laterally, five rhombomeres (r2–r6) can be seen. A first ventricle forms in the dorsal hindbrain (the fourth ventricle). In the trunk, the neural keel develops into a neural rod, albeit still without a cavity.

The *tail* of the developing embryo has also started to elongate and its tip has started to detach from the yolk cell. At its tip, Kupffer's vesicle can be seen. The yolk pinches in and adopts a shape resembling a kidney bean. The caudal-most part of the yolk cell will give rise to the *yolk extension*.

The notochord has now adopted its characteristic 'stack-of-pennies' appearance. In the somites, the myotomes form, giving the somites their characteristic V-shape. When viewed with polarized light, birefringence indicates the formation of myofibrils in the first elongated differentiating muscle pioneer cells.

20 somites (19 hpf)

In the developing eye, the *lens placode* becomes visible. At the same time the otic placode cavitates to form the otic vesicle. The *otoliths* are already present at this stage, but are still so small that they can only be visualized with DIC settings. The yolk extension elongates with the outgrowing tail, whereas the yolk ball has almost halved in size compared to the earlier stages of development.

In the brain, the rudiment of the *cerebellum* can be detected for the first time, using DIC optics. In addition, *ventricles* have formed throughout the developing brain and all seven *rhombomeres* are now recognizable. The neural rod has hollowed into a neural tube almost throughout the entire trunk.

The primordium of the posterior (postotic) *lateral line* can be seen as an ectodermal placode between the presumptive ear and the anterior-most somite. Over the next 20 hours, part of the postotic lateral line primordium will migrate posteriorly. Its position at different times of development will be an important mark for staging during the first part of the pharyngula period (see below).

The vast majority of the somites show the characteristic myotomal V-shape. Beginning with the 17-somite stage, the embryos show the first contractions of the myotomal muscles, i.e. the first *body movements*. As development continues, these movements become progressively more frequent, stronger, and more coordinated.

The anterior cells of the notochord now contain vacuoles. The *pronephric duct* spans the length of the trunk, starting at the anterior first few somites and ending in the region of the prospective anus at the ventral midline. Its termination defines the border between the trunk and the tail.

Day 2

Pharyngula period (24–48 hpf)

See *Figures 1*, 4, and 5.

At the onset of the pharyngula period the embryo has a well-developed, hollow central nervous system with an anterior brain displaying all major subdivisions (five lobes) and somitogenesis is complete. The period derives its name from the seven *pharyngeal arches*, the first two of which give rise to the jaw, and the posterior five form the gills. The pharyngeal arches appear at the start of the period, laterally to the presumptive brain and posterior to the eyes. However, during the early part of the pharyngula period they cannot be discriminated as individual arches.

Another prominent structure during the pharyngula period is the *hatching gland*. It is located in the pericardial region and its cells can be recognized by the brightly refractile cytoplasmic granula.

During the early part of the pharyngula period (up until 31 hpf) the embryo's tail continues to straighten and grow at a rapid rate. The head shortens and becomes more compact in an A–P direction (most easily seen by the shrinking distance between the eyes and ears by a factor of approximately 5) and is straightened out almost completely during the pharyngula period. At 48 hpf the angle between the head and the body is only approximately 50°. The angle between the embryo's head and its trunk—the *head–trunk angle*—can be used to roughly stage the embryos during this period (*Figure 3*) (see ref. 1).

Within the first 28 hpf the primary organization and gross morphology of the *brain* is established and the zebrafish brain has all the typical morphological landmarks of embryonic vertebrate brains.

Additional features that develop during the pharyngula period include *pigment cells* (melanophores (black), xanthophores (yellow), and iridiophores (silverish iridescent)), fins, the cardiovascular system, as well as the first behavioural traits.

The first neural crest-derived *melanocytes* and the *retinal pigment epithelium* begin to appear lightly pigmented at the onset of the pharyngula period. During embryogenesis the melanophores are arranged in characteristic patterns that can be used for staging. The *larval fins* form during the pharyngula period. The rudiments of the paired pectoral fins begin their development and the median fin fold, which runs along the entire tail (trunk) both dorsally and ventrally, forms. During the early phases of this period, the *heart* develops its well-defined chambers, the first heart beats are detectable, and blood can be seen to circulate in

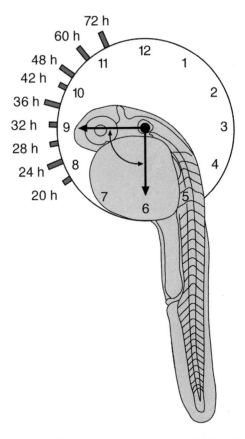

Figure 3 Changes of the head–trunk angle (HTA) during zebrafish development (at 28.5 °C). The HTA is the angle between two imaginary lines. One starting in the centre of the ear and passing through the centre of the eye, and a second line running parallel to the notochord in the midtrunk region (myotomes 5–10). Drawing a correlation between this angle and developmental time is facilitated by imagining these lines as the hands of a clock. For example, as shown in this diagram, at 31 hpf the HTA is at 90° and can easily be imagined as a clock showing quarter to six. The change in HTA is not constant. Starting at 20 hpf, the increments are initially larger (just over 3° per hour). At around 45 hpf they decrease to approx. 1° per hour up until 72 hpf. (Re-drawn after ref. 1).

the closed system of arteries and veins. The isolated contractions of the trunk and tail that appeared late during the segmentation period are replaced by coordinated rhythmic contractions that mimic swimming movements. Moreover, the embryos become sensitive to touch.

Staging pointer

Staging during the pharyngula period is more difficult than during the earlier periods. In the first half (up until 40 hpf) the position of the leading tip of the migrating part of the postotic lateral line primordium in the trunk and tail is the most accurate (albeit slightly cumbersome) method. However, the distance between the eye and ear or the head–trunk angle can also be used to quickly stage embryos.

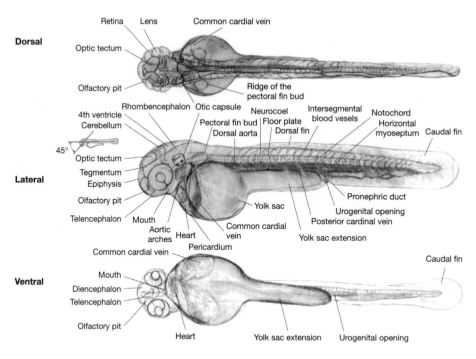

Figure 4 Camera lucida drawing of a zebrafish embryo at 48 hpf (at 28.5 °C), as observed using DIC optics.

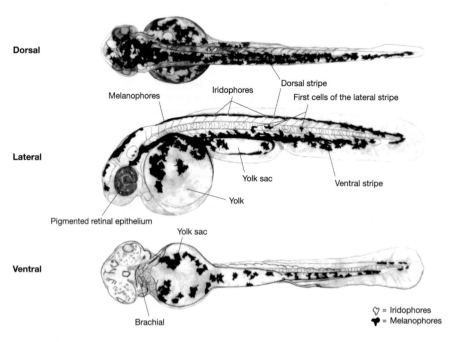

Figure 5 Camera lucida drawing of the pigmentation pattern, as seen at 48 hpf (at 28.5 °C) under DIC optics.

> The movement of the primordia along the body occurs in a linear fashion over time, with the cells migrating in the skin on either side of the embryo at a rate of about 1.7 myotomes per hour (approximately 100 μm/h). The position of the cells with respect to the 30 myotomes defines the primordium or 'prim' stages. Please note that the distance the postotic lateral line primordia have migrated on either side can vary slightly within one individual.

After 40 hpf, the development of the pectoral fin buds provides the best clues as to which stage an embryo is in. Additionally, the head–trunk angle (HTA) provides a fairly accurate means of staging the embryos (see *Table 1*).

Table 1 Head-trunk angle (HTA) and otic vesicle length (OVL) as indicators of developmental stage.

Head-trunk angle (HTA)	OVL	Stage	Age of the embryo (28.5°C)
120°	5	Primula 5	24 hpf
95°	3	Primula 15	30 hpf
75°	1	Primula 25	36 hpf
55°	¾	High pectoral fin	42 hpf
45°	½	Long pectoral fin bud	48 hpf
35°	½	Pectoral fin	60 hpf
25°	½	Protruding mouth	72 hpf

Stages

Primula 5 (24 hpf)

At this stage the developing embryo has around 30 somites and the *median fin fold* can be seen. The *heart* starts beating just prior to the prim-5 stage. Initially, the heart beats at irregular intervals and changing direction of the contractions. At 26 hpf, however, the contractions have become more regular and are now moving from posterior to anterior. *Erythrocytes* can easily be seen on the yolk and in the beating heart, from which they are being pumped into dorsal and anterior regions. Shortly after the prim-5 stage the blood begins circulating.

Melanin synthesis begins, first in the retinal pigment epithelium and later also in the melanophores located in the dorso-lateral skin. The pigmentation at this stage is still very weak and can usually only be seen using a compound microscope.

The most accurate staging criterion at this stage is that the advancing end of the lateral line primordium is located at the fifth somite. The head–trunk angle is 120° and otic vesicle length (OVL) is approximately 5.

Day 3

Hatching period (48–72 hpf)

See *Figures 1* and *6*.

Hatching in the zebrafish is not synchronous. Instead, embryos from a single clutch can hatch at irregular intervals over the entire third day. The time of hatching has no effect on the temporal progression of development. Generally, developing animals are referred to as larvae after hatching. However, since zebrafish do not hatch at a specific time point, and

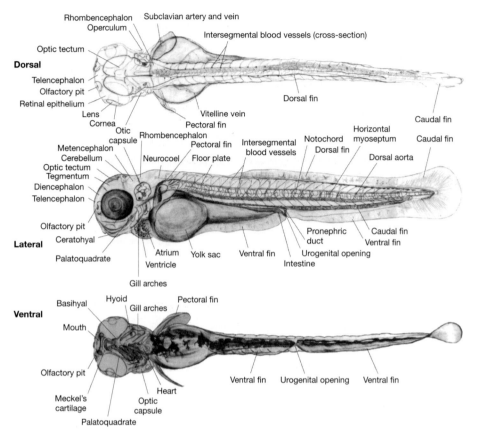

Figure 6 Camera lucida drawing of a zebrafish embryo at 3 dpf (72 hpf, at 28.5 °C), as observed using DIC optics.

since all embryos at the end of day 3 are equally developed irrespective of whether they hatched early that day or late, it has become a general rule to term developing zebrafish as embryos up until the end of day 3 and as larvae thereafter.

The majority of organ rudiments have more or less completed morphogenesis by this stage and their development slows down considerably. However, the development of the jaw, the gill arches, and the pectoral fins are characteristic features of this period.

Development of the jaw and the gill arches

At the onset of the hatching period, the mouth is located midventrally between the eyes. At this stage it can best be visualized with DIC optics. It is small and constitutively open as the jaw cannot yet be moved. During the entire period the mouth migrates anteriorly. However, this migration becomes more rapid during the last 12 hours, i.e. starting at around 60 hpf. Towards the end of the period, the anterior tip of the mouth protrudes beyond the eyes. Arches 3 to 6 will support the gills and arch 7 (the last arch) supports the pharyngeal teeth. Five gill slits form between arches 2–7.

Development of the pectoral fins

At the start of the hatching period, the rudiments of the pectoral fins can be seen as elongated buds caudal to the ear. In their centre they contain the mesenchymal condensations, which will give rise to the girdle cartilages. The distal epithelial fold capping the bud spreads out to form the blade of the pectoral fin. In parallel to the enlarging of the fins, the actinotrichia appear.

Staging pointer

During the early part of the hatching period, the development of the pectoral fins is a useful feature for staging (see also ref. 2). At later stages, the development of the jaw cartilages and the mouth–eye angle yield reliable staging criteria. Once the hatching period has been entered, the otic vesicle length (OVL) no longer changes significantly, from its initial value of around ½. Therefore an OVL of around ½ can be used as a quick way to determine that an embryo is in the hatching period.

Stages

Long pectoral fin bud (48 hpf)

The head–trunk angle (HTA) is now only 45° and the OVL is ½. Prior to the long pectoral fin stage, the *fin buds* stick out at a right angle from the remaining yolk. Beginning at approximately 46–48 hpf, as the fin buds flatten out, they fold onto their dorsal surface, such that the distal tip points towards the posterior (the apical ectodermal ridge being at the point). The pectoral fin buds at this stage are elongated (approximately twice as high as wide) and no longer have the shape of a bud. However, they have not yet fully flattened out and adopted a characteristic fin-like shape.

In the *otic vesicle* the first hair cells start differentiating and the rudiments of the semi-circular canals develop. The *olfactory placodes* are located at the anterior border of the eyes, slightly proximally. In the head and along the posterior lateral line, the first *neuromasts* (including protruding sensory hairs) have formed.

The *notochord* now contains vacuolated differentiating cells all along to its caudal end. When viewed dorsally, *melanophores* cover the head, trunk, and tail as well as the yolk sac. The dorsal melanophore stripe runs along the entire embryo, adjacent to both sides of the median fin fold. In the head, the stripe splits in two and the distribution of melanophores becomes more scattered. Laterally another, at this stage less well-defined, ventral stripe is forming between the myotomes and the yolk sac and yolk extension. At the tip of the tail, the two stripes fuse. Additionally, melanophores migrating ventrally are found scattered over the yolk sac. The first melanophores start populating the horizontal myoseptum, giving rise to the lateral stripe. Reflective *iridiophores* become prominent on the eye, particularly at the edge of the pupil. Some scattered iridiophores are also present in the dorsal tail stripe and the anterior part of the ventral stripe. These last-mentioned iridiophores pre-empt the locations of the 'lateral patches' that partially cover the prospective swim bladder. In the head, trunk, and tail, *xanthophores* can be seen, giving the embryo a weak yellowish appearance that is concentrated dorsally. It should be noted that the intensity of xanthophore pigmentation, and pigmentation in general, can vary considerably with culture conditions.

In side views the remaining *yolk* now has approximately the same size as the developing head. However, when embryos are viewed from the dorsal side, the yolk still protrudes laterally from underneath the head. At around 51 hpf the yolk will have shrunk to about the same width as the head.

As the yolk sac shrinks, the *heart* can be seen more easily. Its atrium is located dorsal to the ventricle and all definitive six pairs of aortic arches have developed. Blood can be seen circulating in the majority of segmental vessels throughout the trunk and tail.

Protruding mouth (72 hpf, 3 dpf)

The head is now nearly in line with the trunk and tail, the HTA being only about 25°. The most obvious characteristic of this stage is that the *mouth* now protrudes anteriorly beyond the eyes. It is still wide open.

Gill slits and the buds of the developing gill filaments can be visualized with DIC optics and the blood can be seen circulating through the pharyngeal arch region in a complex pattern.

The *pectoral fin* now reaches back posteriorly almost to the end of the yolk sac. In lateral views the *gut* can easily be seen along its entire length, beginning at the pharynx and extending to the urogenital opening immediately posterior to the caudal-most part of the yolk extension. The future location of the *swim bladder* is marked by an accumulation of melanocytes overlying the swim bladder rudiment. The *melanophores* have now organized almost completely into the four stripes characteristic for the larval period: the ventral-most yolk sac stripe, the ventral stripe between the trunk and the yolk sac, the lateral stripe along the horizontal myoseptum, and the dorsal stripe, which is forked on the head. Except for the lateral stripe, *iridiophores* can be seen scattered within these domains. The intensity of *xanthophore* pigmentation has increased, rendering the dorsal half of the head and trunk strongly yellow.

Early larval period (3–7 dpf)

See *Figures* 1 and 7–9.

With the end of the embryonic period, most of the morphogenesis in the developing zebrafish is complete and the fish now mainly increase in size. The most characteristic

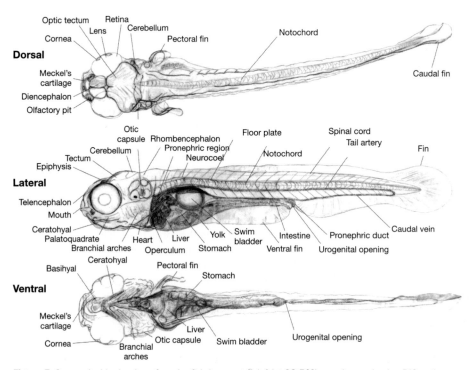

Figure 7 Camera lucida drawing of a zebrafish larvae at 5 dpf (at 28.5 °C), as observed using DIC optics.

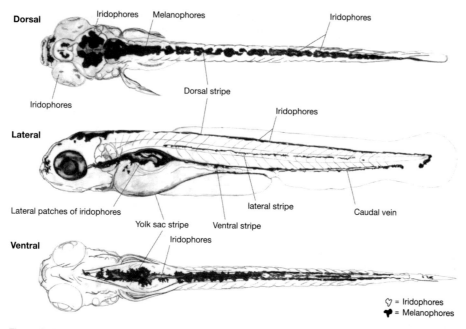

Figure 8 Camera lucida drawing of the pigmentation pattern as seen at 5 dpf (at 28.5 °C) under DIC optics.

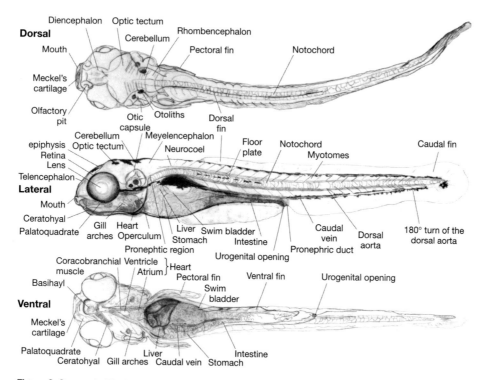

Figure 9 Camera lucida drawing of a zebrafish larvae at 7 dpf (at 28.5 °C), as observed using DIC optics. Note: blood vessels are not indicated in the dorsal view.

change in the developing larvae during day 4 is the inflation of the *swim bladder*, which is visible as a silvery, iridescent oval shape located centrally above the remnants of the yolk. Additionally, the *mouth* continues to grow out more dorsal and anteriorly. Iridiophore pigmentation increases both in surface area covered and in its reflective brightness. The gut relocates more ventrally. The yolk extension is progressively used up. The larvae start to swim actively, display their first *active movements* of the pectoral fins, the jaw, opercular flaps, and the eyes. In addition they now show escape responses, respiration movements, and they start looking for, and eating, food.

For complete descriptions of all larval zebrafish stages, i.e. fish older than 3 dpf (72 hpf), please refer to Chapter 3.

Acknowledgements

I would like to thank Darren Gilmour, Henry Roehl, Scott Holley, Carl Neumann, Marcus Dekens, and Florian Maderspacher for critically reading this manuscript, their comments and helpful discussions. I also thank Karl-Heinz Nill for his help in preparing the figures in this appendix.

References

1. Kimmel, C. B., Ballard, W. W., Kimmel, S. R., Ullmann, B., and Schilling, T. F. (1995). *Dev. Dyn.*, **203**, 253.
2. Grandel, H. and Schulte-Merker, S. (1998). *Mech. Dev.*, **79**, 99.

Appendix 3
Table of zebrafish mutations

Hans Georg Frohnhöfer

Max-Planck-Institut für Entwicklungsbiologie, Abteilung Genetik, Spemannstr. 35, 72076 Tübingen, Germany

The following table provides a summary of mutations that have been described in zebrafish. The list of genes and the corresponding references are based on data from the zfin database of Eugene as per October 2000 (http://zfin.org/ZFIN/). The data were extended by literature searches and include references until April 2002.

Chromosomal aberrations and mutations without a gene name are not included in the table. Mutations from the zfin database without references were included, if a phenotypic description was available online.

Gene

The first two columns of the table give the gene abbreviation and the full name of the locus, respectively. The designations are printed in italics, if anything is at odds with the nomenclature, e.g. the gene name is no longer valid, no 3-letter abbreviation has been defined, or the one chosen is not unique.

The table refers to some unpublished results, that will require a redefinition of the complementation groups fms/pfe/sal (Florian Maderspacher, Tübingen) and ava/hab/weg (Don Kane, Rochester).

Since we consider only mutations with gene names, the table neglects all mutants found in a particular screen designed for gonadal mutations in adults (see Ref. 327).

Description

A short phenotypic description is given for each gene. Unless otherwise stated, the recessive embryonic phenotype visible before the swimming larval stage is described.

ABBREVIATIONS:

h	hours after fertilization
day	days after fertilization
ACh	Acetylcholine
CNS	central nervous system
DRG	dorsal root ganglion
ICM	intermediate cell mass
MFP	medial floorplate
MHB	midbrain-hindbrain boundary

LFP   lateral floorplate

OKR   optokinetic response

RT   retino-tectal

The term *myoseptum* always implies horizontal myoseptum.

The two narrow columns following behind the description indicate whether the mutant shows an embryonic or an adult phenotype. The abbreviations used here are:

e   embryonic phenotype (up to day 7)

a   adult phenotype (recessive)

d   dominant adult phenotype

a, d   distinct recessive and dominant adult phenotypes

m   maternal contribution to zygotic phenotype

M   maternal effect gene

+   adult viable without (obvious) phenotype

(a)   parentheses: adults semiviable or adult phenotype only obtained in hypomorphic alleles

References

The numbers refer to a list of references at the end of this appendix. Entries are sorted according to the year of publication. Reference numbers are printed in italics, if the publication does not use the indicated gene name (e.g. because the mutation was originally published without a gene name).

Gene Product

To date 77 genes have been cloned and published in the zebrafish. The protein products encoded by the corresponding genes are indicated in the last column. Where available, the names (abbreviations) as defined on the zfin website are used. In some cases alternative names are given in square brackets.

Acknowledgements

We thank Judy Sprague for providing us with files containing data on zebrafish mutants from the zfin database.

Table 1 Zebrafish mutants

Gene		Description			References	Gene product
acc	accordion	motility: spasmic contractions of trunk (accordion-like)	e		68	
ace	acerebellar	shape: curled down; no MHB, small ears; abnormal forebrain; mild defects in pectoral fins and paraxial mesoderm	e		59, 110, 157, 171, 205, 225, 241, 307, 308, 316, 364, 382, 385, 389, 396, 409	Fgf8
ache	ache	muscle fibres, Rohon–Beard cells and neuromuscular junctions impaired; embryos become immotile	e		415	Acetylcholine esterase
aei	after eight	no regular somite borders behind ±8th somite	e	+	106, 179, 254, 268, 271, 395, 422	Dld, DeltaD

Table 1 *(continued)*

Gene		Description			References	Gene product
alb	albino	no melanin (embryo and adult)	e	a	1, 69, 82, 277, 321	
alf	another long fin	dominant adult: long fins, finrays irregular		d	69, 105	
ali	alligator	motility reduced; reduced touch response; no voltage-sensitive Na+ current in Rohon–Beard neurons	e		68, 172	
aln	alyron	reduction of premigratory neural crest cells and associated defects (e.g. pigment, DRG, craniofacial cartilage)	e		48, 195	
anh	anhalter	arrest at 24–30 h, small eyes, brain degeneration	e		57	
ant	anty talent	ear does not expand (day 4–5)	e		84	
aoi	aoi	brain necrosis (36 h); forehead, eyes, and branchial arches reduced	e		66	
apa	apparatchik	short head/cartilage reduced: pharyngeal skeleton thick and short; short pectoral fins	e		88	
apo	apollo	CNS degeneration (28 h), enlarged ventricles	e		57	
apt	apo tec	CNS degeneration (tectum, posterior midbrain)	e		Driever Lab	
aqb	aquabat	shape: tail up (day 3); cell death in CNS	e		60	
arc	archie	eyes reduced, lens protruding (day 3), retina: degeneration of ganglion cells	e		118	
arl	arrested lens	small eyes and pupils (day 2), no lens (day 5), cell-autonomous function in lens	e		405, 432	
ase	asterix	adult: fewer and wider melanophore stripes		a,d	69	
asn	astronaut	no balance, circling behaviour; microphonic potential in hair cells present	e		68, 166	
ast	astray	RT-projection: retinal axons may terminate in forebrain, tectum, or hindbrain	e		81, 299, 347, 424	
atl	atlantis	transient extrafolds in hindbrain (day 1)	e	+	78	

239

Table 1 *(continued)*

Gene		Description			References	Gene product
aus	aussicht	defects in forebrain, midbrain, and eyes; upregulation of ace	e		205, 307	
ava	*avalanche*	*ava tm94 is allelic with hab*	e			
awa	awayuki	necrosis throughout CNS (14 h); lysis before 30 h	e		66	
b286	casper	no melanin (body and eyes)	e		76	
b382	colgate	neural crest: Rohon–Beard cells missing in some segments, melanophores reduced, cranial cartilage abnormal	e		76	
bab	babyface	jaw: gaping, abnormal craniofacial elements; short pectoral fins	e		75, 96	
bad	balloonhead	brain abnormally shaped, indistinct neurocoel in trunk, embryo curved	e		60	
baj	bajan	motility reduced, spasmic contractions of trunk (accordion-like)	e		68	
bal	bashful	notochord undifferentiated, blocky somites; disorganized brain with pathfinding errors in RT-projection	e		78, 81, 89, 95, 100, 299	
ban	banshee	arrest at 17 h, no pigment	e		80	
bbh	bubblehead	haemorrhage in brain, brain degeneration	e		99	
bbl	bleached blond	melanin reduced (body and eyes); eyes mottled	e		187	vacuolar ATP synthase
bbr	bubble brain	brain ventricles enlarged (day 1); yolk extension impaired	e		187	60S ribosomal protein L44
bch	blanched	pigments pale, melanophores degenerating; CNS degeneration	e		75, 82, 110	
bcs	bubblicious	shape: curled or sinusoidal; pronephric cysts enlarged, loose glomerulus	e		149	
bea	beamter	no regular somite borders behind 3th somite	e	+	106, 179, 254, 271, 352, 383	
beb	beach boy	CNS degeneration, wavy hindbrain, reduced ventricles, weak circulation	e		57	
bed	bedimmed	iridophores dull; adult: pale, greenish-blue stripes	e	a	69, 82	

240

Table 1 *(continued)*

Gene		Description			References	Gene product
bef	beefeater	liver enlarged and reddish (day 4)	e		91	
bel	belladonna	RT-projection to ipsilateral tectum only; pupils appear big; direction of OKR is reversed	e		81, 218, 299, 309	
bem	beach bum	heart: weak contractility of both chambers; no circulation	e		99	
beo	bandoneon	motility: spasmic contractions of trunk (accordion-like), wavy notochord	e		68	
bge	big ears	otic capsule expanded	e		110	
bib	biber	gastrulation: compressed axis, indistinct somites; unstructured brain	e		71	
bid	big head	head and brain generally enlarged	e	+	78	
bil	bielak	jaw: reduction of mandibular and hyoid structures; no xanthophores; small eyes	e		88	
bks	backstroke	no otoliths; impaired balance, may swim upside down	e		110	
bla	blasen	fin necrosis; pigment gap in tail fin; expanded melanophores	e	+	69, 105	
blc	bleached	pigments pale, melanophores degenerating (body and eyes), no OKR; otoliths reduced	e		66, 75, 82, 110, 218	
ble	big league chew	shape: curled or sinusoidal; pronephric cysts enlarged, loose glomerulus	e		149	
bli	bleich	pigments pale, melanophores degenerating	e		82	
blo	blobbed	notochord curved up in tip of tail, neurocoel ends in a blob	e	+	89	
blp	blowup	shape: curled or sinusoidal; pronephric cysts enlarged, loose glomerulus	e		149	
blr	blurred	melanophores degenerating (body and eyes); no OKR	e		75, 82, 218, 277	
bls	bloodless	erythrocytes and lymphoid cells ± absent in embryos before day 5; haemopoietic precursors reduced and defective; non-cell autonomous	e	+	427	

241

Table 1 *(continued)*

Gene		Description			References	Gene product
bls	blass	melanophores pale (body and eyes), no OKR	e		82, 218	
blu	blumenkohl	RT-projection: nasal retinal axons terminate in enlarged arbors, no OKR	e	(+)	103, 218	
blw	blowout	RT-projection partially ipsilateral; eye cups evaginate into brain, overproliferating retina	e	+	81, 299	
bly	belly ache	small eyes (day 3), neural degeneration; ventral curvature of body	e		Dowling Lab	
bob	bouillabaisse	skin: abnormal cell groups on epidermis, skin falling off	e		105	
bon	bonnie and clyde	heart: cardia bifida, abnormal brain ventricles; endoderm precursors and endoderm tissue reduced	e		99, 279	Mix, Mixer (paired-class homeobox protein)
bot	brain rot	CNS degeneration (spreading), small eyes	e		Driever Lab	
box	boxer	jaw: ventral branchial arches reduced, small pectoral fins; RT-projection: axons misrouted in optic tract (but terminate appropriately); no OKR	e		81, 96, 103, 105, 110	
boz	bozozok	no notochord anlage; floorplate reduced; no myoseptum, blocky somites	e		83, 95, 98, 100, 114, 198, 210, 247, 257, 259, 265, 272, 274, 292, 317, 318, 319, 343, 367, 393, 429, 436	Dharma, (homeobox protein)
brd	braindead	brain necrosis (day 5); RT-projection abnormal	e		66, 103	
bre	breakdance	heart: arrhythmia, ventricle misses every second beat	e	+	61	
bri	brie	xanthophores pale	e	+	82, 90	
brk	brak	melanin severely reduced (body and eyes); reduced touch response; malformed branchial arches	e		88	
brs	brass	melanin reduced; adults yellow	e	a	69	
bru	brudas	no touch response, motility reduced; small eyes: loss of photoreceptors starting from center, no OKR	e		68, 83, 218, 420	

Table 1 *(continued)*

Gene		Description			References	Gene product
bry	brassy	melanin pale; adults: red eyes	e	a	69, 82	
bst	bressot	xanthophores pale	e		82, 90	
bst	barely started	small eyes, early defect in retina development; lens degenerating	e		432	
bub	bubbles	early degeneration (17 h); enlarged cells	e		57	
buf	buzz-off	motility reduced, muscular striation reduced	e		68	
bul	bulldog	short head/cartilage reduced: small chondrocytes; ear: no semicircular canals; small pectoral fins	e		88	
bum	bumper	lens degenerating; partial OKR; adult: expanded melanophores	e	a	69, 75, 218	
bxe	boxed ears	ear: small-uninflated (inner ear is normal)	e		110	
byp	bypass	abnormal blood circulation on yolk; yolk shifted	e		61	
bzj	bazooka joe	shape: curled or sinusoidal; pronephric cysts enlarged, loose glomerulus	e		149	
cab	cabernet	blood cells: decreasing count (day 4), pericardial oedema	e		93	
can	canola	intestinal epithelium degenerates, no exocrine pancreas; processing of phospholipids defective	e		345	
cas	casanova	cardia bifida; immotile (day 2); no endodermal organs; prospective endodermal cells remain mesodermal	e		61, 185, 186, 303, 338, 339, 360, 387, 413	Sox32 (related to subfamily soxF)
cco	choco	xanthophores unpigmented, iridophores dull	e		75, 82	
cdy	chardonnay	blood cells: colourless blood, decreasing count (d2)	e	+	93	Dmt-1, Divalent metal transporter-1 (A. Donovan et al., submitted)
cfs	confused	retina: cell death and lack of differentiation at the end of neuroepithelial proliferation	e		370	
cha	chablis	blood cells: decreasing count (day 2)	e		93	

243

Table 1 (continued)

Gene		Description			References	Gene product
che	cheerios	early degeneration (17 h), often very short, some hatch	e		57, 83	
chg	changeling	notochord differentiates to spherical cells, not fully penetrant	e		100	
chi	chihuahua	abnormality in bone growth and skeletal morphology		d	Fisher Lab	
chn	chinless	jaw: reduction of branchial arches; neural crest derived cartilage cells do not differentiate; secondarily lack of differentiation in muscles	e		97	
cho	choker	myoseptum variably reduced; no lateral melanophore stripe, dorsal and ventral melanophore stripes join behind head	e		60, 82, 106, 299	
cia	chianti	blood cells: hypochromic blood; red blood at day 17	e	+	93	
clb	clear blood	blood cells: colourless blood, abnormal shape of erythrocytes	e		109	
cle	cleopatra	small eyes, no excessive cell death in retina	e		83	
clo	cloche	heart: no endothelial cells; blood cells reduced, precursors of blood cells defective	e		50, 99, 130, 160, 178, 213, 220, 223, 226, 238, 245, 287, 369, 374	
cls	colourless	almost no pigment cells; small ears; affecting all neural crest fates, that are not ectomesenchymal	e		82, 110, 275, 276, 277, 342	Sox10 HMG box transcription factor
clx	clorix	xanthophores pale; small otoliths, reduced motility	e		82, 90, 110	
cob	come back	transient neural degeneration (regional)	e		Driever Lab	
con	chameleon	shape: curled down; thin neural tube, floorplate (LFP) reduced; no myoseptum; no aorta; ipsilateral RT-projections	e		60, 61, 81, 105, 106, 211, 299	
coo	cookie	xanthophores pale, iridophores dull	e	(+)	75, 82	
cos	cosinus	shape: sinusoidal	e		60	
cot	cold-light	melanophores degenerating, xanthophores faint	e		82	

Table 1 (continued)

Gene		Description			References	Gene product
cpt	captain hook	tail shortened and abnormal; *note: definition of complementation group needs reconsideration due to semidominance of mutations*	e		88, 98	
cro	crocodile	cro has been renamed as bru	e			
cru	crusher	short head/cartilage reduced: small chondrocytes; ear: no semicircular canals; small pectoral fins	e		88	
csm	cosmonaut	no balance, but response to vibrational stimuli; microphonic potential in hair cells retained	e	(a)	68, 166	
cst	cyster	pronephric cysts enlarged (day 6)	e		149	
ctd	crash test dummy	notochord undulating	e		89	
cty	city lights	photoreceptors missing	e		Baier Lab	
cud	cudak	CNS degeneration (28 h), enlarged ventricles	e		57	
cup	curly up	shape: curled up	e		60, 117	
cyc	cyclops	shape curled down; reduced prechordal plate, reduced floorplate (MFP), ventral brain defects, fused eyes; pathfinding errors in RT-projection; double mutants cyc; sqt: no hypoblast, mesendodermal tissue reduced except tail mesoderm, no trunk neuroectoderm; expression: blastoderm margin, notochord, prechordal plate, left lateral plate mesoderm, left forebrain	e		12, 13, 14, 15, 17, 23, 26, 27, 28, 32, 34, 36, 46, 47, 54, 56, 60, 65, 70, 74, 81, 83, 95, 98, 100, 101, 104, 107, 116, 117, 121, 122, 123, 139, 148, 150, 158, 170, 173, 186, 192, 201, 215, 216, 229, 232, 235, 236, 237, 244, 245, 247, 251, 256, 258, 263, 269, 279, 285, 289, 294, 299, 311, 317, 319, 333, 343, 346, 364, 368, 372, 390, 409, 410, 414	Ndr2, Nodal-related 2
dad	dappled dead	CNS degeneration; pale xanthophores; branchial arches and pectoral fins reduced; small otoliths	e		Driever Lab	
dak	dackel	jaw: branchial arches short and thick; tiny pectoral fins; inner ear retarded; RT-projection: axons misrouted in optic tract (but terminate appropriately), no OKR	e		81, 96, 103, 105, 110, 261	

Table 1 *(continued)*

Gene		Description			References	Gene product
dan	danish	fins degenerating (median and pectoral)	e		Driever Lab	
dbb	double bubble	shape: curled down; pronephric cysts enlarged	e		149	
ddf	dandruff	skin degenerating, round cells on epidermis	e	+	105	
ded	dead beat	heart: ± silent ventricle, body curved	e		99	
deg	degenerator	jaw: displaced Meckel's; tectal degeneration	e		88	
dem	dead mind	CNS degeneration with defined posterior margin behind the otic vesicle (48 h)	e		Driever Lab	
des	deadly seven	no regular somite borders behind ±8th somite	e	+	106, 179, 254, 271, 349, 422	
dfd	delayed fade	melanophores degenerating; eyes and tectum slightly reduced (day 5)	e		66, 75, 82, 103	
dgn	degenerant	degeneration (20 h), no heart beat	e		57	
dhd	deadhead	CNS degeneration (28 h), wavy hindbrain, reduced ventricles, weak circulation	e		57	
dim	dimmed	iridophores dull	e	+	82	
din	dino	ventralized: ventral/posterior region thickened (10 h), duplicated ventral tail fin	e		72, 73, 105, 117, 119, 126, 138, 144, 153, 199, 230, 238, 244, 247, 250, 259, 295, 350, 400, 410, 426, 429, 434	Chd, Chordin
dis	discontinuous	small eyes (day 5), discontinuities in photoreceptor layer	e		83	
diw	diwanka	motility: spasmic contractions of trunk (accordion-like); required for growth cone guidance of motor axons by adaxial cells			68, 240	
dlA	deltaA	midline: reduced numbers of floorplate and hypochord cells, instead excess of notochord cells; abnormal neurogenesis, including overabundance of some early-specified neurons	e		188, 228, 248	Dla, DeltaA

246

Table 1 *(continued)*

Gene		Description			References	Gene product
dml	daeumling	dominant adult: short body, vertebrae reduced and fused	e	d	69	
dnr	disordered neural retina	small eyes, retinal layers disorganized	e		Cheng Lab	
dns	dirty nose	gastrulation: degeneration of anterior prechordal plate; no hatching gland	e	+	71	
doc	doc	notochord undifferentiated, no myoseptum, blocky somites, short body	e		89	
dog	dog-eared	inner ear: hair cells reduced, cristae absent; lateral line system: fewer neuromasts; jaw retarded	e		110, 363	Eya1, Eyes absent-1
dol	dolphin	jaw: anterior tissue reduced, irregularly aligned cells in anterior arches	e		92	
doo	doolittle	jaw: displaced Meckel's	e		88	
dop	dopey	notochord undifferentiated, blocky somites, short body, degenerate on day 2	e		89, 100	
drb	drobny	notochord and body short (day 3)	e		100	
drc	dracula	fluorescent blood, photosensitive	e		109, 246	Fech, Ferrochelatase
dre	dreumes	small pupils; abnormal ear structure; adults: retarded, no anal fin	e	a	75, 105, 110	
drp	dropje	melanophores expanded, no OKR, abnormal electrophysiology in retina	e	a	69, 82, 218	
dsl	disrupted lens	lens degeneration (day 5), cell-autonomous function in lens	e		405, 432	
dsm	desmodius	fluorescent blood, photosensitive	e		109	
dtr	detour	shape: curled down; floorplate (LFP) reduced; ipsilateral RT-projections; cranial motoneurons reduced			60, 81, 193, 299	
dul	duckbill	jaw: protruding; posterior branchial arches reduced; eyes and brain reduced	e		75, 96	

247

Table 1 (continued)

Gene		Description			References	Gene product
dum	*dumbfish*	adults: reduced response to cocaine, behavioural abnormalities possibly due to abnormal dopamine processing; *the mutant effect is not yet unambiguously attributed to a single locus*			337	
dus	duesentrieb	motility reduced, muscle striation reduced	e		68	
dye	dead eye	CNS necrosis, small eyes			58	Dye (possible nuclear pore protein)
dzz	dizzy gillespie	shape curled down; pronephric cysts enlarged	e		149	
edi	edison	xanthophores pale, blue fluorescence in all cells	e		82, 90	
edw	edawakare	immotile, no muscular striation; abnormal peripheral axons	e		143	
egl	eagle	CNS degeneration (28 h), enlarged ventricles	e		57	
eis	einstein	one otolith	e	+	110	
ele	eisspalte	dent in hindbrain; curved tail; small eyes; generally retarded; abnormal RT-projection and small tectum	e		78, 103	
eli	elipsa	shape: curled down; small eyes: loss of photoreceptors starting from center; pronephric cysts enlarged	e		60, 83, 149, 218, 420	
emp	empty ear	no otoliths	e		84	
end	endeavor	CNS degeneration (28 h), ventricles enlarged	e		57	
enm	enema	enteric neurons reduced; branchial arches reduced	e		Raible Lab	
ent	enterprise	CNS degeneration (28 h), ventricles enlarged	e		57	
enz	end zone	no melanin (body)			76	
eps	ear plugs	ear: pear-like shape, small otoliths, impaired balance	e	+	110	
era	earache	ear: abnormal shape; adults: big pupils (blind?)	e	a	69, 75, 110	

248

Table 1 *(continued)*

Gene		Description			References	Gene product
esa	eraserhead	brain ventricles reduced, heart and circulation affected	e		95	
eso	eselsohr	ear: subtle abnormality of anterior semicircular canal	e	(+)	110	
esr	esrom	xanthophores pale; RT-projection: crossing of midline and outgrowth of axons impaired	e		81, 82, 90, 103	
exi	extraisthmus	malformed MHB, neural degeneration, small eyes, melanin reduced	e		Driever Lab	
exp	expander	motility reduced, spasmic contractions of trunk (accordion-like); body short and curved; degenerating notochord	e		68	
faa	fala	shape: curled down; CNS degeneration (28 h), wavy hindbrain, reduced ventricles	e		57	
fac	facelift	jaw: branchial arches reduced; eyes and brain reduced; flat shape of head	e		75, 96	
fad	fade out	melanophores degenerating (eyes and body); retina degeneration, no OKR	e		82, 218, 277	
fal	falisty	notochord curved, reduced neurocoel	e		100	
fam	fata morgana	melanophores expanded, but normal OKR	e	+	82, 218	
fap	faulpelz	motility reduced, muscle striation reduced	e		68	
far	fakir	motility reduced; touch response and escape response reduced (day 5)	e	+	68	
fau	faust	heart: cardia bifida; endodermal organs defective; endodermal and myocardial precursors reduced	e		61, 227, 229, 387	Gata5, GATA-binding protein 5
fdv	fading vision	melanophores degenerating (eye and body); retina degeneration, no OKR; adult: pale, small eyes	e	a	69, 75, 82, 218	
fel	feelgood	short head/cartilage reduced; reduced pectoral fins; no semicircular canals	e		88	

Table 1 (continued)

Gene		Description			References	Gene product
fet	feta	xanthophores pale	e	+	82, 90	
ffr	fat free	uptake of lipids in gut impaired; biliary emulsification may be defective	e		345	
fla	flat head	branchial arches; eyes and brain reduced; hypertrophy of retinal pigment, no OKR	e		75, 96, 218	
flb	flatbrain	CNS degeneration (28 h), reduced ventricles, no circulation	e		57	
flc	flycatcher	jaw: gaping mouth, ventral branchial elements bent; small eyes	e		88	
flh	floating head	no notochord, cells from dorsal marginal zone form muscle rather than notochord; no hypochord, no dorsal aorta, posterior floorplate patchy; somites blocky and ventrally fused; gene is expressed in notochord-precursors and notochord	e		40, 51, 63, 82, 85, 89, 98, 100, 101, 102, 113, 117, 120, 122, 123, 127, 128, 132, 133, 137, 141, 142, 158, 176, 177, 182, 196, 200, 215, 231, 232, 235, 238, 244, 245, 247, 263, 289, 290, 299, 312, 329, 362, 368, 369	Flh (homeobox protein)
fll	flachland	brain ventricles reduced; abnormal ears, one otolith; heart stretched, no circulation, oedema	e		84, 95	
flo	flotte lotte	small eyes (day 4); stomach thin-walled, no intestinal folds	e		61, 75	
flr	fleer	capillaries in pronephric glomerulus abnormal; wide dorsal aorta; open tissue spaces in liver; loss of photoreceptors	e		149, 420	
fls	finless	adult: no fin-blades		a	69, 105	
flw	falowany	notochord curved, reduced neurocoel; malformed ears with small otoliths	e		100	
fms	panther	pigmentation: xanthophores reduced in embryo and adult's; disruption of adult pigment stripes; gene expressed in lineages of xanthophores, macrophages, and osteoclasts; note: fms is the same gene as sal	e	a,d	301, 353, 378	Csf1r (Colony stimulating factor receptor), [Fms]

250

Table 1 *(continued)*

Gene		Description			References	Gene product
fog	foggy	reduction of dopamine-containing neurons and a corresponding surplus of serotonin-containing neurons in the hypothalamus	e		203, 264, 425	Supt5h, [Spt5] (transcription elongation factor)
fra	fransen	fin necrosis	e	+	69, 105	
frf	frilly fins	fin necrosis	e	a	69, 105	
frk	freckles	melanophores degenerating	e		82	
fro	frozen	motility: twitch only when touched; no muscle striation	e		68	
frs	frascati	blood cells: decreasing count (d2);	e		93	
frx	freixenet	fluorescent blood, photosensitive	e		93	
fsn	fusen	shape: curled down; pronephric cysts enlarged	e		149	
fss	fused somites	no somite borders during segmentation	e	a	106, 179, 254, 268, 313, 352, 383, 395	
fst	frost	early degeneration (14 h)	e		57	
ftp	flat top	jaw: visceral elements reduced	e		76	
fub	fibrils unbundled	immotile, reduced muscular striation; weak heart beat	e		5, 10, 68	
ful	fullbrain	mild cyclopia; collapsed ventricles; rough skin; retarded, not hatching, curved shape	e		78, 82, 95, 110	
fyd	frayed	fin necrosis, short body	e	+	69, 105	
gal	galileo	CNS degeneration (28 h), ventricles enlarged	e		57	
gam	gammler	liver necrosis	e		61	
gap	*gaping mouth*	*gap has been renamed as bab*				
gdr	goodyear	small eyes (day 3); degenerating retina and brain, pycnotic nuclei	e		432	
gec	grainy tec	neural degeneration (midbrain), small eyes	e		Driever Lab	
gef	good effort	small eyes, early defect in retina development	e		432	
gem	gemini	no balance, circling behaviour; vibration insensitive; no microphonic potential in hair cells	e		68, 166	

251

Table 1 *(continued)*

Gene		Description			References	Gene product
ger	gossamer	melanophore shape	e		82, 96	
get	geist	short head/cartilage reduced: chondrocytes irregularly aligned; cartilage matrix with changed staining properties	e		92	
ghl	ghoul	arrest at 20 h, no pigment; skinny tail	e		80	
gin	ginger	heart: ventricle becomes silent	e		99	
gin10	genomic instability 10	generates mosaicism in eyes			Cheng Lab	
git	gitolo	notochord short with variable defects	e		100	
glc	glaca	brain ventricles reduced; weak heart and reduced circulation	e		95	
glo	glass onion	brain opaque (17 h); abnormal somite borders; abnormal brain and retina; few blood cells	e		83, 384	
gna	gnarled	RT-projection error of nasodorsal axons	e		103	
gno	gnome	notochord undifferentiated	e		100	
gob	gobbler	jaw: cartilage elements abnormally shaped	e		76	
gol	golden	melanin reduced	e	a	1, 4, 16, 69, 82	
gon	goner	early degeneration preferentially in CNS and tail (14 h), no heart beat	e		57	
gor	gorp	early degeneration, preferentially in CNS and tail	e		57	
gos	golas	ears small and abnormally shaped (2d), abnormal semicircular canals, otoliths small; melanocytes not visible	e		84	
gre	grenache	blood cells: decreasing count (day 4)	e		93	
gri	*grinch*	*gri has been renamed as laf*				
grl	gridlock	no circulation in trunk, local defects in aorta; early expression in lateral plate mesoderm, possibly affecting arterial–venous distinction between angioblasts	e		55, 99, 324, 412	Hey2, (Hairy/Enhancer of split family of bHLH proteins)

Table 1 *(continued)*

Gene		Description		References	Gene product
gro	grossmaul	jaw: gaping mouth, ventral branchial elements are bent	e	88	
gsp	goosepimples	skin degenerating, round cells on epidermis; reduced head–trunk angle	e	105	
gts	*goody-two-shoes*	abnormal response to cocaine; *the mutant effect is not yet unambiguously attributed to a single locus*	d	337	
gul	gulliver	notochord folded until day 3, but on day 5 elongated beyond normal; small head	e	100	
gup	grumpy	notochord undifferentiated, blocky somites; disorganized brain with pathfinding errors in RT-projection, no OKR	e	78, 81, 89, 95, 100, 218, 299	
guw	gumowy	CNS degeneration (28 h), ventricles enlarged	e	57	
hab	half baked	epiboly stops at about 50% coverage	e	79	
hag	hagoromo	adult pigmentation: disorganized stripe patterns	d	273	F-box/WD40-repeat protein
hai	hai	jaw: dorsally displaced Meckel's; small eyes; flat head	e	88	
hal	hal	heart: beat of ventricle very weak, atrium enlarged	e	99	
ham	hammerhead	short head/cartilage reduced: misaligned cells; short body; reduced pectoral fins	e	92, 105	
han	hands off	heart and fin patterning abnormal; heart and fin field fail to get expanded in lateral plate mesoderm	e	322, 433	Hand2 (bHLH-protein)
hap	happy	notochord undifferentiated, short body; blocky somites; degenerate on day 2	e	89	
has	heart and soul	brain ventricles collapsed; patchy eye pigmentation, retina disorganized; heart distended	e	83, 95, 99, 239, 354, 381	Prkci, Protein kinase C / iota
hat	heart attack	immotile; enlarged heart cavity	e	61, 68, 110	
hba	hubba bubba	shape: curled or sinusoidal; pronephric cysts enlarged	e	149	

Table 1 (continued)

Gene		Description			References	Gene product
hdl	headless	zygotic: eyes reduced; maternal-zygotic: complete loss of eyes, forebrain and anterior midbrain; region of MHB and Wnt signalling rostrally expanded	e	m	280	Tcf3, Transcription factor tcf3
heg	heart of glass	heart: walls grossly extended	e		99	
hei	heirloom	iridophores dull, number may be decreased	e	(+)	82	
hek	helter skelter	ear: no otoliths, transient abnormalities in ear formation	e		84	
hel	herzschlag	immotile; weak heartbeat, small brain ventricles	e		61, 68	
hem	hermes	motility reduced, muscle striation reduced	e		68	
hen	head on	jaw/cartilage: anterior parts missing; small chondrocytes	e		92, 105	
her	hertz	motility: uncoordinated, tail flips	e		68	
hez	helderziend	lens colourless; short head, reduced body length	e	(+)	75	
hht	huli hutu	retinal disorganization	e		Cheng Lab	
hip	hiphop	heart: weak irregular beat (atrium : ventricle = 3 : 1); retrograde blood flow	e		61	
hkn	haken	shape: curled down (day ominant)	e		60	
hlx	helix	jaw: neurocranial elements missing	e		Eisen Lab	
hmp	humpback	shape: curled down or sinusoidal; adult: malformed vertebra and shape; phenotype co-segregates with mutation in Hmp	e	a	209	Hmp (membrane-associated guanylate kinase)
hoe	hoernle	abrupt developmental delay (17 h); increased cell death (22 h); no heart beat or pigment	e		57	
hoo	hoover	jaw: anterior two pharyngeal arches reduced	e		92	
hos	haos	otoliths missing or reduced and occasionally in ectopic positions	e		84	
hot	hanging out	jaw: open mouth	e		92	
hph	headphones	ear: expanded otic capsules (day 5)	e	+	110	

Table 1 (continued)

Gene		Description			References	Gene product
hps	hypersensitive	extra neuromasts in lateral line system; adult: pigmentation defects	e	a	110	
hrp	harpy	arrest at 20 h, no pigment; skinny tail	e		80	
hst	half stoned	otoliths missing or reduced	e		110	
hub	hubble	CNS degeneration (28 h), ventricles enlarged	e		57	
hup	heads up	notochord thin; small head; uncoordinated movement	e		100	
hyd	mr. hyde	cartilage differentiation: poor outline of rods, small chondrocytes; small ears, no semicircular canals; short pectoral fins	e		88	
ich	ichabod	maternal effect: embryonic shield almost missing; dorsal and anterior structures lacking; ventral/posterior structures enlarged; no beta-catenin in yolk syncytial layer	e	M	274	
ifl	inflated	shape: curled or sinusoidal; pronephric cysts enlarged, loose glomerulus	e		149	
igu	iguana	shape: curled down; floor plate (LFP) reduced, myoseptum absent posteriorly; ipsilateral RT-projections	e		60, 81, 117, 282, 299	
ika	ikarus	pectoral fins tiny or missing (embryo and adult)	e	a	69, 105	
ind	indigested	yolk: remnants of the yolk with dark brown colour (day 5)	e	m	71	
isl	island beat	heart: only isolated twitching	e		99, 391, 431	Cacna1 c (Calcium channel α1C subunit)
itf	interface	small eyes, melanophores and iridophores reduced in eyes	e		432	
itr	interrail	CNS degeneration (28 h), ventricles enlarged	e		57	
ivy	ivory	melanophores degenerating (eyes and body); retina degeneration, no OKR	e		75, 82, 218	
jag	jaguar	adult: wider and fewer pigment stripes		a,d	378	

Table 1 *(continued)*

Gene		Description			References	Gene product
jam	jam	motility reduced, muscle striation reduced; blood not passing from heart	e		61, 68	
jan	janus	blastoderm splits, axis duplication (maternal effect with variable frequency)	e	M	25, 111	
jef	jellyfish	short head/cartilage reduced: cartilage elements severely reduced	e		92, 105	
jek	jekyll	heart: no valve, no atrioventricular border cells; short head/reduced cartilage: cartilage elements severely reduced; cartilage matrix with altered staining properties; ear: no semicircular canals; small pectoral fins	e		88, 99, 406	Ugdh, UDP-glucose dehydrogenase (related to Sugarless)
jnr	junior	shape: curled down; pronephric cysts enlarged, loose glomerulus	e		149	
jpy	*jumpy*	abnormal response to cocaine; *the mutant effect is not yet unambiguously attributed to a single locus*			337	
kap	kasper	arrest at 17 h; lyse day 2	e		80	
kas	knollnase	dorsal midline in anterior forebrain absent (day 1), abnormal head shape; body slightly curled up; general necrosis on day 3	e		75	
kby	kirby	ballooned cells in gut	e		Cheng Lab	
kef	kefir	xanthophores pale	e	+	82, 90	
kei	keinstein	no otoliths	e		110	
kep	kepler	CNS degeneration (28 h), ventricles enlarged	e		57	
kgg	kugelig	gastrulation: short tail, dense tail somites, reduced posterior yolk extension	e		71	
kik	kinks	notochord wavy	e	+	89	
kim	kimble	short head/reduced cartilage: pharyngeal skeleton thick and short; short pectoral fins	e		88	
klu	kluska	tail curved ventrally; ventral fin fold malformations	e		98	

256

Table 1 *(continued)*

Gene		Description			References	Gene product
kny	knypek	gastrulation: reduced convergence and extension; short body; partial cyclopia	e		98, 100, 163, 238, 267, 269, 329, 352, 401	Kny (heparan sulfate proteoglycan)
koi	knorrig	short head/reduced cartilage: tumour-like outgrowths of chondrocytes at jaw and neurocranium	e		92	
kon	korken	notochord lacks sheath cells, notochord and body short	e		89	
kor	korinthe	small lenses (day 5), lens degeneration	e	(+)	69, 75	
krm	krom	curled pectoral fins (day 3); adult: no pectoral fins, anal fins reduced	e	a	69, 105	
krt	krenty	small eyes (day 5), discontinuities in photoreceptor layer	e		83	
kuk	kuehler kopf	brain ventricles reduced; slow heart, no circulation in head	e		95	
kus	kurzschluss	circulation restricted to heart and head	e		61	
laf	lost-a-fin	dorsalized: ventral tail fin reduced	e	m	87, 98, 105, 117, 194, 326, 376, 380	Alk8, Activin receptor-like kinase 8
lak	lakritz	melanophores expanded; partial OKR; retinal ganglion cells reduced, cell numbers in inner nuclear layer increased	e		82, 214, 218, 359	Ath5, Atonal homolog 5 (HLH transcription factor)
lam	laughing man	neural crest: DRG neurons displaced dorsally	e		76	
lau	lauscher	ear: expanded otic capsules (day 4); abnormal semicircular canals	e	+	110	
law	lawine	epiboly: round cells at the edges of the neural tube in trunk and tail	e		79	
laz	lazy susan	heart: weak beat (atrium and ventricle); no circulation	e		99	
lcl	lacklustre	iridophores dull	e	+	82	
led	lead	melanin pale	e		82	
lee	lichee	small eyes (2d), degenerating cells in retina, abnormal brain	e		Driever Lab	

Table 1 *(continued)*

Gene		Description			References	Gene product
leg	legong	heart: thin (day 1), later tube-like (day 2); ventricle ± silent, atrium fibrillating	e		61	
leh	leaky heart	haemorrhage in pericardial area	e		99	
leo	leopard	adult: spotted pigmentation		a,d	41, 42, 69, 190	
lep	leprechaun	rhombic eyes, small pupils; abnormal inner ears; big pectoral fins	e	(+)	75, 105, 110	
let	less tec	neural degeneration (spreading); retina and pectoral fins affected	e		Driever Lab	
lev	leviathan	notochord folded	e		100	
lil	liliput	dominant adult: short body, normal skeleton		d	69	
lit	little richard	ears small and pear shaped, no semicircular canals; jaw: ventral branchial arches reduced	e		84, 88	
ljw	long jaw	jaw: displaced Meckel's	e		88	
lnd	landing	melanophore number reduced, pale melanophores	e		82	
lnf	landfill	brain ventricles reduced; slow heart, no circulation in head	e		95	
loa	lonely atrium	heart: no ventricle, enlarged atrium	e		61	
loc	low octane	heart: weak beats of ventricle	e		99	
loe	loose ends	posterior somite boundaries irregular; disorganized motor axons; die around d9	e		Beattie Lab	
lof	long fin	adult: long fins		d	37, 69, 105, 270	
log	logelei	brain ventricles reduced; slow heart, no circulation in head	e		95	
lok	locke	shape: curled down; pronephric cysts enlarged	e		60, 61, 117	
lop	lens opaque	opaque lenses, fibres disorganized	e		432	
los	lost trail	melanophore pattern changed (reduced on yolk sac); adult: incomplete pattern in tail	e	d	82	
lot	lost and found	CNS degeneration (28 h), reduced ventricles	e		57	

Table 1 *(continued)*

Gene		Description			References	Gene product
low	lockjaw	jaw: pharyngeal arches reduced, lower jaw hanging; melanophores reduced	e		96	
lte	little ears	small ears; poor balance, may swim upside down	e		110	
luc	lucky	notochord degenerating, short body	e	+	89	
lum	lumpazi	liver necrosis	e		61	
lzr	lazarus	jaw: fusion of branchial elements; abnormal segmentation of branchial region	e		305, 408	Lzr (Pbx/Exd family of homeodomain)
m385	hybernaied	motility: weak touch response (day 2 onwards)	e		Driever Lab	
mac	melancholic	melanophores expanded	e		82	
mah	mach two	motility reduced, muscle striation reduced; swimming improves (day 5)	e	+	68	
maj	major tom	CNS degeneration (28 h), ventricles enlarged	e		57	
mao	macho	motility: no touch response; no voltage-sensitive Na$^+$ current in Rohon–Beard neurons; RT-projection with enlarged arborization of nasodorsal axons on tectum; no OKR	e		68, 103, 172, 218, 348, 399	
mar	mariner	no balance, circling behaviour; vibration insensitive; ear: hair bundles abnormal	e		68, 166, 233, 255	Myosin VIIA
mat	matt	iridophores dull	e	+	82	
mbl	masterblind	no optic vesicles, ± no telencephalon (17 h); anterior head structures missing; enlarged epiphysis	e		75, 132, 351, 402	Axin 1
mcr	massacre	small eyes (day 3); degenerating retina and brain, pycnotic nuclei	e		432	
med	microwaved	fin necrosis	e	+	105	
men	menhir	one otolith	e	+	110	
mes	*mercedes*	*mes has been renamed as ogo*				
mfb	mush for brains	haemorrhage in anterior brain, brain degeneration	e		99	

259

Table 1 *(continued)*

Gene		Description			References	Gene product
mfn	mini fin	dorsalized: ventral tail fin reduced	e		87, 105, 194	Tolloid (metalloprotease, related to BMP1)
mgt	maggot	notochord small and cells have spherical shape; short head/cartilage reduced: poorly delineated cartilage elements	e		88, 100	
mia	mid granola	neural degeneration (regional)	e		Driever Lab	
mib	mind bomb	somite boundaries abnormal; brain morphology disturbed; enlarged numbers of various neuronal cell types; floor plate and hypochord reduced; melanophores reduced	e		*78, 82,* 84, 95, 100, *106,* 154, *179,* 188, 200, 204, 228, 271, 302, 356, 361, 365, 423	related to Delta
mic	microps	small eyes (day 1); expanded melanophores, but normal OKR	e	(a)	75, 218	
mig	migraine	haemorrhage in brain, brain degeneration, general degeneration	e		99	
mik	mikry	notochord undifferentiated, short body; blocky somites; degenerate on day 3	e		100	
mil	miles apart	heart: cardia bifida; bubbly tail fin; non-cell autonomous failure of heart precursors to migrate to the midline	e		61, 99, 283	Edg5 (lysosphingolipid G-protein-coupled receptor)
mio	miro	small eyes, head and jaw; abnormal RT-Projection onto a reduced tectum	e		103	
mir	mirage	melanophor number decreased (body and eyes); no OKR	e		82, 218	
miz	mizerny	small eyes (day 3), degenerating cells in retina with loss of pigmentation; degenerating melanophores, small otoliths	e		83	
mlk	milky	xanthophores faint; head, eyes, and branchial arches reduced	e		75, 82	
mlt	meltdown	posterior intestine abnormal with expanded mesenchym	e		91	
mnl	monolith	one otolith	e	+	94, 136, 310	
mob	mont blanc	jaw: second anterior branchial arch and ventral arch elements reduced; eyes reduced	e		88	

Table 1 (continued)

Gene		Description			References	Gene product
moe	mosaic eyes	retina: lamination, localization of cell divisions and pigmented epithelium abnormal	e		357	
mok	mikre oko	small eyes (day 5), photoreceptor cell layer absent	e		83, 340	
mol	monorail	shape: curled down; floorplate reduced	e		60, 299	
mom	momo	no notochord and reduced floorplate in trunk, no myoseptum, blocky somites, but normal in tail; cyclopia; variable penetrance	e		82, 89, 117, 244, 247, 299	
mon	moonshine	no blood cells (day 1); iridophore number increased; *note: vmp is the same gene as mon*	e		82, 93, 105	
mos	mother superior	jaw: anterior branchial arches reduced; eyes reduced; supernummary neuromast organs	e		88	
mot	merlot	blood cells: decreasing count (day 2)	e		93	
mot	motionless	dopaminergic and noradrenergic neurons affected in number and differentiation	e		203	
mpt	mal profit	yolk: remnants of the yolk with dark brown colour (day 5); small adults	e	(+)	71	
mrb	mr bubble	shape: curled or sinusoidal; pronephric cysts enlarged	e		149	
mrc	mercury	no balance, circling behaviour; vibration insensitive; no microphonic potential in hair cells	e		166, 233	
mrd	mustard	no melanin	e	a	69, 82	
mre	marginal eye	small eyes (day 4); retina: cell death in marginal proliferative zone and scattered in inner nuclear layer	e		118	
msk	mask	lens not tightly associated with optic cup (24 h); epithelial cells cover lens (45 h)	e		118	
msq	main squeeze	heart: silent ventricle; body curved laterally; faint pigmentation	e		99	
mtc	microtic	small ears; fins and jaw retarded; large diameter of retina	e		110	

Table 1 *(continued)*

Gene		Description			References	Gene product
mur	murasaki	CNS necrosis (20 h), MHB abnormal; abnormal olfactory epithelium; eyes and jaw reduced	e		66	
nac	nacre	lack of melanophores accompanied by increased iridophores number; gene product induces melanophores ectopically	e	a	212, 253, 296, 371	Mitfa, (microphthalmia-associated transcription factor)
nag	nagel	fin necrosis	e	+	69, 105	
nan	nearly normal	no air in swim bladder, partially lethal, survivors have reduced body size	e	(a)	187	translation factor-EF1
nar	no arches	pharyngeal arches reduced	e		67	polyadenylation factor/RNase
nat	natter	brain ventricles collapsed; abnormal heart	e		61, 78	
nba	night blindness a	small eyes (day 5); dominant: age-dependent retinal degeneration	e	d	129	
nbb	night blindness b	CNS degeneration; dominant: visual sensitivity fluctuates, when dark-adapted; age-dependent postembryonic abnormalities in retina	e	d	284	
ndv	nosedive	neural crest: DRG neurons absent, but appear to be replaced by ectopic melanophores	e		76	
neb	nebel	turbid cytoplasm, cell adhesion defects; lyse during 3–36 h	e	M	224, 281	
ned	neural degeneration	massive cell death in CNS	e		2	
nel	neil	neural degeneration, enlarged brain ventricles; no circulation	e		57	
nev	nevermind	RT-projection: dorsal retinal axons terminate dorsally and ventrally on tectum; motility reduced, poor balance; thin neural tube	e		68, 103	
nic	nicotinic receptor	immotile; ACh receptors are blocked	e		8, 13, 31, 68, 175	Chma1 (cholinergic receptor)
nie	niezerka	eyes reduced (day 5), loss of photoreceptors, no OKR	e		83, 420	
nir	noir	poor balance, lazy; melanophores expanded, no OKR, abnormal electrophysiology in retina	e		82, 218	

Table 1 *(continued)*

Gene		Description			References	Gene product
niv	nirvana	arrest at 17 h, no pigment; immotile, no heartbeat	e		80	
nkl	nickel	melanophores pale	e	+	82	
nls	neckless	mesoderm lacking between rhombomere 5 and somite 1; midline defects, no pectoral fins; expressed in early paraxial mesoderm of trunk	e		328	Aldh1a2, [Raldh2] (retin-aldehyde dehydrogenase, synthesis of retinoic acid),
noa	no optokinetic response a	visual behaviour: no OKR, retina appears normal	e		35, 146	
nob	non blond	xanthophores pale; small otoliths, generally retarded	e		82, 90, 110	
noc	nocyrano	jaw: abnormal neurocranium (deformed trabeculae)	e		88	
noi	no isthmus	no MHB, tiny tectum, no cerebellum; abnormal optic stalk and postoptic commissure; abnormal ear; no pronephric tubules	e		59, 61, 103, 131, 157, 161, 168, 171, 205, 228, 238, 291, 307, 364	Pax2a, [Pax2.1]
nok	nagie oko	brain ventricles collapsed; patchy eye pigmentation, retina disorganized; circulation affected	e		83, 95	
nok	no knack	oedema, poor circulation	e		187	protein-sorting factor-αNAC
nor	no relief	intestinal epithelium degenerating, no exocrine pancreas; branchial arches reduced	e		91	
nos	no soul	no noradrenergic neurons in branchial region	e		203	
not	nototod	notochord and somites degenerating from day 2	e		100	
nrb	no optokinetic response b	visual behaviour: no OKR; retina appears normal	e		146	
nrc	no optokinetic response c	visual behaviour: no OKR; abnormal layers in retina	e		146, 325, 403	
nrd	narrowminded	no Rohon–Beard neurons, formation of neural crest impaired	e		189, 423	
nrf	not really finished	retina: photoreceptors reduced	e		145	Nrf1, Nuclear respiratory factor 1
nrs	not really started	yolk degenerates	e		437	Spinl, Spinster-like

Table 1 (continued)

Gene		Description			References	Gene product
ntl	no tail	narrow notochord primordium, undifferentiated notochord; enlarged floorplate; no tail forming; no myoseptum, blocky somites; expression in mesendodermal precursors (germring), tailbud, and notochord	e		20, 21, 22, 24, 29, 30, 33, 38, 39, 44, 45, 49, 52, 62, 63, 64, 65, 70, 85, 89, 98, 100, 101, 108, 111, 114, 117, 120, 122, 123, 125, 128, 133, 137, 141, 151, 152, 155, 158, 162, 192, 196, 215, 229, 232, 244, 247, 256, 296, 299, 366, 369	Ntl
ntn	no tectal neuron	no tectal neurons, reduced eyes	e		143	
obd	out of bounds	small eyes (day 6); degeneration at margin of retina	e		418, 432	
obe	obelix	adult: wider and fewer pigment stripes		a,d	69	
obs	obscure	abnormal melanophore shape; adult: pinkish color	e	a	69, 82	
oep	one eyed pinhead	zygotic: prechordal plate, floorplate (MFP), and endoderm reduced; no anterior forebrain, cyclopia; maternal-zygotic: additionally reduction of mesoderm	e	m	60, 71, 83, 95, 98, 100, 137, 140, 153, 167, 184, 186, 201, 215, 232, 234, 263, 265, 269, 278, 296, 299, 303, 311, 331, 362, 364, 367, 375, 388, 390, 394, 410, 436	Oep (EGF-CFC family)
ogo	ogon	ventralized: excess cells below tailbud (day 0), duplicated ventral tail fin (day 3)	e	m,a	69, 72, 82, 98, 105, 144, 217	
ogr	ogre	early necrosis (13 h) all over the embryo	e		80	
olp	overlooped	heart: abnormal position of chambers	e		61	
ome	oko meduzy	brain ventricles collapsed; patchy eye pigmentation, retina disorganized; circulation affected	e		83, 95, 214	
orb	orbiter	no balance, circling behaviour; vibration insensitive; no microphonic potential in hair cells	e	(a)	68, 166, 233	
ori	orient express	early degeneration (CNS, tail)	e		57	

Table 1 *(continued)*

Gene		Description			References	Gene product
ott	otter	collapsed brain ventricles; eyes and ears small, no inner ear structures	e		78, 110	
out	out of sight	eyes reduced, neural lamina present	e		83	
ovl	oval	shape: curly down; pronephric cysts enlarged; loss of photoreceptors starting from centre	e		60, 420	
pac	parachute	hindbrain irregular (day 1); abnormal growth of epidermis above hindbrain	e		78	
paf	passing fancy	CNS degeneration	e		57	
pal	pale tec	body slightly smaller (day 2), degenerating (day 5)	e		Driever Lab	
pan	pandora	heart: ± no ventricle; pleiotropic effects on tail, brain, otic vesicles, retina, and pigment	e		57, 83, 84, 99, 239, 262, 425	Spt6 (transcription elongation factor)
pap	pao pao tang	pronephric cysts enlarged (day 6)	e		149	
pat	pinatubo	embryos disintegrate (19 h), periderm defect	e		57	
paw	pale and wan	blood cells: decrease in number	e		109	
pca	photo-receptors absent	no optokinetic response; abnormal eyes	e		118	
pch	pech	pigments pale (body and retina); melanophores degenerating; retarded	e		82	
pde	parade	melanophore and iridophore migration defective	e	+	82	
pec	platy tec	degeneration in brain and retina	e		Driever Lab	
pek	pekinese	short head/cartilage reduced: anterior tissue missing; cartilaginous elements kinked; reduced pectoral fins	e		92, 105	
pel	pelican	jaw: gaping mouth, ventral branchial elements bent	e		88	
pen	penner	skin: abnormal cell growth especially around pectoral fins and branchial arches	e		105	

Table 1 (continued)

Gene		Description			References	Gene product
pep	pepita	melanophores degenerating	e		82	
pes	pescadillo	eyes, brain, pharyngeal arches, and gut are reduced; reduced organs correlate with sites of strong gene expression	e		58	Pes (possibly involved in cell cycle control)
pet	petroglyph	melanophore shape spindly	e		82	
pew	pewter	melanin pale	e	+	82	
pfe	pfeffer	pfe is renamed as sal				
pgu	piegus	eyes reduced, cell death and loss of pigment in retina; degenerating melanophores; altered brain shape	e		83	
pgy	piggytail	dorsalized: somites laterally extended, ventral tail fin reduced	e		69, 87, 105, 117, 194	
pic	pinscher	jaw: branchial arches reduced; short body; RT-projection: axons misrouted in optic tract (but terminate appropriately); partial OKR	e		81, 96, 103	
pie	piebold	intestinal epithelium degenerates, no exocrine pancreas; processing of phospholipids defective	e		91, 345	
pif	pinfin	fins reduced (embryonic and adult)	e	a	69, 105	
pik	pickwick	heart: contractility reduced in both chambers	e		99, 435	Titin (sarcomeric protein)
pio	pistachio	xanthophores unpigmented; brain, eyes, and jaw reduced	e		75, 82, 96	
pip	pipe heart	heart: weak beat (day 2), ventricle ± silent; no circulation; heart stretched (day 3)	e		61	
pir	pirueta	shape: curled up	e		60	
pkt	punktata	small eyes, cell death, and loss of pigment in retina; body melanophores degenerating	e		83	
plk	polka	heart: no beat, isolated contractions in atrium; small ventricle, large atrium	e		61	
plt	poltergeist	early necrosis (15 h), lyse at 20 h	e		80	
plx	perplexed	retina: cell death and lack of differentiation at the end of neuroepithelial proliferation	e		370	

Table 1 *(continued)*

Gene		Description			References	Gene product
png	ping pong	heart: retrograde blood flow	e		99	
pnt	pinotage	blood cells: reduced count (day 2), colourless	e	(+)	93	
poa	partial optokinetic response a	visual behaviour: OKR small and quick; retina appears normal	e		35, 146	
pob	partial optokinetic response b	visual behaviour: OKR response under red light reduced; red cones missing in retina	e		115, 146	
pod	podgladacz	small eyes, no extensive cell death in retina	e		83	
pol	polished	melanophores degenerating, xanthophores pale	e		82	
pos	postdoc	short head/cartilage reduced: individual cartilage bars are reduced in size; small pectoral fins	e		88	
ppa	photoreceptor patches	photoreceptor layer abnormal	e		Cheng Lab	
ppd	pipe dream	heart: contractility reduced, ventricle enlarged	e		99	
ppl	pipeline	heart: contractility reduced, no circulation	e		99	
ppt	pipe tail	gastrulation abnormal: somites short and laterally extended, notochord expanded, trunk shortened; outgrowth of tail abnormal; cartilage in head affected	e		71, 92, 135	Wnt5
pro	proteus	notochord differentiation: local breaks and degeneration; not fully penetrant	e		100	
ptw	pan twardowski	CNS degeneration (28 h), ventricles enlarged	e		57	
pun	punkt	pale melanin, melanophores degenerating; iridophores dull; dominant: cloudy yolk	e		82, 89, 277	
pup	pile-up	pale melanin, melanophores degenerating; iridophores dull	e		82	
puz	puzzle	abnormal pigment cells; reduced jaw; abnormal ears	e		82, 96, 110	

Table 1 *(continued)*

Gene		Description			References	Gene product
pyr	pyry	CNS degeneration (28 h); small eyes, plexiform layers in retina abnormal	e		83	
qam	quasimodo	notochord undulating; few and pale melanophores (body and eyes); no OKR	e		82, 89, 218	
qua	quark	xanthophores pale	e	+	82, 90	
que	quetschkom mode	motility reduced, spasmic contractions of trunk (accordion-like)	e		68	
quh	quiet heart	heart: nearly silent	e		61	
qui	quitter	early degeneration (14 h)	e		57	
quo	quadro	jaw: second anterior branchial arch and ventral arch elements reduced; otic placode small or split into two fields; branchial arches reduced	e		84, 88	
rde	round eye	retina: cell death in marginal proliferative zone and scattered in inner nuclear layer; variable degeneration of tectum and forebrain	e		118	
rdy	red eye	small eyes (day 3), degenerating cells in retina	e		Driever Lab	
rec	recover	CNS necrosis: tectum (day 4), recovering, adult viable	e	+	66	
red	relaxed	immotile; internal Ca^{++} release defective in skeletal muscles	e		68, 377	
reg	reggae	heart: spasmic beat (silent ventricle, atrial fibrilations?)	e		99	
reg5	reg5	fin regeneration fails		a	43, 207	
reg6	reg6	fin regeneration with formation of abnormal bumps; embryo: skin-defect and fluid filled sacs in ventral fin fold	e	a	43, 207	
ret	retsina	blood cells: decreasing count; proerythroblasts abnormal; haemoglobin present	e		93, 379	Slc4a1 (solute carrier family 4, anion exchanger)
rfl	rafels	fin necrosis	e	+	69, 105	
rhe	rheostat	heart: weak beat of ventricle, atrium enlarged	e		Driever Lab	

Table 1 *(continued)*

Gene		Description			References	Gene product
ric	ricotta	xanthophores pale	e	+	82, 90	
ris	riesling	blood cells: decreasing count (day 3), erythrocytes fail to achieve terminal differentiation	e		93, 286	Sptb, (erythrocytic β-Spectrin)
rix	rieux	short head/cartilage reduced: pharyngeal skeleton thick and short; short pectoral fins	e		88	
rnd	round	short head/cartilage reduced: small chondrocytes; ear: no semicircular canals; small pectoral fins	e		88	
rne	rosine	small lenses (day 4 and adult)	e	a	69, 75	
roc	roller coaster	motility: uncoordinated; no touch sensitivity	e		68	
roh	rochen	jaw: dorsally displaced Meckel's; small eyes; flat head	e		88	
rot	rotten	neuronal degeneration (midbrain, later spreading)	e		Driever Lab	
roy	roy orbison	no iridophores (including eyes)	e	+	Dowling Lab	
rse	rose	adult: adult pink colour; gene required for late melanocyte population		a	41, 42, 69, 300, 378	Edndrb1 (Endothelin receptor)
rst	rolling stones	ectopic otoliths, sometimes loose as well	e	+	110	
ruz	runzel	motility reduced, muscle degeneration	e		68	
sad	stop and drop	arrest at 17 h, no heart beat	e		57	
sah	sahne	pigments pale, melanophores degenerating; generally retarded	e		75, 82, 110	
sal	salz	xanthophores reduced (embryonic and adult); *note: sal is the same gene as fms*	e	a,d	69, 82, 90	
sam	samson	gastrulation: short body, thin notochord, wide somites	e		71	
san	santa	heart: both chambers become enlarged, single myocardium layer in ventricle	e		61, 99	
sap	sapje	motility reduced, muscle degeneration	e		68	

Table 1 *(continued)*

Gene		Description			References	Gene product
sas	stars-and-stripes	melanophore shape	e	+	82	
sat	saturn	CNS degeneration (28 h), ventricles enlarged	e		57	
sau	sauternes	blood cells: colourless blood, decreasing count (day 2); abnormal globin gene expression	e	(+)	93, 147	Alas2 (δ-aminolevulinate synthase, haem-synthesis)
sbd	scabland	hindbrain abnormal (day 1), midbrain and jaw abnormal	e		78	
sbl	sunbleached	pigments pale, melanophores degenerating (body and eyes); retina degeneration, no OKR; jaw reduced	e		75, 82, 96, 218, 277	
sbn	somitabun	dorsalized (dominant maternal): oval embryo (day 0); somites extend laterally; posterior trunk twisted; ventral elements (blood, ventral ectoderm) reduced; antimorphic mutation	e	m	87, 165, 191, 194, 206, 250, 292, 298, 326, 400, 429	Madh5, MAD homolog 5, [Smad5]
sch	sancho panza	gastrulation: short axis, blocky somites, undifferentiated cells at tip of tail	e		80	
sco	scotch tape	heart: matrix on endocardium and endothelia in tail disrupted	e	(+)	61	
sdp	sensory deprived	no dorsal root ganglia; later severe oedema	e		Raible Lab	
sdy	sandy	no melanin (eyes and body); no OKR	e	a	69, 82, 218, 277	
sed	straight edge	no otoliths	e		84	
sen	sense	shape: curled down; small eyes	e		60	
ser	screamer	jaw: posterior branchial arches reduced; eyes and brain reduced; flat shape of head	e		75, 96	
sfy	stuffy	cell differentiation: increased luminal mucous and mucous cell patches	e		Cheng Lab	
shd	shady	iridophores reduced (embryonic and adult)	e	a	69, 82	
she	schmerle	jaw: anterior two arches reduced; mouth points ventrally on day 3	e		92, 105, 156	

Table 1 (continued)

Gene		Description			References	Gene product
shf	short fin	fins have reduced length	a		270	
sho	shocked	motility: uncoordinated	e		68	
shr	shrunken head	body axis is curved (day 2); branchial arches and eyes reduced; cell degeneration in retina	e		249	
sic	sickle	shape: curled down	e		60	
sid	sinusoida	small eyes, discontinuities in photoreceptor layer	e		83	
sih	silent heart	no heartbeat	e		61	
sil	silent partner	heart: silent ventricle	e		88, 99	
sin	sinus	shape: curled down or sinusoidal	e	(+)	60	
sis	sisyphus	heart: weak beat of ventricle, atrium enlarged	e		99	
sit	stop it	developmental delay: 12 somites, when siblings already have 19	e		57	
skb	skylab	no balance, circling behaviour; degeneration of hair cells in ear	e		68, 233	
skm	skid mark	developmental arrest at 12 h; degenerative loss of eyes and ears	e		57	
sky	slinky	motility reduced, muscle striation reduced; circulation defects	e		61, 68	
sla	schlaffi	motility: reduced; somitomeres short and contracted	e		68	
slb	silberblick	gastrulation: only partial convergent extension; short shield; cyclopia; hanging jaw	e	(+)	75, 124, 266, 436	Wnt11
sli	slip jig	heart: silent ventricle, weak atrium, no circulation	e		61	
slj	slim jim	intestinal epithelium degenerates anteriorly; no exocrine pancreas; branchial arches reduced; processing of phospholipids defective	e		91, 345	
slk	sparse-like	melanophore number reduced (anterior head region, more gradual in trunk)	e		82	
sll	sallow	pigments pale, melanophores degenerating; CNS degeneration	e		75, 82	

271

Table 1 *(continued)*

Gene		Description			References	Gene product
slm	slumber	motility reduced, curved laterally	e		68	
sln	slow tan	melanophores pale	e		82	
slo	sloth	immotile, no muscle striation	e		68	
slp	slop	motility reduced, muscle striation reduced; heart: atrium enlarged	e		61, 68	
slt	saltarin	tail variably curved; neurocoel indistinct	e		60	
slw	slow motion	motility reduced, muscle striation reduced	e		68	
sly	sleepy	notochord undifferentiated; blocky somites; disorganized brain with pathfinding errors in RT-projection, no OKR	e		78, 81, 89, 95, 100, 218, 299	
smf	smurf	dominant adult: short body		d	69	
smh	schmalhans	shape: curled down; floorplate reduced	e		60, 117, 299	
sml	schwammerl	motility reduced, muscle degeneration	e		68	
smo	slow mo	heart: slow rate (day 2–3), recovering	e	+	99, 112, 407	
smu	slow-muscle-omitted	no slow muscle fibres or muscle pioneers; cyclopia, jaw and fin growth abnormal; hedgehog signalling disrupted	e		242, 332, 368, 390, 404	Smoh, Smoothened homolog
snd	stop and die	early degeneration	e		57	
sne	schnecke	motility reduced, muscle striation reduced	e		68	
snh	snailhouse	dorsalized: oval embryo (day 0); somites extend laterally; posterior trunk twisted like snail shell; ventral elements (blood, ventral ectoderm) reduced	e		87, 117, 165, 194, 250, 269, 292, 298, 314, 326, 350, 410, 434	Bmp7
snk	snakehead	collapsed brain ventricles, small eyes, small ears	e		78, 84, 95, 110	
sno	snorri	notochord short and with local lesions	e		100	
snp	stolen pearls	iridophores: decreased count (regional reduction)	e	+	82	

Table 1 *(continued)*

Gene		Description			References	Gene product
snt	schnitter	shape: curled down; small eyes, brain necrosis	e		60	
snw	snow white	notochord: spherical cells, short	e		100	
sny	sneezy	notochord undifferentiated, short body; blocky somites; degenerate on day 2	e		89, 100, 336	coatomer alpha protein
sof	softy	motility reduced, muscle degeneration	e		68	
sol	solaris	CNS degeneration (28 h), ventricles enlarged	e		57	
sop	sofa potato	immotile; postsynaptic complex defective in skeletal muscles	e		68, 377	
sou	soulless	no catecholaminergic neurons in locus coeruleus and branchial region	e		202, 203	Arix (aristaless homeobox)
spa	sparse	melanophores reduced on head and ventrally; adult: mottled stripes; required for larval and early-adult melanocytes	e	a	1, 42, 69, 82, 222, 277, 296, 386	Kit, Kit receptor (type III receptor tyrosine kinase)
spb	speed bump	epiboly: early necrosis, lyse on day 2	e		80	
spc	space cadet	motility: uncoordinated contractions, flicks of tail; defects in commissural hindbrain neurons (spiral fibre neurons)	e		68, 373	
spg	spiel ohne grenzen	MHB largely missing; defects in midbrain and hindbrain	e		95, 417, 421, 430	Pou2, POU domain gene 2
spi	spirale	shape: curled down	e		60	
spk	spock	ear: pear shaped, tiny otoliths	e		110	
spl	spitzmaul	jaw: displaced Meckel's; lower jaw appears elongated	e		Driever Lab	
spm	space man	CNS degeneration (28 h), ventricles enlarged	e		57	
spo	spaced out	motility: uncoordinated contractions, flicks of tail	e	(+)	68	
spr	specter	early necrosis (17 h), small necrotic head (day 1), lyse day 2	e		80	
sps	spine-stein	spongeous material in neurocoel; adult: short body	e	a	60, 69	

Table 1 *(continued)*

Gene		Description			References	Gene product
spt	spadetail	reduced paraxial mesoderm in trunk, somites irregular; thickened spade-like tail; no pectoral fins; compaction and convergent movement of lateral mesoderm cells fails	e		3, 6, 7, 9, 11, 13, 18, 19, 22, 24, 47, 52, 53, 71, 77, 98, 102, 108, 117, 134, 142, 152, 176, 178, 181, 182, 196, 220, 231, 232, 238, 244, 247, 256, 329, 366, 374, 400	Tbx16, T-box gene 16
spu	sputnik	no balance, circling behaviour; vibration insensitive; ear: hair bundles splayed	e	(a)	68, 166, 233, 397	cadherin-related protein
spw	space cowboy	CNS degeneration (28 h), ventricles enlarged	e		57	
spy	spy eye	small eyes (day 5), no excessive cell death in retina	e		83	
spz	spazz	CNS degeneration (28 h), ventricles enlarged	e		57	
sqt	squint	embryonic shield not evident; later phenotype with variable penetrance: reduced prechordal plate, cyclopia	e		150, 186, 192, 201, 216, 229, 245, 258, 263, 265, 274, 279, 317, 318, 319, 333, 343, 415, 436	Ndr1, Nodal-related 1
sri	sunrise	lens protruding, cornea irregular (day 4)	e	+	75	
sst	straight shot	intestinal epithelium degenerating; exocrine pancreas abnormal; branchial arches reduced	e		91	
ssw	sideswipe	early degeneration (18 h), no brain ventricles, no heart beat	e		57	
stb	saltimbanqui	tail curled up (day 3); floor plate indistinct; small eyes	e		60	
std	stem degen	transient, widespread brain degeneration	e		Driever Lab	
ste	steifftier	motility reduced; touch response reduced; expanded melanophores, no OKR; no voltage-sensitive Na$^+$ current in Rohon–Beard neurons	e		68, 172, 218	
stf	stuffed	yolk: dark brownish material ventral of gut	e		61	
sth	still heart	heart: silent; reduced motility	e		61	

Table 1 *(continued)*

Gene		Description			References	Gene product
sti	sticky blood	blood cells do not ripen	e		109	
stn	strangelove	short head/cartilage reduced; reduced pectoral fins; no semicircular canals	e		88	
sto	stomp	pectoral fins necrosing	e	+	105	
stp	stoepsel	*abbreviation has been changed to sts*				
stp	stumpf	short head/cartilage reduced; reduced pectoral fins	e		88	
str	stretched	heart beats, but there is no circulation (day 1); weak beat (day 2)	e		61	
sts	stoepsel	dominant adult: short body, vertebrae shortened		d	69, 438	
stu	sturgeon	jaw: anterior two arches reduced; gaping mouth	e		66, 92, 156	
stw	stone-washed	pigments pale, melanophores degenerating; retarded	e		82, 110	
sty	stumpy	abnormal branching and pathfinding in primary and secondary motorneurons	e	+	243	
sub	stein und bein	otoliths reduced; adult: dorsal and pelvic fin reduced	e	a	69, 105, 110	
suc	sucker	jaw: anterior two pharyngeal arches reduced, ventral arch specification abnormal	e		92, 156, 293	Edn1, [Et-1] Endothelin 1
sug	superglue	thin matrix between myocardium and endocardium	e		61, 75	
sum	submarine	poor balance; melanophores expanded, but normal OKR	e		82, 218	
sur	schmalspur	shape: curled down; floorplate (MFP) and posterior prechordal plate reduced	e	m	60, 83, 95, 98, 100, 117, 244, 256, 299, 304, 320	Foxh1, Forkhead box H1
swr	swirl	dorsalized: oval embryo (day 0); somites extend around circumference; embryo bursts (day 1) due to mechanical constriction; ventral primordia reduced in hypoblast and epiblast	e		69, 73, 87, 105, 126, 159, 165, 191, 194, 197, 217, 221, 230, 238, 250, 260, 292, 295, 298, 314, 326, 350, 388, 410	Bmp2b

HANS GEORG FROHNHÖFER

Table 1 *(continued)*

Gene		Description			References	Gene product
syu	sonic-you	shape: curled down; floorplate (LFP) reduced; no myoseptum; no aorta; reduced pectoral fins	e		60, 61, 105, 106, 148, 158, 174, 211, 219, 289, 290, 294, 297, 299, 330, 335, 341, 344, 362, 368, 390, 420	Shh, Sonic hedgehog
tan	tango	heart: weak ventricle, atrium enlarged (day 2)	e		61	
tar	tartar	xanthophores pale; fluorescent; adults: pale melanophore stripes	e	a	69, 82, 90	
tbo	the boss	jaw: displaced Meckel's	e		88	
tbr	thunderbird	blood cells: decreasing count (day 3)	e	+	93	
tdo	touchdown	melanophore number reduced; touch response reduced; adult: short and kinked at head	e	(a)	75, 82, 172, 277	
tel	tell tale heart	heart: nearly silent	e		99	
tem	tec melt	CNS degeneration (day 2), dorsal curvature, branchial arches reduced	e		Driever Lab	
ten	tennis match	heart: retrograde blood flow	e		99	
tep	tec perdu	CNS degeneration (regional day 2, scattered day 4)	e		Driever Lab	
tew	tec wreck	small eyes (day 3), degenerating cells in retina, brain degeneration	e		Driever Lab	
tfu	too few	dopaminergic neurons reduced in hypothalamus	e		203	
thc	thanotec	neural degeneration, pale pigment, small eyes (day 5)	e		Driever Lab	
thi	thinner	jaw: ventral branchial arches reduced; small eyes	e		88	
tig	tiger	shape: curled down; melanophore migration abnormal	e		60, 82	
til	tilsit	xanthophores pale; RT-projection: crossing of midline and outgrowth of axons impaired	e		81, 82, 90, 103	
tin	tinte	melanophore shape abnormal, degenerate until day 4	e		82, 103	
tip	tippelbruder	red blood cells accumulate in liver	e		61	

276

Table 1 *(continued)*

Gene		Description			References	Gene product
tnd	tarnished	iridophores dull	e		82	
tnt	techno trousers	motility: uncoordinated, crazy movements, spasmic contractions	e		68	
tod	toned-down	iridophores dull	e	+	82	
tof	tofu	xanthophores pale; RT-projection: crossing of midline and outgrowth of axons impaired	e	+	81, 82, 90, 103	
top	topped	abnormal motorneurons	e		Beattie Lab	
tra	transparent	adult: iridophores reduced, black eyes and transparent trunk		a	69	
tre	tremblor	heart: weak (day 1), fibrillating irregular beat (day 2)	e		61, 99	
tri	trilobite	gastrulation: short axis, broad somites; reduced convergent extension	e		71, 98, 100, 163, 267, 315, 352, 416	
trl	troll	arrest at 17 h; no heart visible	e		80	
trp	tramp	liver necrosis, small gut without internal folds	e		61	
tst	toast	neural crest derivatives ± missing	e		311	
tub	turbulent	small eyes (day 3), degenerating cells in retina, abnormal plexiform layers	e		83	
tun	turned on	hindbrain ventricle reduced; head tilted (day 5), body with dorsal curvature	e		95	
tur	turtle	motility reduced, muscular striation reduced	e		68	
tut	tutu	gaps in tail fin and pectoral fins	e	+	105	
twi	twitch twice	motility: uncoordinated, flicks of tail	e		68	
twn	turned down	brain ventricles reduced; heart and circulation affected	e		95	
two	twitch once	motility: uncoordinated	e		68	
ubo	U-boot	no myoseptum, blocky somites; necrotic tail fin	e		82, 105, 106, 211, 299, 392	
udu	ugly duckling	gastrulation: short axis; indistinct myoseptum; short tail	e		71	

Table 1 *(continued)*

Gene		Description			References	Gene product
uhi	uchu hikoushi	CNS degeneration (28 h), enlarged ventricles; no circulation; tail curved	e		57	
uki	ukkie	small pupils; retarded ears; pectoral fin broadened; adult: no anal fins	e	a	75, 105, 110	
uml	umleitung	shape: curled down; floorplate (LFP) reduced; ipsilateral RT-projection	e		81, 299	
unf	*uncle freddy*	*unf has been renamed as sur*				
uni	union jack	melanophore shape abnormal	e	+	82	
unp	unplugged	motility: nearly immotile (day 1), motility improving later; pathfinding errors of pioneering motor axons	e		68, 323, 411	
uns	unsaddled	melanophore pattern altered	e		82	
val	valentino	hindbrain segmentation fails behind rhombomeres 3/4; val is required in rhombomeres 5 and 6	e		86, 116, 164, 169, 183, 215, 306, 334, 398	Val (bZip transcription factor)
van	vanille	xanthophores pale, iridophores dull, brain necrosis, small eyes	e		66, 75, 82	
vgo	van gogh	ears very small, no semicircular canals; no segmentation of branchial region	e		92, 110, 303	
vhnf1	vhnf1	enlarged pronephric cyst, reductions in gut region (pancreas, liver) and ear; expression in pronephros, gut, and hindbrain; involved in subdividing the primordium of the organs in each of these three regions	e		398	homeobox related to vHnf1 and TCF2
vic	vicious cycle	shape: curled down, sinusoidal or tail up	e		60	
vik	viking	CNS degeneration (28 h), enlarged ventricles	e		57	
vip	viper	retarded brain (day 1); silent heart	e		61, 78	
vis	visionary	small eyes (day 5), no excessive cell death	e		83	
vlt	vlad tepes	no blood cells (day 1); by 26 h expression of haematopoietic stem cell markers is lost apart from residual staining in posterior ICM	e		109, 428	Gata1

Table 1 *(continued)*

Gene		Description			References	Gene product
vmp	vampire	no blood cells (day 1); severe reduction in blood cell precursors and accumulation of debris in the ICM; *note: vmp is the same gene as mon*	e		109, 358	
vol	volcano	epiboly: slow vegetal expansion, early lysis	e		98	
vos	vestigial outer segments	photoreceptors: no visible outer segments	e		Cheng Lab	
vox	vox	dorsalized with low penetrance; but strongly dorsalized, when combined with a deletion of the related gene vent	e		355	vox/vent class homeobox protein
voy	voyager	CNS degeneration (28 h), enlarged ventricles	e		57	
vtn	valentine	heart: walls grossly extended	e		99	
wan	wanda	dominant adult: abnormal shape of body and fins, vertebrae reduced	d		69, 105	
wat	wavy tail	notochord undulating in tail, short body	e	+	89	
way	wavy	motility: reduced, may swim upside down	e		68, 218	
wde	wide eye	small eyes (day 4), cell death scattered across inner nuclear and photoreceptor layer	e		118	
wea	weak atrium	heart: atrium ± silent	e		61, 99	
web	weak beat	heart: weak beat; ± no circulation	e		61	
weg	weg	*weg is allelic with hab*	e		79	
weh	weissherbst	blood cells: hypochromic blood, decreasing count; cells not fully differentiated; may transport iron from yolk into circulation	e		93, 252	Slc39a1 (solute carrier family 39), [Ferroportin1]
wei	weiss	pigments pale, melanophores degenerating; CNS degeneration, retarded	e		82, 110, 277	
whz	weiches herz	heart: weak beat of ventricle	e		99	
wib	whitebread	no xanthophores, no iridophores; small eyes (day 5); reduced lower jaw	e		88	

Table 1 *(continued)*

Gene		Description			References	Gene product
win	wicked brain	hindbrain abnormal, thin MHB (day 1); small eyes (day 3)	e		78	
wir	wirbel	shape: curled down; motility: touch response reduced	e		60	
wis	white snake	immotile; neural tube thinner, brain necrosis; no melanin; no circulation	e		78, 95	
wit	*whitetail*	*wit has been renamed as mib*				
woe	who-cares	RT-projection: dorsal retinal axons terminate dorsally and ventrally on tectum	e		103	
wsh	washed-out	pigments pale, melanophores degenerating	e		82	
wup	whats up?	no otoliths; impaired balance, may swim upside down	e		110	
yhd	yellowhead	small eyes and brain; degeneration in all retinal layers; pericardial oedema	e		249	
yng	young	small eyes; choroid fissure open; body curved	e		288	
yob	yobo	xanthophores pale; adult: small eyes; maternal effect	e	a,m	69, 82, 90	
yoc	yocca	xanthophores pale, adults normal	e	+	82, 90	
yol	yol	eyes lack ciliary marginal zone of the retina	e		Baier Lab	
yot	you-too	shape: curled down; floorplate (LFP) reduced; no myoseptum, no dorsal aorta; ipsilateral RT-projections; ectopic lens formation	e		60, 61, 81, 106, 177, 179, 208, 211, 282, 289, 290, 299, 341	Gli2 (zinc finger transcription factor)
you	U shaped somites	shape: curled down; no myoseptum, blocky somites	e		61, 106, 148, 211, 232, 299	Urod (Uroporphyrinogen decarboxylase)
yqe	yquem	fluorescent blood, photosensitive; pericardial oedema	e		93, 180	
yug	yugiri	CNS necrosis (36 h); abnormal neurocoel; small eyes and small head	e		66	66
yur	yura	CNS degeneration (28 h), ventricles enlarged	e		57	
yyo	yoyo	heart: retrograde blood flow	e		99	

Table 1 *(continued)*

Gene		Description			References	Gene product
zem	ziemniok	small eyes (day 3), degenerating cells in retina, abnormal plexiform layers	e		83	
zez	zezem	CNS degeneration (28 h), ventricles enlarged	e		57	
zhi	zhivago	short head/cartilage reduced; reduced pectoral fins; no semicircular canals	e		88	
zim	ziehharmonika	motility: spasmic contractions of trunk (accordion-like), wavy notochord	e		68	
zin	zinfandel	blood cells: colourless blood, decreasing count	e	(+)	93	
ziz	zickzack	notochord undulating	e		89	
zja	zwangsjacke	hatching defective	e	+	71	
znk	znikam	ear: small otoliths, transient abnormalities in ear differentiation	e		84	
zny	zimny	small eyes (day 2), degenerating cells in retina; later general neural degeneration	e		83	
zom	zombie	early necrosis, arrested before somitogenesis	e		80	
zon	zonderzen	no pectoral fins; brain ventricles collapsed (transient); variable defects of jaw and motility	e		95	
zwa	zwart	melanophores expanded; inactive, frequently lying on the side	e		82	

References

1. Streisinger G, Singer F, Walker C, Knauber D, Dower N (1986). *Genetics* 112, 311.
2. Grunwald DJ, Kimmel CB, Westerfield M, Walker C, Streisinger G (1988). *Dev Biol* 126, 115.
3. Kimmel CB, Kane DA, Walker C, Warga RM, Rothman MB (1989). *Nature* 337, 358.
4. Streisinger G, Coale F, Taggart C, Walker C, Grunwald DJ (1989). *Dev Biol* 131, 60.
5. Felsenfeld AL, Walker C, Westerfield M, Kimmel C, Streisinger G (1990). *Development* 108, 443.
6. Ho RK, Kane DA (1990). *Nature* 348, 728.
7. Molven A, Wright CV, Bremiller R, De Robertis EM, Kimmel CB (1990). *Development* 109, 279.
8. Westerfield M, Liu DW, Kimmel CB, Walker C (1990). *Neuron* 4, 867.
9. Eisen JS, Pike SH (1991). *Neuron* 6, 767.
10. Felsenfeld AL, Curry M, Kimmel CB (1991). *Dev Biol* 148, 23.
11. Hatta K, Bremiller R, Westerfield M, Kimmel CB (1991). *Development* 112, 821.
12. Hatta K, Kimmel CB, Ho RK, Walker C (1991). *Nature* 350, 339.
13. Kimmel CB, Hatta K, Eisen JS (1991). *Development* Suppl, 47.

14. Bernhardt RR, Nguyen N, Kuwada JY (1992). *Neuron* 8, 869.

15. Bernhardt RR, Patel CK, Wilson SW, Kuwada JY (1992). *J Comp Neurol* 326, 263.

16. Grunwald DJ, Streisinger G (1992). *Gen Res* 59, 103.

17. Hatta K (1992). *Neuron* 9, 629.

18. Ho RK (1992). *Development Supp* 65, 65.

19. Westerfield M, Wegner J, Jegalian BG, De Robertis EM, Puschel AW (1992). *Gen Dev* 6, 591.

20. Halpern ME, Ho RK, Walker C, Kimmel CB (1993). *Cell* 75, 99.

21. Hammerschmidt M, Nusslein-Volhard C (1993). *Development* 119, 1107.

22. Joly JS, Joly C, Schulte-Merker S, Boulekbache H, Condamine H (1993). *Development* 119, 1261.

23. Strahle U, Blader P, Henrique D, Ingham PW (1993). *Gen Dev* 7, 1436.

24. Thisse C, Thisse B, Schilling TF, Postlethwait JH (1993). *Development* 119, 1203.

25. Abdelilah S, Solnica-Krezel L, Stainier DY, Driever W (1994). *Nature* 370, 468.

26. Allende ML, Weinberg ES (1994). *Dev Biol* 166, 509.

27. Hatta K, Puschel AW, Kimmel CB (1994). *PNAS* 91, 2061.

28. Patel CK, Rodriguez LC, Kuwada JY (1994). *J Neurobiol* 25, 345.

29. Schulte-Merker S, Hammerschmidt M, Beuchle D, Cho KW, De Robertis EM, Nusslein-Volhard C (1994). *Development* 120, 843.

30. Schulte-Merker S, van Eeden FJ, Halpern ME, Kimmel CB, Nusslein-Volhard C (1994). *Development* 120, 1009.

31. Sepich DS, Ho RK, Westerfield M (1994). *Dev Biol* 161, 84.

32. Thisse C, Thisse B, Halpern ME, Postlethwait JH (1994). *Dev Biol* 164, 420.

33. Xu Q, Holder N, Patient R, Wilson SW (1994). *Development* 120, 287.

34. Barth KA, Wilson SW (1995). *Development* 121, 1755.

35. Brockerhoff SE, Hurley JB, Janssen-Bienhold U, Neuhauss SC, Driever W, Dowling JE (1995). *PNAS* 92, 10545.

36. Ekker SC, Ungar AR, Greenstein P, von Kessler DP, Porter JA, Moon RT, Beachy PA (1995). *Curr Biol* 5, 944.

37. Geraudie J, Monnot MJ, Brulfert A, Ferretti P (1995). *Int J Dev Biol* 39, 373.

38. Greenspoon S, Patel CK, Hashmi S, Bernhardt RR, Kuwada JY (1995). *J Neurosci* 15, 5956.

39. Griffin K, Patient R, Holder N (1995). *Development* 121, 2983.

40. Halpern ME, Thisse C, Ho RK, Thisse B, Riggleman B, Trevarrow B, Weinberg ES, Postlethwait JH, Kimmel CB (1995). *Development* 121, 4257.

41. Johnson SL, Africa D, Horne S, Postlethwait JH (1995). *Genetics* 139, 1727.

42. Johnson SL, Africa D, Walker C, Weston JA (1995). *Dev Biol* 167, 27.

43. Johnson SL, Weston JA (1995). *Genetics* 141, 1583.

44. Kelly GM, Erezyilmaz DF, Moon RT (1995). *Mech Dev* 53, 261.

45. Kelly GM, Greenstein P, Erezyilmaz DF, Moon RT (1995). *Development* 121, 1787.

46. Macdonald R, Barth KA, Xu Q, Holder N, Mikkola I, Wilson SW (1995). *Development* 121, 3267.

47. Neave B, Rodaway A, Wilson SW, Patient R, Holder N (1995). *Mech Dev* 51, 169.

48. Riley BB, Grunwald DJ (1995). *PNAS* 92, 5997.

49. Rissi M, Wittbrodt J, Delot E, Naegeli M, Rosa FM (1995). *Mech Dev* 49, 223.

50. Stainier DY, Weinstein BM, Detrich HW 3rd, Zon LI, Fishman MC (1995). *Development* 121, 3141.

51. Talbot WS, Trevarrow B, Halpern ME, Melby AE, Farr G, Postlethwait JH, Jowett T, Kimmel CB, Kimelman D (1995). *Nature* 378, 150.

52. Thisse C, Thisse B, Postlethwait JH (1995). *Dev Biol* 172, 86.

53. Tokumoto M, Gong Z, Tsubokawa T, Hew CL, Uyemura K, Hotta Y, Okamoto H (1995). *Dev Biol* 171, 578.

54. Toyama R, Curtiss PE, Otani H, Kimura M, Dawid IB, Taira M (1995). *Dev Biol* 170, 583.

55. Weinstein BM, Stemple DL, Driever W, Fishman MC (1995). *Nat Med* 1, 1143.

56. Yan YL, Hatta K, Riggleman B, Postlethwait JH (1995). *Dev Dyn* 203, 363.

57. Abdelilah S, Mountcastle-Shah E, Harvey M, Solnica-Krezel L, Schier AF, Stemple DL, Malicki J, Neuhauss SC, Zwartkruis F, Stainier DY, Rangini Z, Driever W (1996). *Development* 123, 217.

58. Allende ML, Amsterdam A, Becker T, Kawakami K, Gaiano N, Hopkins N (1996). *Gen Dev* 10, 3141.

59. Brand M, Heisenberg CP, Jiang YJ, Beuchle D, Lun K, Furutani-Seiki M, Granato M, Haffter P, Hammerschmidt M, Kane DA, Kelsh RN, Mullins MC, Odenthal J, van Eeden FJ, Nusslein-Volhard C (1996). *Development* 123, 179.

60. Brand M, Heisenberg CP, Warga RM, Pelegri F, Karlstrom RO, Beuchle D, Picker A, Jiang YJ, Furutani-Seiki M, van Eeden FJ, Granato M, Haffter P, Hammerschmidt M, Kane DA, Kelsh RN, Mullins MC, Odenthal J, Nusslein-Volhard C (1996). *Development* 123, 129.

61. Chen JN, Haffter P, Odenthal J, Vogelsang E, Brand M, van Eeden FJ, Furutani-Seiki M, Granato M, Hammerschmidt M, Heisenberg CP, Jiang YJ, Kane DA, Kelsh RN, Mullins MC, Nusslein-Volhard C (1996). *Development* 123, 293.

62. Conlon FL, Sedgwick SG, Weston KM, Smith JC (1996). *Development* 122, 2427.

63. Danos MC, Yost HJ (1996). *Dev Biol* 177, 96.

64. Essner JJ, Laing JG, Beyer EC, Johnson RG, Hackett PB Jr (1996). *Dev Biol* 177, 449.

65. Franklin JI, Sargent TD (1996). *Dev Dyn* 206, 121.

66. Furutani-Seiki M, Jiang YJ, Brand M, Heisenberg CP, Houart C, Beuchle D, van Eeden FJ, Granato M, Haffter P, Hammerschmidt M, Kane DA, Kelsh RN, Mullins MC, Odenthal J, Nusslein-Volhard C (1996). *Development* 123, 229.

67. Gaiano N, Amsterdam A, Kawakami K, Allende M, Becker T, Hopkins N (1996). *Nature* 383, 829.

68. Granato M, van Eeden FJ, Schach U, Trowe T, Brand M, Furutani-Seiki M, Haffter P, Hammerschmidt M, Heisenberg CP, Jiang YJ, Kane DA, Kelsh RN, Mullins MC, Odenthal J, Nusslein-Volhard C (1996). *Development* 123, 399.

69. Haffter, P., Odenthal, J., Mullins, M.C., Lin, S., Farrell, M.J., Vogelsang, E., Haas, F., Brand, M., van Eeden, F.J.M., Furutani-Seiki, M., Granato, M., Hammerschmidt, M., Heisenberg, C.P., Jiang, Y.J., Kane, D.A., Kelsh, R.N., Hopkins, N., Nüsslein-Volhard, C. (1996). *Dev Genes Evol* 206, 260.

70. Hammerschmidt M, Bitgood MJ, McMahon AP (1996). *Gen Dev* 10, 647.

71. Hammerschmidt M, Pelegri F, Mullins MC, Kane DA, Brand M, van Eeden FJ, Furutani-Seiki M, Granato M, Haffter P, Heisenberg CP, Jiang YJ, Kelsh RN, Odenthal J, Warga RM, Nusslein-Volhard C (1996). *Development* 123, 143.

72. Hammerschmidt M, Pelegri F, Mullins MC, Kane DA, van Eeden FJ, Granato M, Brand M, Furutani-Seiki M, Haffter P, Heisenberg CP, Jiang YJ, Kelsh RN, Odenthal J, Warga RM, Nusslein-Volhard C (1996). *Development* 123, 95.

73. Hammerschmidt M, Serbedzija GN, McMahon AP (1996). *Gen Dev* 10, 2452.

74. Hauptmann G, Gerster T (1996). *Development* 122, 1769.

75. Heisenberg CP, Brand M, Jiang YJ, Warga RM, Beuchle D, van Eeden FJ, Furutani-Seiki M, Granato M, Haffter P, Hammerschmidt M, Kane DA, Kelsh RN, Mullins MC, Odenthal J, Nusslein-Volhard C (1996). *Development* 123, 191.

76. Henion PD, Raible DW, Beattie CE, Stoesser KL, Weston JA, Eisen JS (1996). *Dev Gen* 18, 11.

77. Jesuthasan S (1996). *Development* 122, 381.

78. Jiang YJ, Brand M, Heisenberg CP, Beuchle D, Furutani-Seiki M, Kelsh RN, Warga RM, Granato M, Haffter P, Hammerschmidt M, Kane DA, Mullins MC, Odenthal J, van Eeden FJ, Nusslein-Volhard C (1996). *Development* 123, 205.

79. Kane DA, Hammerschmidt M, Mullins MC, Maischein HM, Brand M, van Eeden FJ, Furutani-Seiki M, Granato M, Haffter P, Heisenberg CP, Jiang YJ, Kelsh RN, Odenthal J, Warga RM, Nusslein-Volhard C (1996). *Development* 123, 47.

80. Kane DA, Maischein HM, Brand M, van Eeden FJ, Furutani-Seiki M, Granato M, Haffter P, Hammerschmidt M, Heisenberg CP, Jiang YJ, Kelsh RN, Mullins MC, Odenthal J, Warga RM, Nusslein-Volhard C (1996). *Development* 123, 57.

81. Karlstrom RO, Trowe T, Klostermann S, Baier H, Brand M, Crawford AD, Grunewald B, Haffter P, Hoffmann H, Meyer SU, Muller BK, Richter S, van Eeden FJ, Nusslein-Volhard C, Bonhoeffer F (1996). *Development* 123, 427.

82. Kelsh RN, Brand M, Jiang YJ, Heisenberg CP, Lin S, Haffter P, Odenthal J, Mullins MC, van Eeden FJ, Furutani-Seiki M, Granato M, Hammerschmidt M, Kane DA, Warga RM, Beuchle D, Vogelsang L, Nusslein-Volhard C (1996). *Development* 123, 369.

83. Malicki J, Neuhauss SC, Schier AF, Solnica-Krezel L, Stemple DL, Stainier DY, Abdelilah S, Zwartkruis F, Rangini Z, Driever W (1996). *Development* 123, 263.

84. Malicki J, Schier AF, Solnica-Krezel L, Stemple DL, Neuhauss SC, Stainier DY, Abdelilah S, Rangini Z, Zwartkruis F, Driever W (1996). *Development* 123, 275.

85. Melby AE, Warga RM, Kimmel CB (1996). *Development* 122, 2225.

86. Moens CB, Yan YL, Appel B, Force AG, Kimmel CB (1996). *Development* 122, 3981.

87. Mullins MC, Hammerschmidt M, Kane DA, Odenthal J, Brand M, van Eeden FJ, Furutani-Seiki M, Granato M, Haffter P, Heisenberg CP, Jiang YJ, Kelsh RN, Nusslein-Volhard C (1996). *Development* 123, 81.

88. Neuhauss SC, Solnica-Krezel L, Schier AF, Zwartkruis F, Stemple DL, Malicki J, Abdelilah S, Stainier DY, Driever W (1996). *Development* 123, 357.

89. Odenthal J, Haffter P, Vogelsang E, Brand M, van Eeden FJ, Furutani-Seiki M, Granato M, Hammerschmidt M, Heisenberg CP, Jiang YJ, Kane DA, Kelsh RN, Mullins MC, Warga RM, Allende ML, Weinberg ES, Nusslein-Volhard C (1996). *Development* 123, 103.

90. Odenthal J, Rossnagel K, Haffter P, Kelsh RN, Vogelsang E, Brand M, van Eeden FJ, Furutani-Seiki M, Granato M, Hammerschmidt M, Heisenberg CP, Jiang YJ, Kane DA, Mullins MC, Nusslein-Volhard C (1996). *Development* 123, 391.

91. Pack M, Solnica-Krezel L, Malicki J, Neuhauss SC, Schier AF, Stemple DL, Driever W, Fishman MC (1996). *Development* 123, 321.

92. Piotrowski T, Schilling TF, Brand M, Jiang YJ, Heisenberg CP, Beuchle D, Grandel H, van Eeden FJ, Furutani-Seiki M, Granato M, Haffter P, Hammerschmidt M, Kane DA, Kelsh RN, Mullins MC, Odenthal J, Warga RM, Nusslein-Volhard C (1996). *Development* 123, 345.

93. Ransom DG, Haffter P, Odenthal J, Brownlie A, Vogelsang E, Kelsh RN, Brand M, van Eeden FJ, Furutani-Seiki M, Granato M, Hammerschmidt M, Heisenberg CP, Jiang YJ, Kane DA, Mullins MC, Nusslein-Volhard C (1996). *Development* 123, 311.

94. Riley BB, Grunwald DJ (1996). *Dev Biol* 179, 427.

95. Schier AF, Neuhauss SC, Harvey M, Malicki J, Solnica-Krezel L, Stainier DY, Zwartkruis F, Abdelilah S, Stemple DL, Rangini Z, Yang H, Driever W (1996). *Development* 123, 165.

96. Schilling TF, Piotrowski T, Grandel H, Brand M, Heisenberg CP, Jiang YJ, Beuchle D, Hammerschmidt M, Kane DA, Mullins MC, van Eeden FJ, Kelsh RN, Furutani-Seiki M, Granato M, Haffter P, Odenthal J, Warga RM, Trowe T, Nusslein-Volhard C (1996). *Development* 123, 329.

97. Schilling TF, Walker C, Kimmel CB (1996). *Development* 122, 1417.

98. Solnica-Krezel L, Stemple DL, Mountcastle-Shah E, Rangini Z, Neuhauss SC, Malicki J, Schier AF, Stainier DY, Zwartkruis F, Abdelilah S, Driever W (1996). *Development* 123, 67.

99. Stainier DY, Fouquet B, Chen JN, Warren KS, Weinstein BM, Meiler SE, Mohideen MA, Neuhauss SC, Solnica-Krezel L, Schier AF, Zwartkruis F, Stemple DL, Malicki J, Driever W, Fishman MC (1996). *Development* 123, 285.

100. Stemple DL, Solnica-Krezel L, Zwartkruis F, Neuhauss SC, Schier AF, Malicki J, Stainier DY, Abdelilah S, Rangini Z, Mountcastle-Shah E, Driever W (1996). *Development* 123, 117.

101. Strahle U, Blader P, Ingham PW (1996). *Int J Dev Biol* 40, 929.

102. Ticho BS, Stainier DY, Fishman MC, Breitbart RE (1996). *Mech Dev* 59, 205.

103. Trowe T, Klostermann S, Baier H, Granato M, Crawford AD, Grunewald B, Hoffmann H, Karlstrom RO, Meyer SU, Muller B, Richter S, Nusslein-Volhard C, Bonhoeffer F (1996). *Development* 123, 439.

104. Ungar AR, Moon RT (1996). *Dev Biol* 178, 186.

105. van Eeden FJ, Granato M, Schach U, Brand M, Furutani-Seiki M, Haffter P, Hammerschmidt M, Heisenberg CP, Jiang YJ, Kane DA, Kelsh RN, Mullins MC, Odenthal J, Warga RM, Nusslein-Volhard C (1996). *Development* 123, 255.

106. van Eeden FJ, Granato M, Schach U, Brand M, Furutani-Seiki M, Haffter P, Hammerschmidt M, Heisenberg CP, Jiang YJ, Kane DA, Kelsh RN, Mullins MC, Odenthal J, Warga RM, Allende ML, Weinberg ES, Nusslein-Volhard C (1996). *Development* 123, 153.

107. Vriz S, Joly C, Boulekbache H, Condamine H (1996). *Brain Res Bull* 40, 221.

108. Weinberg ES, Allende ML, Kelly CS, Abdelhamid A, Murakami T, Andermann P, Doerre OG, Grunwald DJ, Riggleman B (1996). *Development* 122, 271.

109. Weinstein BM, Schier AF, Abdelilah S, Malicki J, Solnica-Krezel L, Stemple DL, Stainier DY, Zwartkruis F, Driever W, Fishman MC (1996). *Development* 123, 303.

110. Whitfield TT, Granato M, van Eeden FJ, Schach U, Brand M, Furutani-Seiki M, Haffter P, Hammerschmidt M, Heisenberg CP, Jiang YJ, Kane DA, Kelsh RN, Mullins MC, Odenthal J, Nusslein-Volhard C (1996). *Development* 123, 241.

111. Abdelilah S, Driever W (1997). *Dev Biol* 184, 70.

112. Baker K, Warren KS, Yellen G, Fishman MC (1997). *PNAS* 94, 4554.

113. Beattie CE, Eisen JS (1997). *Development* 124, 713.

114. Blagden CS, Currie PD, Ingham PW, Hughes SM (1997). 11, 2163.

115. Brockerhoff SE, Hurley JB, Niemi GA, Dowling JE (1997). *J Neurosci* 17, 4236.

116. Chandrasekhar A, Moens CB, Warren JT Jr, Kimmel CB, Kuwada JY (1997). *Development* 124, 2633.

117. Chen JN, van Eeden, FJ, Warren, KS, Chin, A., Nusslein-Volhard C, Haffter P, Fishman MC (1997). *Development* 124, 4373.

118. Fadool, J. M., Brockerhoff, S. E., Hyatt, G. A., Dowling, J. E. (1997). *Dev Gen* 20, 288.

119. Fisher S, Amacher SL, Halpern ME (1997). *Development* 124, 1301.

120. Fouquet B, Weinstein BM, Serluca FC, Fishman MC (1997). *Dev Biol* 183, 37.

121. Fulwiler C, Schmitt EA, Kim JM, Dowling JE (1997). *J Comp Neurol* 381, 449.

122. Glasgow E, Karavanov AA, Dawid IB (1997). *Dev Biol* 192, 405.

123. Halpern ME, Hatta K, Amacher SL, Talbot WS, Yan YL, Thisse B, Thisse C, Postlethwait JH, Kimmel CB (1997). *Dev Biol* 187, 154.

124. Heisenberg CP, Nusslein-Volhard C (1997). *Dev Biol* 184, 85.

125. Hug B, Walter V, Grunwald DJ (1997). *Dev Biol* 183, 61.

126. Kishimoto Y, Lee KH, Zon L, Hammerschmidt M, Schulte-Merker S (1997). *Development* 124, 4457.

127. Lauderdale JD, Davis NM, Kuwada JY (1997). *Mol Cell Neurosci* 9, 293.

128. Lele Z, Krone PH (1997). *Mech Dev* 61, 89.

129. Li L, Dowling JE (1997). *PNAS* 94, 11645.

130. Liao W, Bisgrove BW, Sawyer H, Hug B, Bell B, Peters K, Grunwald DJ, Stainier DY (1997). *Development* 124, 381.

131. Macdonald R, Scholes J, Strahle U, Brennan C, Holder N, Brand M, Wilson SW (1997). *Development* 124, 2397.

132. Masai I, Heisenberg CP, Barth KA, Macdonald R, Adamek S, Wilson SW (1997). *Neuron* 18, 43.

133. Melby AE, Kimelman D, Kimmel CB (1997). *Dev Dyn* 209, 156.

134. Miller-Bertoglio VE, Fisher S, Sanchez A, Mullins MC, Halpern ME (1997). *Dev Biol* 192, 537.

135. Rauch GJ, Hammerschmidt M, Blader P, Schauerte HE, Strahle U, Ingham PW, McMahon AP, Haffter P (1997). *CSHS Quant Biol* 62, 227.

136. Riley BB, Zhu C, Janetopoulos C, Aufderheide KJ (1997). *Dev Biol* 191, 191.

137. Schier AF, Neuhauss SC, Helde KA, Talbot WS, Driever W (1997). *Development* 124, 327.

138. Schulte-Merker S, Lee KJ, McMahon AP, Hammerschmidt M (1997). *Nature* 387, 862.

139. Strahle U, Fischer N, Blader P (1997). *Mech Dev* 62, 147.

140. Strahle U, Jesuthasan S, Blader P, Garcia-Villalba P, Hatta K, Ingham PW (1997). *Genes Function* 1, 131.

141. Sumoy L, Keasey JB, Dittman TD, Kimelman D (1997). *Mech Dev* 63, 15.

142. Amacher SL, Kimmel CB (1998). *Development* 125, 1397.

143. Ando H, Mishina M (1998). *Neurosci Lett* 244, 81.

144. Bauer H, Meier A, Hild M, Stachel S, Economides A, Hazelett D, Harland RM, Hammerschmidt M (1998). *Dev Biol* 204, 488.

145. Becker TS, Burgess SM, Amsterdam AH, Allende ML, Hopkins N (1998). *Development* 125, 4369.

146. Brockerhoff SE, Dowling JE, Hurley JB (1998). *Vision Res* 38, 1335.

147. Brownlie A, Donovan A, Pratt SJ, Paw BH, Oates AC, Brugnara C, Witkowska HE, Sassa S, Zon LI (1998). *Nat Genet* 20, 244.

148. Chandrasekhar A, Warren JT Jr, Takahashi K, Schauerte HE, van Eeden FJ, Haffter P, Kuwada JY (1998). *Mech Dev* 76, 101.

149. Drummond IA, Majumdar A, Hentschel H, Elger M, Solnica-Krezel L, Schier AF, Neuhauss SC, Stemple DL, Zwartkruis F, Rangini Z, Driever W, Fishman MC (1998). *Development* 125, 4655.

150. Feldman B, Gates MA, Egan ES, Dougan ST, Rennebeck G, Sirotkin HI, Schier AF, Talbot WS (1998). *Nature* 395, 181.

151. Goldstein AM, Fishman MC (1998). *Dev Biol* 201, 247.

152. Griffin KJ, Amacher SL, Kimmel CB, Kimelman D (1998). *Development* 125, 3379.

153. Grinblat Y, Gamse J, Patel M, Sive H (1998). *Development* 125, 4403.

154. Haddon C, Jiang YJ, Smithers L, Lewis J (1998). *Development* 125, 4637.

155. Joore J, van de Water S, Betist M, van den Eijnden-van Raaij A, Zivkovic D (1998). *Mech Dev* 79, 5.

156. Kimmel CB, Miller CT, Kruze G, Ullmann B, BreMiller RA, Larison KD, Snyder HC (1998). *Dev Biol* 203, 245.

157. Korzh V, Sleptsova I, Liao J, He J, Gong Z (1998). *Dev Dyn* 213, 92.

158. Lauderdale JD, Pasquali SK, Fazel R, van Eeden FJ, Schauerte HE, Haffter P, Kuwada JY (1998). *Mol Cell Neurosci* 11, 194.

159. Lee KH, Marden JJ, Thompson MS, MacLennan H, Kishimoto Y, Pratt SJ, Schulte-Merker S, Hammerschmidt M, Johnson SL, Postlethwaite JH, Beier DC, Zon LI (1998). *Dev Gen* 23, 97.

160. Liao EC, Paw BH, Oates AC, Pratt SJ, Postlethwait JH, Zon LI (1998). *Gen Dev* 12, 621.

161. Lun K, Brand M (1998). *Development* 125, 3049.

162. Makita R, Mizuno T, Koshida S, Kuroiwa A, Takeda H (1998). *Mech Dev* 71, 165.

163. Marlow F, Zwartkruis F, Malicki J, Neuhauss SC, Abbas L, Weaver M, Driever W, Solnica-Krezel L (1998). *Dev Biol* 203, 382.

164. Moens CB, Cordes SP, Giorgianni MW, Barsh GS, Kimmel CB (1998). *Development* 125, 381.

165. Nguyen VH, Schmid B, Trout J, Connors SA, Ekker M, Mullins MC (1998). *Dev Biol* 199, 93.

166. Nicolson T, Rusch A, Friedrich RW, Granato M, Ruppersberg JP, Nusslein-Volhard C (1998). *Neuron* 20, 271.

167. Peyrieras N, Strahle U, Rosa F (1998). *Curr Biol* 8, 783.

168. Pfeffer PL, Gerster T, Lun K, Brand M, Busslinger M (1998). *Development* 125, 3063.

169. Prince VE, Moens CB, Kimmel CB, Ho RK (1998). *Development* 125, 393.

170. Rebagliati MR, Toyama R, Haffter P, Dawid IB (1998). *PNAS* 95, 9932.

171. Reifers F, Bohli H, Walsh EC, Crossley PH, Stainier DY, Brand M (1998). *Development* 125, 2381.

172. Ribera AB, Nusslein-Volhard C (1998). *J Neurosci* 18, 9181.

173. Sampath K, Rubinstein AL, Cheng AM, Liang JO, Fekany K, Solnica-Krezel L, Korzh V, Halpern ME, Wright CV (1998). *Nature* 395, 185.

174. Schauerte HE, van Eeden FJ, Fricke C, Odenthal J, Strahle U, Haffter P (1998). *Development* 125, 2983.

175. Sepich DS, Wegner J, O'Shea S, Westerfield M (1998). *Genetics* 148, 361.

176. Shanmugalingam S, Wilson SW (1998). *Mech Dev* 78, 85.

177. Shoji W, Yee CS, Kuwada JY (1998). *Development* 125, 1275.

178. Thompson MA, Ransom DG, Pratt SJ, MacLennan H, Kieran MW, Detrich HW 3rd, Vail B, Huber TL, Paw B, Brownlie AJ, Oates AC, Fritz A, Gates MA, Amores A, Bahary N, Talbot WS, Her H, Beier DR, Postlethwait JH, Zon LI (1998). *Dev Biol* 197, 248.

179. van Eeden FJ, Holley SA, Haffter P, Nusslein-Volhard C (1998). *Dev Gen* 23, 65.

180. Wang H, Long Q, Marty SD, Sassa S, Lin S (1998). *Nat Genet* 20, 239.

181. Warga RM, Nusslein-volhard C (1998). *Dev Biol* 203, 116.

182. Yamamoto A, Amacher SL, Kim SH, Geissert D, Kimmel CB, De Robertis EM (1998). *Development* 125, 3389.

183. Yan YL, Jowett T, Postlethwait JH (1998). *Dev Dyn* 213, 370.

184. Zhang J, Talbot WS, Schier AF (1998). *Cell* 92, 241.

185. Alexander J, Rothenberg M, Henry GL, Stainier DY (1999). *Dev Biol* 215, 343.

186. Alexander J, Stainier DY (1999). *Curr Biol* 9, 1147.

187. Amsterdam A, Burgess S, Golling G, Chen W, Sun Z, Townsend K, Farrington S, Haldi M, Hopkins N (1999). *Gen Dev* 13, 2713.

188. Appel B, Fritz A, Westerfield M, Grunwald DJ, Eisen JS, Riley BB (1999). *Curr Biol* 9, 247.

189. Artinger KB, Chitnis AB, Mercola M, Driever W (1999). *Development* 126, 3969.

190. Asai R, Taguchi E, Kume Y, Saito M, Kondo S (1999). *Mech Dev* 89, 87.

191. Barth KA, Kishimoto Y, Rohr KB, Seydler C, Schulte-Merker S, Wilson SW (1999). *Development* 126, 4977.

192. Bisgrove BW, Essner JJ, Yost HJ (1999). *Development* 126, 3253.

193. Chandrasekhar A, Schauerte HE, Haffter P, Kuwada JY (1999). *Development* 126, 2727.

194. Connors SA, Trout J, Ekker M, Mullins MC (1999). *Development* 126, 3119.

195. Cretekos CJ, Grunwald DJ (1999). *Dev Biol* 210, 322.

196. Dheen T, Sleptsova-Friedrich I, Xu Y, Clark M, Lehrach H, Gong Z, Korzh V (1999). *Development* 126, 2703.

197. Dick A, Meier A, Hammerschmidt M (1999). *Dev Dyn* 216, 285.

198. Fekany K, Yamanaka Y, Leung T, Sirotkin HI, Topczewski J, Gates MA, Hibi M, Renucci A, Stemple D, Radbill A, Schier AF, Driever W, Hirano T, Talbot WS, Solnica-Krezel L (1999). *Development* 126, 1427.

199. Fisher S, Halpern ME (1999). *Nat Genet* 23, 442.

200. Gothilf Y, Coon SL, Toyama R, Chitnis A, Namboodiri MA, Klein DC (1999). *Endocrinology* 140, 4895.

201. Gritsman K, Zhang J, Cheng S, Heckscher E, Talbot WS, Schier AF (1999). *Cell* 97, 121.

202. Guo S, Brush J, Teraoka H, Goddard A, Wilson SW, Mullins MC, Rosenthal A (1999). *Neuron* 24, 555.

203. Guo S, Wilson SW, Cooke S, Chitnis AB, Driever W, Rosenthal A (1999). *Dev Biol* 208, 473.

204. Haddon C, Mowbray C, Whitfield T, Jones D, Gschmeissner S, Lewis J (1999). *J Neurocytol* 28, 837.

205. Heisenberg CP, Brennan C, Wilson SW (1999). *Development* 126, 2129.

206. Hild M, Dick A, Rauch GJ, Meier A, Bouwmeester T, Haffter P, Hammerschmidt M (1999). *Development* 126, 2149.

207. Johnson SL, Bennett P (1999). *Meth Cell Biol* 59, 301.

208. Karlstrom RO, Talbot WS, Schier AF (1999). *Gen Dev* 13, 388.

209. Konig C, Yan YL, Postlethwait J, Wendler S, Campos-Ortega JA (1999). *Mech Dev* 86, 17.

210. Koos DS, Ho RK (1999). *Dev Biol* 215, 190.

211. Lewis KE, Currie PD, Roy S, Schauerte H, Haffter P, Ingham PW (1999). *Dev Biol* 216, 469.

212. Lister JA, Robertson CP, Lepage T, Johnson SL, Raible DW (1999). *Development* 126, 3757.

213. Majumdar A, Drummond IA (1999). *Dev Gen* 24, 220.

214. Malicki J, Driever W (1999). *Development* 126, 1235.

215. Mendonsa ES, Riley BB (1999). *Dev Biol* 206, 100.

216. Meno C, Gritsman K, Ohishi S, Ohfuji Y, Heckscher E, Mochida K, Shimono A, Kondoh H, Talbot WS, Robertson EJ, Schier AF, Hamada H (1999). *Mol Cell* 4, 287.

217. Miller-Bertoglio V, Carmany-Rampey A, Furthauer M, Gonzalez EM, Thisse C, Thisse B, Halpern ME, Solnica-Krezel L (1999). *Dev Biol* 214, 72.

218. Neuhauss SC, Biehlmaier O, Seeliger MW, Das T, Kohler K, Harris WA, Baier H (1999). *J Neurosci* 19, 8603.

219. Neumann CJ, Grandel H, Gaffield W, Schulte-Merker S, Nusslein-Volhard C (1999). *Development* 126, 4817.

220. Oates AC, Brownlie A, Pratt SJ, Irvine DV, Liao EC, Paw BH, Dorian KJ, Johnson SL, Postlethwait JH, Zon LI, Wilks AF (1999). *Blood* 94, 2622.

221. Ober EA, Schulte-Merker S (1999). *Dev Biol* 215, 167.

222. Parichy DM, Rawls JF, Pratt SJ, Whitfield TT, Johnson SL (1999). *Development* 126, 3425.

223. Parker L, Stainier DY (1999). *Development* 126, 2643.

224. Pelegri F, Knaut H, Maischein HM, Schulte-Merker S, Nusslein-Volhard C (1999). *Curr Biol* 9, 1431.

225. Picker A, Brennan C, Reifers F, Clarke JD, Holder N, Brand M (1999). *Development* 126, 2967.

226. Porcher C, Liao EC, Fujiwara Y, Zon LI, Orkin SH (1999). *Development* 126, 4603.

227. Reiter JF, Alexander J, Rodaway A, Yelon D, Patient R, Holder N, Stainier DY (1999). *Gen Dev* 13, 2983.

228. Riley BB, Chiang M, Farmer L, Heck R (1999). *Development* 126, 5669.

229. Rodaway A, Takeda H, Koshida S, Broadbent J, Price B, Smith JC, Patient R, Holder N (1999). *Development* 126, 3067.

230. Rohr KB, Schulte-Merker S, Tautz D (1999). *Mech Dev* 85, 147.

231. Sass JB, Martin CC, Krone PH (1999). *Int J Dev Biol* 43, 835.

232. Schilling TF, Concordet JP, Ingham PW (1999). *Dev Biol* 210, 277.

233. Seiler C, Nicolson T (1999). *J Neurobiol* 41, 424.

234. Shinya M, Furutani-Seiki M, Kuroiwa A, Takeda H (1999). *Dev Growth Diff* 41, 135.

235. Tongiorgi E (1999). *Blood Cells Mol. Dis.* 48, 79.

236. Varga ZM, Wegner J, Westerfield M (1999). *Development* 126, 5533.

237. Warga RM, Nusslein-Volhard C (1999). *Development* 126, 827.
238. Weidinger G, Wolke U, Koprunner M, Klinger M, Raz E (1999). *Development* 126, 5295.
239. Yelon D, Horne SA, Stainier DY (1999). *Dev Biol* 214, 23.
240. Zeller J, Granato M (1999). *Development* 126, 3461.
241. Adamska M, Leger S, Brand M, Hadrys T, Braun T, Bober E (2000). 97, 161.
242. Barresi MJ, Stickney HL, Devoto SH (2000). *Development* 127, 2189.
243. Beattie CE, Melancon E, Eisen JS (2000). *Development* 127, 2653.
244. Bisgrove BW, Essner JJ, Yost HJ (2000). *Development* 127, 3567.
245. Brown LA, Rodaway AR, Schilling TF, Jowett T, Ingham PW, Patient RK, Sharrocks AD (2000). *Mech Dev* 90, 237.
246. Childs S, Weinstein BM, Mohideen MA, Donohue S, Bonkovsky H, Fishman MC (2000). *Curr Biol* 10, 1001.
247. Chin AJ, Tsang M, Weinberg ES (2000). *Dev Biol* 227, 403.
248. Cornell RA, Eisen JS (2000). *Development* 127, 2873.
249. Daly FJ, Sandell JH (2000). *Anat Rec* 258, 145.
250. Dick A, Hild M, Bauer H, Imai Y, Maifeld H, Schier AF, Talbot WS, Bouwmeester T, Hammerschmidt M (2000). *Development* 127, 343.
251. Dick A, Mayr T, Bauer H, Meier A, Hammerschmidt M (2000). *Gene* 246, 69.
252. Donovan A, Brownlie A, Zhou Y, Shepard J, Pratt SJ, Moynihan J, Paw BH, Drejer A, Barut B, Zapata A, Law TC, Brugnara C, Lux SE, Pinkus GS, Pinkus JL, Kingsley PD, Palis J, Fleming MD, Andrews NC, Zon LI (2000). *Nature* 403, 776.
253. Dorsky RI, Raible DW, Moon RT (2000). *Gen Dev* 14, 158.
254. Durbin L, Sordino P, Barrios A, Gering M, Thisse C, Thisse B, Brennan C, Green A, Wilson S, Holder N (2000). *Development* 127, 1703.
255. Ernest S, Rauch GJ, Haffter P, Geisler R, Petit C, Nicolson T (2000). *Hum Mol Genet* 9, 2189.
256. Essner JJ, Branford WW, Zhang J, Yost HJ (2000). *Development* 127, 1081.
257. Fekany-Lee K, Gonzalez E, Miller-Bertoglio V, Solnica-Krezel L (2000). *Development* 127, 2333.
258. Feldman B, Dougan ST, Schier AF, Talbot WS (2000). *Curr Biol* 10, 531.
259. Gonzalez EM, Fekany-Lee K, Carmany-Rampey A, Erter C, Topczewski J, Wright CV, Solnica-Krezel L (2000). *Genes Dev* 14, 3087.
260. Goutel C, Kishimoto Y, Schulte-Merker S, Rosa F (2000). *Mech Dev* 99, 15.
261. Grandel H, Draper BW, Schulte-Merker S (2000). *Development* 127, 4169.
262. Griffin KJ, Stoller J, Gibson M, Chen S, Yelon D, Stainier DY, Kimelman D (2000). *Dev Biol* 218, 235.
263. Gritsman K, Talbot WS, Schier AF (2000). *Development* 127, 921.
264. Guo S, Yamaguchi Y, Schilbach S, Wada T, Lee J, Goddard A, French D, Handa H, Rosenthal A (2000). *Nature* 408, 366.
265. Hashimoto H, Itoh M, Yamanaka Y, Yamashita S, Shimizu T, Solnica-Krezel L, Hibi M, Hirano T (2000). *Dev Biol* 217, 138.
266. Heisenberg CP, Tada M, Rauch GJ, Saude L, Concha ML, Geisler R, Stemple DL, Smith JC, Wilson SW (2000). *Nature* 405, 76.
267. Henry CA, Hall LA, Burr Hille M, Solnica-Krezel L, Cooper MS (2000). *Curr Biol* 10, 1063.
268. Holley SA, Geisler R, Nusslein-Volhard C (2000). *Genes Dev* 14, 1678.
269. Imai Y, Feldman B, Schier AF, Talbot WS (2000). *Genetics* 155, 261.
270. Iovine MK, Johnson SL (2000). *Genetics* 155, 1321.
271. Jiang YJ, Aerne BL, Smithers L, Haddon C, Ish-Horowicz D, Lewis J (2000). *Nature* 408, 475.
272. Kawahara A, Wilm T, Solnica-Krezel L, Dawid IB (2000). *PNAS* 97, 12121.
273. Kawakami K, Amsterdam A, Shimoda N, Becker T, Mugg J, Shima A, Hopkins N (2000). *Curr Biol* 10, 463.
274. Kelly C, Chin AJ, Leatherman JL, Kozlowski DJ, Weinberg ES (2000). *Development* 127, 3899.
275. Kelsh RN, Dutton K, Medlin J, Eisen JS (2000). *Mech Dev* 93, 161.
276. Kelsh RN, Eisen JS (2000). *Development* 127, 515.
277. Kelsh RN, Schmid B, Eisen JS (2000). *Dev Biol* 225, 277.
278. Kiecker C, Muller F, Wu W, Glinka A, Strahle U, Niehrs C (2000). *Mech Dev* 94, 37.

279. Kikuchi Y, Trinh LA, Reiter JF, Alexander J, Yelon D, Stainier DY (2000). *Gen Dev* 14, 1279.

280. Kim CH, Oda T, Itoh M, Jiang D, Artinger KB, Chandrasekharappa SC, Driever W, Chitnis AB (2000). *Nature* 407, 913.

281. Knaut H, Pelegri F, Bohmann K, Schwarz H, Nusslein-Volhard C (2000). *J Cell Biol* 149, 875.

282. Kondoh H, Uchikawa M, Yoda H, Takeda H, Furutani-Seiki M, Karlstrom RO (2000). *Mech Dev* 96, 165.

283. Kupperman E, An S, Osborne N, Waldron S, Stainier DY (2000). *Nature* 406, 192.

284. Li L, Dowling JE (2000). *J Neurosci* 20, 1883.

285. Liang JO, Etheridge A, Hantsoo L, Rubinstein AL, Nowak SJ, Izpisua Belmonte JC, Halpern ME (2000). *Development* 127, 5101.

286. Liao EC, Paw BH, Peters LL, Zapata A, Pratt SJ, Do CP, Lieschke G, Zon LI (2000). *Development* 127, 5123.

287. Liao W, Ho CY, Yan YL, Postlethwait J, Stainier DY (2000). *Development* 127, 4303.

288. Link BA, Fadool JM, Malicki J, Dowling JE (2000). *Development* 127, 2177.

289. Liu A, Majumdar A, Schauerte HE, Haffter P, Drummond IA (2000). *Mech Dev* 91, 409.

290. Majumdar A, Drummond IA (2000). *Dev Biol* 222, 147.

291. Majumdar A, Lun K, Brand M, Drummond IA (2000). *Development* 127, 2089.

292. Melby AE, Beach C, Mullins M, Kimelman D (2000). *Dev Biol* 224, 275.

293. Miller CT, Schilling TF, Lee K, Parker J, Kimmel CB (2000). *Development* 127, 3815.

294. Muller F, Albert S, Blader P, Fischer N, Hallonet M, Strahle U (2000). *Development* 127, 3889.

295. Muraoka O, Ichikawa H, Shi H, Okumura S, Taira E, Higuchi H, Hirano T, Hibi M, Miki N (2000). *Dev Biol* 228, 29.

296. Nasevicius A, Ekker SC (2000). *Nat Genet* 26, 216.

297. Neumann CJ, Nuesslein-Volhard C (2000). *Science* 289, 2137.

298. Nguyen VH, Trout J, Connors SA, Andermann P, Weinberg E, Mullins MC (2000). *Development* 127, 1209.

299. Odenthal J, van Eeden FJ, Haffter P, Ingham PW, Nusslein-Volhard C (2000). *Dev Biol* 219, 350.

300. Parichy DM, Mellgren EM, Rawls JF, Lopes SS, Kelsh RN, Johnson SL (2000). *Dev Biol* 227, 294.

301. Parichy DM, Ransom DG, Paw B, Zon LI, Johnson SL (2000). *Development* 127, 3031.

302. Park HC, Kim CH, Bae YK, Yeo SY, Kim SH, Hong SK, Shin J, Yoo KW, Hibi M, Hirano T, Miki N, Chitnis AB, Huh TL (2000). *Dev Biol* 227, 279.

303. Piotrowski T, Nusslein-Volhard C (2000). *Dev Biol* 225, 339.

304. Pogoda HM, Solnica-Krezel L, Driever W, Meyer D (2000). *Curr Biol* 10, 1041.

305. Popperl H, Rikhof H, Chang H, Haffter P, Kimmel CB, Moens CB (2000). *Mol Cell* 6, 255.

306. Raible DW, Kruse GJ (2000). *J Comp Neurol* 421, 189.

307. Reifers F, Adams J, Mason IJ, Schulte-Merker S, Brand M (2000). *Mech Dev* 99, 39.

308. Reifers F, Walsh EC, Leger S, Stainier DY, Brand M (2000). *Development* 127, 225.

309. Rick JM, Horschke I, Neuhauss SC (2000). *Curr Biol* 10, 595.

310. Riley BB, Moorman SJ (2000). *J Neurobiol* 43, 329.

311. Rubinstein AL, Lee D, Luo R, Henion PD, Halpern ME (2000). *Genesis J Gen Dev* 26, 86.

312. Saude L, Woolley K, Martin P, Driever W, Stemple DL (2000). *Development* 127, 3407.

313. Sawada A, Fritz A, Jiang Y, Yamamoto A, Yamasu K, Kuroiwa A, Saga Y, Takeda H (2000). *Development* 127, 1691.

314. Schmid B, Furthauer M, Connors SA, Trout J, Thisse B, Thisse C, Mullins MC (2000). *Development* 127, 957.

315. Sepich DS, Myers DC, Short R, Topczewski J, Marlow F, Solnica-Krezel L (2000). *Genesis* 27, 159.

316. Shanmugalingam S, Houart C, Picker A, Reifers F, Macdonald R, Barth A, Griffin K, Brand M, Wilson SW (2000). *Development* 127, 2549.

317. Shimizu T, Yamanaka Y, Ryu SL, Hashimoto H, Yabe T, Hirata T, Bae YK, Hibi M, Hirano T (2000). *Mech Dev* 91, 293.

318. Shinya M, Eschbach C, Clark M, Lehrach H, Furutani-Seiki M (2000). *Mech Dev* 98, 3.

319. Sirotkin HI, Dougan ST, Schier AF, Talbot WS (2000). *Development* 127, 2583.

320. Sirotkin HI, Gates MA, Kelly PD, Schier AF, Talbot WS (2000). *Curr Biol* 10, 1051.

321. Vihtelic TS, Hyde DR (2000). *J Neurobiol* 44, 289.

322. Yelon D, Ticho B, Halpern ME, Ruvinsky I, Ho RK, Silver LM, Stainier DY (2000). *Development* 127, 2573.

323. Zhang J, Granato M (2000). *Development* 127, 2099.

324. Zhong TP, Rosenberg M, Mohideen MA, Weinstein B, Fishman MC (2000). *Science* 287, 1820.

325. Allwardt BA, Lall AB, Brockerhoff SE, Dowling JE (2001). *J Neurosci* 21, 2330.

326. Bauer H, Lele Z, Rauch GJ, Geisler R, Hammerschmidt M (2001). *Development* 128, 849.

327. Bauer MP, Goetz FW (2001). *Biol Reprod* 64, 548.

328. Begemann G, Schilling TF, Rauch GJ, Geisler R, Ingham PW (2001). *Development* 128, 3081.

329. Biemar F, Argenton F, Schmidtke R, Epperlein S, Peers B, Driever W (2001). *Dev Biol* 230, 189.

330. Bingham S, Nasevicius A, Ekker SC, Chandrasekhar A (2001). *Genesis* 30, 170.

331. Carmany-Rampey A, Schier AF (2001). *Curr Biol* 11, 1261.

332. Chen W, Burgess S, Hopkins N (2001). *Development* 128, 2385.

333. Chen Y, Schier AF (2001). *Nature* 411, 607.

334. Cooke J, Moens C, Roth L, Durbin L, Shiomi K, Brennan C, Kimmel C, Wilson S, Holder N (2001). *Development* 128, 571.

335. Coutelle O, Blagden CS, Hampson R, Halai C, Rigby PW, Hughes SM (2001). *Dev Biol* 236, 136.

336. Coutinho P, Saúde L, Stemple D (2001). *Sec Europ Conf ZF Gen & Dev* 9999, 5.

337. Darland T, Dowling JE (2001). *Proc Natl Acad Sci U S A* 98, 11691.

338. David NB, Rosa FM (2001). *Development* 128, 3937.

339. Dickmeis T, Mourrain P, Saint-Etienne L, Fischer N, Aanstad P, Clark M, Strahle U, Rosa F (2001). *Genes Dev* 15, 1487.

340. Doerre G, Malicki J (2001). *J Neurosci* 21, 6745.

341. Du SJ, Dienhart M (2001). *Differentiation* 67, 84.

342. Dutton KA, Pauliny A, Lopes SS, Elworthy S, Carney TJ, Rauch J, Geisler R, Haffter P, Kelsh RN (2001). *Development* 128, 4113.

343. Erter CE, Wilm TP, Basler N, Wright CV, Solnica-Krezel L (2001). *Development* 128, 3571.

344. Etheridge LA, Wu T, Liang JO, Ekker SC, Halpern ME (2001). *Genesis* 30, 164.

345. Farber SA, Pack M, Ho SY, Johnson ID, Wagner DS, Dosch R, Mullins MC, Hendrickson HS, Hendrickson EK, Halpern ME (2001). *Science* 292, 1385.

346. Fernandez-Llebrez P, Hernandez S, Andrades JA (2001). *Cell Tissue Res* 305, 115.

347. Fricke C, Lee JS, Geiger-Rudolph S, Bonhoeffer F, Chien CB (2001). *Science* 292, 507.

348. Gnuegge L, Schmid S, Neuhauss SC (2001). *J Neurosci* 21, 3542.

349. Gray M, Moens CB, Amacher SL, Eisen JS, Beattie CE (2001). *Dev Biol* 237, 306.

350. Grinblat Y, Sive H (2001). *Dev Dyn* 222, 688.

351. Heisenberg CP, Houart C, Take-Uchi M, Rauch GJ, Young N, Coutinho P, Masai I, Caneparo L, Concha ML, Geisler R, Dale TC, Wilson SW, Stemple DL (2001). *Genes Dev* 15, 1427.

352. Henry CA, Crawford BD, Yan YL, Postlethwait J, Cooper MS, Hille MB (2001). *Dev Biol* 240, 474.

353. Herbomel P, Thisse B, Thisse C (2001). 238, 274.

354. Horne-Badovinac S, Lin D, Waldron S, Schwarz M, Mbamalu G, Pawson T, Jan Y, Stainier DY, Abdelilah-Seyfried S (2001). *Curr Biol* 11, 1492.

355. Imai Y, Gates MA, Melby AE, Kimelman D, Schier AF, Talbot WS (2001). *Development* 128, 2407.

356. Itoh M, Chitnis AB (2001). *Mech Dev* 102, 263.

357. Jensen AM, Walker C, Westerfield M (2001). *Development* 128, 95.

358. Kawahara A, Dawid IB (2001). 11, 1353.

359. Kay JN, Finger-Baier KC, Roeser T, Staub W, Baier H (2001). *Neuron* 30, 725.

360. Kikuchi Y, Agathon A, Alexander J, Thisse C, Waldron S, Yelon D, Thisse B, Stainier DY (2001). *Genes Dev* 15, 1493.

361. Kim CH, Palardy G, Oda T, Itoh M, Jiang YJ, Lewis J, Chandrasekharappa S, Chitnis AB (2001). *Sec Europ Conf ZF Gen & Dev* 9999, 16.

362. Korzh S, Emelyanov A, Korzh V (2001). *Mech Dev* 103, 137.

363. Kozlowsky DJ, Whitfield TT, Hukriede NA, Weinberg SE (2001). *Sec Europ Conf ZF Gen & Dev* 9999, 9.

364. Kudoh T, Dawid IB (2001). *Mech Dev* 109, 95.

365. Lawson ND, Scheer N, Pham VN, Kim CH, Chitnis AB, Campos-Ortega JA, Weinstein BM (2001). *Development* 128, 3675.

366. Lekven AC, Thorpe CJ, Waxman JS, Moon RT (2001). *Dev Cell* 1, 103.

367. Lele Z, Nowak M, Hammerschmidt M (2001). *Dev Dyn* 222, 681.

368. Lewis KE, Eisen JS (2001). *Development* 128, 3485.

369. Liang D, Chang JR, Chin AJ, Smith A, Kelly C, Weinberg ES, Ge R (2001). *Mech Dev* 108, 29.

370. Link BA, Kainz PM, Ryou T, Dowling JE (2001). *Dev Biol* 236, 436.

371. Lister JA, Close J, Raible DW (2001). *Dev Biol* 237, 333.

372. Liu Q, Marrs JA, Chuang JC, Raymond PA (2001). *Brain Res Dev Brain Res* 131, 17.

373. Lorent K, Liu KS, Fetcho JR, Granato M (2001). *Development* 128, 2131.

374. Lyons SE, Shue BC, Oates AC, Zon LI, Liu PP (2001). *Blood* 97, 2611.

375. Minchiotti G, Manco G, Parisi S, Lago CT, Rosa F, Persico MG (2001). *Development* 128, 4501.

376. Mintzer KA, Lee MA, Runke G, Trout J, Whitman M, Mullins MC (2001). *Development* 128, 859.

377. Ono F, Higashijima S, Shcherbatko A, Fetcho JR, Brehm P (2001). *J Neurosci* 21, 5439.

378. Parichy DM, Johnson SL (2001). *Dev Genes Evol* 211, 319.

379. Paw BH. (2001). *Blood Cells Mol Dis* 27, 62.

380. Payne TL, Postlethwait JH, Yelick PC (2001). *Mech Dev* 100, 275.

381. Peterson RT, Mably JD, Chen JN, Fishman MC (2001). *Curr Biol* 11, 1481.

382. Phillips BT, Bolding K, Riley BB (2001). *Dev Biol* 235, 351.

383. Prince VE, Holley SA, Bally-Cuif L, Prabhakaran B, Oates AC, Ho RK, Vogt TF (2001). *Mech Dev* 105, 175.

384. Pujic Z, Malicki J (2001). *Dev Biol* 234, 454.

385. Raible F, Brand M (2001). *Mech Dev* 107, 105.

386. Rawls JF, Johnson SL (2001). *Development* 128, 1943.

387. Reiter JF, Kikuchi Y, Stainier DYR (2001). *Development* 128, 125.

388. Reiter JF, Verkade H, Stainier DY (2001). *Dev Biol* 234, 330.

389. Roehl H, Nusslein-Volhard C (2001). *Curr Biol* 11, 503.

390. Rohr KB, Barth KA, Varga ZM, Wilson SW (2001). *Neuron* 29, 341.

391. Rottbauer W, Baker K, Wo ZG, Mohideen MA, Cantiello HF, Fishman MC (2001). *Dev Cell* 1, 265.

392. Roy S, Wolff C, Ingham PW (2001). *Genes Dev* 15, 1563.

393. Ryu SL, Fujii R, Yamanaka Y, Shimizu T, Yabe T, Hirata T, Hibi M, Hirano T (2001). *Dev Biol* 231, 397.

394. Sakaguchi T, Kuroiwa A, Takeda H (2001). 107, 25.

395. Sawada A, Shinya M, Jiang YJ, Kawakami A, Kuroiwa A, Takeda H (2001). *Development* 128, 4873.

396. Shinya M, Koshida S, Sawada A, Kuroiwa A, Takeda H (2001). *Development* 128, 4153.

397. Söllner C, Rauch GJ, Schuster SC, Geisler R, Nicolson T (2001). *Sec Europ Conf ZF Gen & Dev* 9999, 10.

398. Sun Z, Hopkins N (2001). *Genes Dev* 15, 3217.

399. Svoboda KR, Linares AE, Ribera AB (2001). *Development* 128, 3511.

400. Topczewska JM, Topczewski J, Solnica-Krezel L, Hogan BLM (2001). *Mech Dev* 100, 343.

401. Topczewski J, Sepich DS, Myers DC, Walker C, Amores A, Lele Z, Hammerschmidt M, Postlethwait J, Solnica-Krezel L (2001). *Dev Cell* 1, 251.

402. van de Water S, van de Wetering M, Joore J, Esseling J, Bink R, Clevers H, Zivkovic D (2001). *Development* 128, 3877.

403. Van Epps HA, Yim CM, Hurley JB, Brockerhoff SE (2001). *Invest Ophthalmol Vis Sci* 42, 868.

404. Varga ZM, Amores A, Lewis KE, Yan YL, Postlethwait JH, Eisen JS, Westerfield M (2001). *Development* 128, 3497.

405. Vihtelic TS, Yamamoto Y, Sweeney MT, Jeffery WR, Hyde DR (2001). *Dev Dyn* 222, 625.

406. Walsh EC, Stainier DY (2001). *Science* 293, 1670.

407. Warren KS, Baker K, Fishman MC (2001). *Am J Physiol Heart Circ Physiol* 281, H1711.

408. Waskiewicz AJ, Rikhof HA, Hernandez RE, Moens CB (2001). *Development* 128, 4139.

409. Winkler C, Moon RT (2001). *Dev Biol* 229, 102.

410. Yang ZG, Liu NG, Lin S (2001). *Dev Biol* 231, 138.

411. Zhang J, Malayaman S, Davis C, Granato M (2001). *Dev Biol* 240, 560.

412. Zhong TP, Childs S, Leu JP, Fishman MC (2001). *Nature* 414, 216.

413. Aoki TO, David NB, Minchiotti G, Saint-Etienne L, Dickmeis T, Persico GM, Strahle U, Mourrain P, Rosa FM (2002). *Development* 129, 275.

414. Aoki TO, Mathieu J, Saint-Etienne L, Rebagliati MR, Peyrieras N, Rosa FM (2002). *Dev Biol* 241, 273.

415. Behra M, Cousin X, Bertrand C, Vonesch JL, Biellmann D, Chatonnet A, Strahle U (2002). *Nat Neurosci* 5, 111.

416. Bingham S, Higashijima S, Okamoto H, Chandrasekhar A (2002). *Dev Biol* 242, 149.

417. Burgess S, Reim G, Chen W, Hopkins N, Brand M (2002). *Development* 129, 905.

418. Childs S, Chen JN, Garrity DM, Fishman MC (2002). *Development* 129, 973.

419. diIorio PJ, Moss JB, Sbrogna JL, Karlstrom RO, Moss LG (2002). *Dev Biol* 244, 75.

420. Doerre G, Malicki J (2002). *Mech Dev* 110, 125.

421. Hauptmann G, Belting HG, Wolke U, Lunde K, Soll I, Abdelilah-Seyfried S, Prince V, Driever W (2002). *Development* 129, 1645.

422. Holley SA, Julich D, Rauch GJ, Geisler R, Nusslein-Volhard C (2002). *Development* 129, 1175.

423. Hong SK, Kim CH, Yoo KW, Kim HS, Kudoh T, Dawid IB, Huh TL (2002). *Mech Dev* 112, 199.

424. Hutson LD, Chien CB (2002). *Neuron* 33, 205.

425. Keegan BR, Feldman JL, Lee DH, Koos DS, Ho RK, Stainier DY, Yelon D (2002). *Development* 129, 1623.

426. Koshida S, Shinya M, Nikaido M, Ueno N, Schulte-Merker S, Kuroiwa A, Takeda H (2002). *Dev Biol* 244, 9.

427. Liao EC, Trede NS, Ransom D, Zapata A, Kieran M, Zon LI (2002). *Development* 129, 649.

428. Lyons SE, Lawson ND, Lei L, Bennett PE, Weinstein BM, Liu PP (2002). 99, 5454.

429. Myers DC, Sepich DS, Solnica-Krezel L (2002). *Dev Biol* 243, 81.

430. Reim G, Brand M (2002). *Development* 129, 917.

431. Serluca FC, Drummond IA, Fishman MC (2002). 12, 492.

432. Vihtelic TS, Hyde DR (2002). *Vision Res* 42, 535.

433. Weidinger G, Wolke U, Koprunner M, Thisse C, Thisse B, Raz E (2002). *Development* 129, 25.

434. Willot V, Mathieu J, Lu Y, Schmid B, Sidi S, Yan YL, Postlethwait JH, Mullins M, Rosa F, Peyrieras N (2002). *Dev Biol* 241, 59.

435. Xu X, Meiler SE, Zhong TP, Mohideen M, Crossley DA, Burggren WW, Fishman MC (2002). *Nat Genet* 30, 205.

436. Yamashita S, Miyagi C, Carmany-Rampey A, Shimizu T, Fujii R, Schier AF, Hirano T (2002). *Dev Cell* 2, 363.

437. Young RM, Marty S, Nakano Y, Wang H, Yamamoto D, Lin S, Allende ML (2002). *Dev Dyn* 223, 298.

438. Zhang Y, Cui FZ, Wang XM, Feng QL, Zhu XD (2002). *Bone* 30, 541.

Index